LIQUID INTERFACIAL SYSTEMS

SURFACTANT SCIENCE SERIES

1. Nonionic Surfactants, *edited by Martin J. Schick* (see also Volumes 19, 23, and 60)
2. Solvent Properties of Surfactant Solutions, *edited by Kozo Shinoda* (see Volume 55)
3. Surfactant Biodegradation, *R. D. Swisher* (see Volume 18)
4. Cationic Surfactants, *edited by Eric Jungermann* (see also Volumes 34, 37, and 53)
5. Detergency: Theory and Test Methods (in three parts), *edited by W. G. Cutler and R. C. Davis* (see also Volume 20)
6. Emulsions and Emulsion Technology (in three parts), *edited by Kenneth J. Lissant*
7. Anionic Surfactants (in two parts), *edited by Warner M. Linfield* (see Volume 56)
8. Anionic Surfactants: Chemical Analysis, *edited by John Cross*
9. Stabilization of Colloidal Dispersions by Polymer Adsorption, *Tatsuo Sato and Richard Ruch*
10. Anionic Surfactants: Biochemistry, Toxicology, Dermatology, *edited by Christian Gloxhuber* (see Volume 43)
11. Anionic Surfactants: Physical Chemistry of Surfactant Action, *edited by E. H. Lucassen-Reynders*
12. Amphoteric Surfactants, *edited by B. R. Bluestein and Clifford L. Hilton* (see Volume 59)
13. Demulsification: Industrial Applications, *Kenneth J. Lissant*
14. Surfactants in Textile Processing, *Arved Datyner*
15. Electrical Phenomena at Interfaces: Fundamentals, Measurements, and Applications, *edited by Ayao Kitahara and Akira Watanabe*
16. Surfactants in Cosmetics, *edited by Martin M. Rieger* (see Volume 68)
17. Interfacial Phenomena: Equilibrium and Dynamic Effects, *Clarence A. Miller and P. Neogi*
18. Surfactant Biodegradation: Second Edition, Revised and Expanded, *R. D. Swisher*
19. Nonionic Surfactants: Chemical Analysis, *edited by John Cross*
20. Detergency: Theory and Technology, *edited by W. Gale Cutler and Erik Kissa*
21. Interfacial Phenomena in Apolar Media, *edited by Hans-Friedrich Eicke and Geoffrey D. Parfitt*
22. Surfactant Solutions: New Methods of Investigation, *edited by Raoul Zana*
23. Nonionic Surfactants: Physical Chemistry, *edited by Martin J. Schick*
24. Microemulsion Systems, *edited by Henri L. Rosano and Marc Clausse*
25. Biosurfactants and Biotechnology, *edited by Naim Kosaric, W. L. Cairns, and Neil C. C. Gray*
26. Surfactants in Emerging Technologies, *edited by Milton J. Rosen*
27. Reagents in Mineral Technology, *edited by P. Somasundaran and Brij M. Moudgil*
28. Surfactants in Chemical/Process Engineering, *edited by Darsh T. Wasan, Martin E. Ginn, and Dinesh O. Shah*
29. Thin Liquid Films, *edited by I. B. Ivanov*
30. Microemulsions and Related Systems: Formulation, Solvency, and Physical Properties, *edited by Maurice Bourrel and Robert S. Schechter*
31. Crystallization and Polymorphism of Fats and Fatty Acids, *edited by Nissim Garti and Kiyotaka Sato*

32. Interfacial Phenomena in Coal Technology, *edited by Gregory D. Botsaris and Yuli M. Glazman*
33. Surfactant-Based Separation Processes, *edited by John F. Scamehorn and Jeffrey H. Harwell*
34. Cationic Surfactants: Organic Chemistry, *edited by James M. Richmond*
35. Alkylene Oxides and Their Polymers, *F. E. Bailey, Jr., and Joseph V. Koleske*
36. Interfacial Phenomena in Petroleum Recovery, *edited by Norman R. Morrow*
37. Cationic Surfactants: Physical Chemistry, *edited by Donn N. Rubingh and Paul M. Holland*
38. Kinetics and Catalysis in Microheterogeneous Systems, *edited by M. Grätzel and K. Kalyanasundaram*
39. Interfacial Phenomena in Biological Systems, *edited by Max Bender*
40. Analysis of Surfactants, *Thomas M. Schmitt* (see Volume 96)
41. Light Scattering by Liquid Surfaces and Complementary Techniques, *edited by Dominique Langevin*
42. Polymeric Surfactants, *Irja Piirma*
43. Anionic Surfactants: Biochemistry, Toxicology, Dermatology. Second Edition, Revised and Expanded, *edited by Christian Gloxhuber and Klaus Künstler*
44. Organized Solutions: Surfactants in Science and Technology, *edited by Stig E. Friberg and Björn Lindman*
45. Defoaming: Theory and Industrial Applications, *edited by P. R. Garrett*
46. Mixed Surfactant Systems, *edited by Keizo Ogino and Masahiko Abe*
47. Coagulation and Flocculation: Theory and Applications, *edited by Bohuslav Dobiáš*
48. Biosurfactants: Production • Properties • Applications, *edited by Naim Kosaric*
49. Wettability, *edited by John C. Berg*
50. Fluorinated Surfactants: Synthesis • Properties • Applications, *Erik Kissa*
51. Surface and Colloid Chemistry in Advanced Ceramics Processing, *edited by Robert J. Pugh and Lennart Bergström*
52. Technological Applications of Dispersions, *edited by Robert B. McKay*
53. Cationic Surfactants: Analytical and Biological Evaluation, *edited by John Cross and Edward J. Singer*
54. Surfactants in Agrochemicals, *Tharwat F. Tadros*
55. Solubilization in Surfactant Aggregates, *edited by Sherril D. Christian and John F. Scamehorn*
56. Anionic Surfactants: Organic Chemistry, *edited by Helmut W. Stache*
57. Foams: Theory, Measurements, and Applications, *edited by Robert K. Prud'homme and Saad A. Khan*
58. The Preparation of Dispersions in Liquids, *H. N. Stein*
59. Amphoteric Surfactants: Second Edition, *edited by Eric G. Lomax*
60. Nonionic Surfactants: Polyoxyalkylene Block Copolymers, *edited by Vaughn M. Nace*
61. Emulsions and Emulsion Stability, *edited by Johan Sjöblom*
62. Vesicles, *edited by Morton Rosoff*
63. Applied Surface Thermodynamics, *edited by A. W. Neumann and Jan K. Spelt*
64. Surfactants in Solution, *edited by Arun K. Chattopadhyay and K. L. Mittal*
65. Detergents in the Environment, *edited by Milan Johann Schwuger*

66. Industrial Applications of Microemulsions, *edited by Conxita Solans and Hironobu Kunieda*
67. Liquid Detergents, *edited by Kuo-Yann Lai*
68. Surfactants in Cosmetics: Second Edition, Revised and Expanded, *edited by Martin M. Rieger and Linda D. Rhein*
69. Enzymes in Detergency, *edited by Jan H. van Ee, Onno Misset, and Erik J. Baas*
70. Structure–Performance Relationships in Surfactants, *edited by Kunio Esumi and Minoru Ueno*
71. Powdered Detergents, *edited by Michael S. Showell*
72. Nonionic Surfactants: Organic Chemistry, *edited by Nico M. van Os*
73. Anionic Surfactants: Analytical Chemistry, Second Edition, Revised and Expanded, *edited by John Cross*
74. Novel Surfactants: Preparation, Applications, and Biodegradability, *edited by Krister Holmberg*
75. Biopolymers at Interfaces, *edited by Martin Malmsten*
76. Electrical Phenomena at Interfaces: Fundamentals, Measurements, and Applications, Second Edition, Revised and Expanded, *edited by Hiroyuki Ohshima and Kunio Furusawa*
77. Polymer-Surfactant Systems, *edited by Jan C. T. Kwak*
78. Surfaces of Nanoparticles and Porous Materials, *edited by James A. Schwarz and Cristian I. Contescu*
79. Surface Chemistry and Electrochemistry of Membranes, *edited by Torben Smith Sørensen*
80. Interfacial Phenomena in Chromatography, *edited by Emile Pefferkorn*
81. Solid–Liquid Dispersions, *Bohuslav Dobiáš, Xueping Qiu, and Wolfgang von Rybinski*
82. Handbook of Detergents, *editor in chief: Uri Zoller*
    Part A: Properties, *edited by Guy Broze*
83. Modern Characterization Methods of Surfactant Systems, *edited by Bernard P. Binks*
84. Dispersions: Characterization, Testing, and Measurement, *Erik Kissa*
85. Interfacial Forces and Fields: Theory and Applications, *edited by Jyh-Ping Hsu*
86. Silicone Surfactants, *edited by Randal M. Hill*
87. Surface Characterization Methods: Principles, Techniques, and Applications, *edited by Andrew J. Milling*
88. Interfacial Dynamics, *edited by Nikola Kallay*
89. Computational Methods in Surface and Colloid Science, *edited by Małgorzata Borówko*
90. Adsorption on Silica Surfaces, *edited by Eugène Papirer*
91. Nonionic Surfactants: Alkyl Polyglucosides, *edited by Dieter Balzer and Harald Lüders*
92. Fine Particles: Synthesis, Characterization, and Mechanisms of Growth, *edited by Tadao Sugimoto*
93. Thermal Behavior of Dispersed Systems, *edited by Nissim Garti*
94. Surface Characteristics of Fibers and Textiles, *edited by Christopher M. Pastore and Paul Kiekens*
95. Liquid Interfaces in Chemical, Biological, and Pharmaceutical Applications, *edited by Alexander G. Volkov*

96. Analysis of Surfactants: Second Edition, Revised and Expanded, *Thomas M. Schmitt*
97. Fluorinated Surfactants and Repellents: Second Edition, Revised and Expanded, *Erik Kissa*
98. Detergency of Specialty Surfactants, *edited by Floyd E. Friedli*
99. Physical Chemistry of Polyelectrolytes, *edited by Tsetska Radeva*
100. Reactions and Synthesis in Surfactant Systems, *edited by John Texter*
101. Protein-Based Surfactants: Synthesis, Physicochemical Properties, and Applications, *edited by Ifendu A. Nnanna and Jiding Xia*
102. Chemical Properties of Material Surfaces, *Marek Kosmulski*
103. Oxide Surfaces, *edited by James A. Wingrave*
104. Polymers in Particulate Systems: Properties and Applications, *edited by Vincent A. Hackley, P. Somasundaran, and Jennifer A. Lewis*
105. Colloid and Surface Properties of Clays and Related Minerals, *Rossman F. Giese and Carel J. van Oss*
106. Interfacial Electrokinetics and Electrophoresis, *edited by Ángel V. Delgado*
107. Adsorption: Theory, Modeling, and Analysis, *edited by József Tóth*
108. Interfacial Applications in Environmental Engineering, *edited by Mark A. Keane*
109. Adsorption and Aggregation of Surfactants in Solution, *edited by K. L. Mittal and Dinesh O. Shah*
110. Biopolymers at Interfaces: Second Edition, Revised and Expanded, *edited by Martin Malmsten*
111. Biomolecular Films: Design, Function, and Applications, *edited by James F. Rusling*
112. Structure–Performance Relationships in Surfactants: Second Edition, Revised and Expanded, *edited by Kunio Esumi and Minoru Ueno*
113. Liquid Interfacial Systems: Oscillations and Instability, *Rudolph V. Birikh, Vladimir A. Briskman, Manuel G. Velarde, and Jean-Claude Legros*

ADDITIONAL VOLUMES IN PREPARATION

Novel Surfactants: Preparation, Applications, and Biodegradability: Second Edition, Revised and Expanded, *edited by Krister Holmberg*

Colloidal Polymers: Preparation and Biomedical Applications, *edited by Abdelhamid Elaissari*

Gemini Surfactants: Interfacial and Solution-Phase Behavior, *edited by Raoul Zana and Jiding Xia*

Colloidal Science of Flotation, *Anh V. Nguyen and Hans Joachim Schulze*

LIQUID INTERFACIAL SYSTEMS

Oscillations and Instability

Rudolph V. Birikh

Perm State Pedagogical University
Perm, Russia

Vladimir A. Briskman

Russian Academy of Sciences
Perm, Russia

Manuel G. Velarde

Instituto Pluridisciplinar
Universidad Complutense of Madrid
Madrid, Spain

Jean-Claude Legros

Université Libre de Bruxelles
Brussels, Belgium

MARCEL DEKKER

MARCEL DEKKER, INC. NEW YORK • BASEL

Library of Congress Cataloging-in-Publication Data
A catalog record for this book is available from the Library of Congress.

ISBN: 0-8247-4295-8

This book is printed on acid-free paper.

Headquarters
Marcel Dekker, Inc., 270 Madison Avenue, New York, NY 10016, U.S.A.
tel: 212-696-9000; fax: 212-685-4540

Distribution and Customer Service
Marcel Dekker, Inc., Cimarron Road, Monticello, New York 12701, U.S.A.
tel: 800-228-1160; fax: 845-796-1772

Eastern Hemisphere Distribution
Marcel Dekker AG, Hutgasse 4, Postfach 812, CH-4001 Basel, Switzerland
tel: 41-61-260-6300; fax: 41-61-260-6333

World Wide Web
http://www.dekker.com

The publisher offers discounts on this book when ordered in bulk quantities. For more information, write to Special Sales/Professional Marketing at the headquarters address above.

Current printing (last digit):

10 9 8 7 6 5 4 3 2 1

PRINTED IN THE UNITED STATES OF AMERICA

Preface

It has long been recognized that a large number of seemingly diverse phenomena derive from the unique properties and behavior of fluids at or near interfaces. Incidentally, 1 cm^3 of water sprayed into spherical droplets 10^{-6} cm in diameter provides a liquid–air interface of 6 million cm^2.

Interfacial phenomena are widely seen in nature and play a vital role in many biological and industrial processes. They are of importance in fluid–phase separation and other processes such as flotation (by bubbling gas through a liquid, it is possible to separate solid matter from the liquid because of its tendency to adsorb onto the bubble surfaces), coating (fluids containing molecular species readily adsorb to the surface of a spreading fluid layer as in traditional photography), laser irradiation of metals and semiconductors, laser and electron beam welding, liquid extraction, gas adsorption accompanied or not by surface chemical reactions (as the O_2/CO_2 exchange in the lungs; in the healthy lungs of an adult, with a surface area about 100 m^2, surface tension is naturally regulated, ensuring that minimal effort is required when expanding the liquid-lined airways during inspiration, a case quite different from that of prematurely born babies or adults suffering from respiratory distress syndrome), anti-icing film spreading on grounded aircraft and de-icing of harbors, drying of solvent-containing paints, dewetting silicon wafers for microchip manufacture, boiling heat and mass transfer, retardation of evaporation from reservoirs, jet decay, corrosion mechanisms as in salt corrosion of aeroturbine blades, flame spreading, distillation, engineering operations involving foaming, dispersions, spraying or liquid atomization (ink-jet printing), metallurgy due to complex mixing hydrodynamics in furnaces containing melts (in particular, in the steel-making industry, in which erosion of the walls of ceramic crucibles occurs), continuous casting of lubricating films, spreading of lubricants, tertiary oil recovery, certain coal treatments, and pollution abatement. Surface tension gradients have also been used for crystal growth to produce semiconductors by imposing large temperature gradients along a molten material in the low-effective-gravity environment (termed microgravity) of a space craft.

Cleaning a liquid surface is not the same as cleaning a liquid. If a kilogram of a liquid contains 1 mg of an impurity and this impurity is uniformly distributed in the volume, then the concentration is one part per million and the bulk properties of the material (such as density or viscosity) may remain almost unaffected by the contamination. However, if 1 mg of the foreign matter is concentrated in the surface layer, it may form a skin 10 nm thick. Because this thickness is greater than the radius of the sphere of molecular action, the surface will behave as if the main ingredient were absent. Another estimate may complete the perspective. If a molecule of solute occupies 0.4 nm^2, then 2.5×10^{14} molecules are needed to cover a surface area of 1 cm^2 (keep in mind the surface of the liquid solvent). Such a number of molecules corresponds to a 4×10^{-7} molar bulk concentration when dissolved in 1 cm^3 of solvent.

Capillary properties and surface stresses play a significant role in wave generation and motion. In 1774 (from a letter dated November 7, 1773), Benjamin Franklin read a paper to the Royal Society (London) in which he reported on an experiment on the large pond on the common at Clapham:

> the oil, though not more than a tea spoonful, produced an instant calm over a space several yards square, which spread amazingly, and extended itself gradually till it reached the lee side making all that quarter of the pond, perhaps half an acre, as smooth as a looking glass.

An estimate of the thickness of Franklin's oil film is 2.5 nm, which is of the order of molecular size (monolayer). Franklin was not the first to make such an observation—he referred to Pliny the Elder, and we know that Plutarch and others mentioned a similar finding by Mediterranean sailors. In 1982, Lucassen gave a reappraisal on the damping of (gravity) waves in a liquid surface using adequate limited amounts of useful pollutants. In particular, he commented on the case of a stormy sea and offered advice about using oil mixtures for air–sea or shore–sea rescue operations in which human lives are involved.

Other scientists were less fortunate in relating capillarity to the force underlying some apparent interfacial phenomena. In about 1900, Fritz Haber entertained the idea that impulse transmission in nerves was due to sudden large variations in surface tension. Experiments conducted by W. D. Harkins (1910) proved that Haber was not correct. It was not until the publications of A. L. Hodgkin and A. F. Huxley, around 1950, that we were given a clear, unambiguous picture of action potential/impulse neural transmission.

SURFace ACTive AgeNTS (in short, surfactants) are materials and chemicals that tend to accumulate at surfaces or interfaces, causing changes in their behavior. Surfactant molecules are generally amphiphilics in that their structure has a hydrophilic head (ionic or nonionic) with affinity for the water solution and a hydrophobic tail (a hydrocarbon group) that is repelled by an aqueous phase. Both soluble and insoluble surfactants tend to form monolayers at interfaces. They are used in engineering (some processes mentioned earlier depend on the use of natural or artificial surfactants) and at home under various names, such as soap (sodium salt of fatty acids), detergent (sodium sulfate or sulfonate of long-chain hydrocarbons), wetting agent, emulsifier, demulsifier, or dispersing agent, depending on their function. Fatty acids, alcohols, and some proteins (natural polymers) are typical surfactants. Surfactants are primary components of cell walls, as lining material in the lungs, of the tear film and artificial tears (used to cope with the dry eye syndrome), and of bile salts, which play a major role in digestive processes.

Adsorbed materials acting on the surface tension, temperature gradients, electric fields, or inhomogeneous electric charge distribution lead to nonuniform surface or interfacial tension and hence give rise to forces resulting in fluid motion. If the surface tension varies along the liquid–air interface, a tangential force per unit area develops on the interface, given by $f_s = \nabla_s \sigma$, where σ denotes the surface or interfacial tension and ∇ indicates the gradient operator; the subscript s restricts the corresponding vector to its surface component (e.g., water $dT/dx = 10^2$ K/m, $d\sigma/dT = -0.15$ mN/mK, $d\sigma/dx = 15$ mN/m^2; mN/m = dyne/cm). The liquid tends to move from lower to higher surface tension, the Marangoni effect.

Great scientists have contributed to our understanding of interfacial phenomena: Leonardo da Vinci (1490), Young (1805), Laplace (1806), Gauss (1830), Plateau, Marangoni, Dupre de Rennes, Van der Mensbrugghe, Lord Kelvin and his brother J. Thomson, Maxwell, Gibbs, Lord Rayleigh, N. Bohr (presumably his early work—1906—on the surface tension of water later led him to the drop model of the nucleus), Langmuir, Derjaguin, Levich, Defay, Prigogine, Scriven, Lucassen, Linde, Ostrach, Acrivos, Brenner, Wasan, de Gennes, and others. For an account of the work of the pioneers we take pleasure in referring the reader to the book by Plateau (1873) and to two scholarly written papers by Scriven and Sternling (1960) and by Ross and Becher (1992), respectively. References, with no pretense of being complete, are provided in Chapter 1 with the understanding that the authors' bias toward some areas and ignorance in others will no doubt be reflected. Apologies are made to contributors whose work is not cited.

A scientist of particular relevance to the topics treated in this book is Carlo Marangoni, whose work for well over 40 years (1869–1910) paved the way to our understanding of surface tension gradient (Marangoni) driven flows. He seems to have had lively exchanges with Plateau, as shown by the following quotation from a paper published in 1878 (p. 54):

> *Ma pur troppo vi è un proverbio che dice: Chi vuol provar troppo, non prova nulla! e così avuenne a me che, volendo enumerare troppi e svariati fatti, ho trascurato l'idea fondamentale ed ho lasciata troppo facile la via ad una lunga critica quale mi mosse l'ilustre Plateau.*
>
> *Peccato confessato, mezzo perdonato. L'altra metà del perdono me la saprò meritare colla seguente difesa . . .*

and then follows a long paper in three parts.*

The list above of interfacial phenomena appears vast indeed. Hence we should alert the reader that this is far from a complete treatise. However, we hope that by discussing just a few cases, paradigmatic in our opinion, the reader will be introduced to methodologies for the study of the Marangoni effect in the dynamics of interfacial phenomena. Some model problems are treated redundantly in the hope of being useful (alas!, impenitent sinners). Now and again throughout the book, the surface tension nonuniformity is shown to be an active agent either triggering or enhancing interfacial instability, with subsequent bulk flow and other consequences, or trying to suppress instability when competing with other destabilizing mechanisms. Thus we can say that, thanks to the Marangoni effect, free surfaces or interfaces become active in driving flow or instability in liquid layers, films, or drops, and bubbles. Heuristic arguments are complemented with analytical and numerical (computerized) results, and in various cases these are confronted with experimental data. Needless to say we provide references to works in which further and more complete results on interfacial phenomena can be found and, besides, we expect that the reader will easily realize how ample the territory appears for future work.

In Chapter 1 (Introduction), we provide, in general terms, the basic equations and boundary conditions to be used throughout the book. However, later in each chapter, for the sake of comprehensive treatment, these equations are recalled to the extent needed for the particular problem treated. Dimensionless groups are also defined in the Introduction.

The Marangoni effect is defined and its role in triggering flow motions and instability is highlighted in Chapter 2, where a discussion of various types of instabilities driven by the Marangoni effect is provided, which should help the reader in subsequent chapters. Thus we provide: 1) a simplified description of the onset of steady patterned convection (Benard cells), 2) a physically appealing harmonic oscillator analysis of surface oscillations, transverse (capillary-gravity)

---

* "But, unfortunately, as the proverb says: He who wants to prove too much, proves nothing! And this is what has happened to me: wanting to enumerate too many different facts, I have neglected the fundamental ideal, and left myself wide open to a long criticism such as that which the illustrious Plateau has brought against me. A sin shriven, half forgiven. I shall know to gain the other half of the pardon with the following defense. . ."

waves and longitudinal (dilational) waves excited and eventually sustained by the Marangoni effect, and 3) for the sake of completeness, although the subject is not really treated further in the book, a similar description of oscillatory surface phenomena occurring when either a tangential or an orthogonal electric field is applied to the surface of a liquid layer. Threshold values are given for the onset of gravito-electro-capillary waves due to the action of the electric field, very much as the Marangoni effect does for mass or heat transfer processes under thermal constraints. The above-mentioned flows, steady or oscillatory, will be designated under the general term surface tension gradient (Marangoni) driven convection or, alternatively: thermocapillary or solutocapillary flows.

Chapter 3 deals with the parametric excitation of an isothermal liquid layer leading to surface waves by means of forced oscillations of the bulk of the liquid as a parametric liquid pendulum and by a variable electric field applied orthogonally to the open surface. In Chapter 4 we discuss the suppression or delay of a possible Rayleigh–Taylor instability by means of high-frequency vibrations in an isothermal system. Vibrations are taken as either uniform harmonic oscillations of the container or tangential along the surface of the liquid layer.

Chapter 5 deals again with the role, disregarding buoyancy, of the Marangoni effect in triggering steady patterned convection in a shallow liquid layer heated from below, a paradigmatic model problem. Then, in the case of a liquid layer with deformable open surface, we treat oscillatory convection (overstability) that offers various types of surface wave behavior when heating from above. We also discuss different instabilities of a liquid film, or a liquid layer with two open boundaries, subjected to a transverse temperature gradient or internal heat sources. Finally, the effect of the combined action of thermocapillarity and buoyancy is also considered.

In Chapter 6 we discuss the influence of a permeable partition or divider in a film subjected to heat transfer and the Marangoni effect. Chapter 7 examines the thermocapillary instability of two-layer systems with the Marangoni effect, with and without buoyancy effects. Chapter 8 deals with multi- (three-) layer systems, with particular attention paid to the role of finite lateral extent and variable effective gravity (microgravity) in both monotonic instability and overstability. In Chapter 9 we consider one more case of surface tension gradient (Marangoni) driven convection when the surface is restricted by solid parts, as in experiments with liquid zones having solid guards on their liquid surface.

Chapter 10 is devoted to the problem of drop-and-bubble migration due to the Marangoni effect. Both passive migration and instability, and eventual self-propulsion, due to the Marangoni effect in *active* drops are considered. Chapter 11 deals with spreading layers and their eventual breakdown (deformation and rupture), all due to the Marangoni effect and in Chapter 12 we study the parametric excitation of thermocapillary instability and oscillations in nonisothermal liquid layers. Chapter 13 is devoted to a study of the combined action of high-frequency vibrations and the Marangoni effect on the interface between two fluid layers.

When writing this book we had in mind graduate and postgraduate (Ph.D. and postdoctoral) students and researchers from physical chemistry, physico-chemical hydrodynamics, materials science and engineering, chemical, petroleum and mechanical engineering, applied mathematics, and physics. We have done our best to make the contents self-contained. This book is part of the training effort produced by the authors in the framework of the Interfacial Convection and Phase Change (ICOPAC) Network sponsored by the European Union.

Rudolph V. Birikh
Vladimir A. Briskman
Manuel G. Velarde
Jean-Claude Legros

Acknowledgments

We wish to express our gratitude to the following colleagues, from whom we have learned much (apologies are made to all for our not discriminating the degree of gratitude and academic rank): Andreas (Andy) Acrivos, Ramaswamy Balasubramanian, Pierre Berge (deceased), Nikolai A. Bezdenezhnykh, Svetlana V. Boushoueba, Jean Bragard, Yurii K. Bratukhin, Howard Brenner, John W. Bush, Manuel Castans, Antonio Castellanos, Anatoly A. Cherepanov, Vladimir I. Chernatynsky, Xiao-Lin Chu, Pierre Colinet, Steve H. Davis, Pierre-Gilles de Gennes, Alain de Ryck, Monique Dubois, Aleksandr B. Ezersky, Grigory Z. Gershuni (deceased), Ramon Gonzalez-Rubio, Benoit Goyeau, Etienne Guyon, Marcel Hennenberg, Georges (Bud) Homsy, Ivan B. Ivanov, Serafim Kalliadasis, E. Lothar Koschmieder, Serguei Kosvintsev, Vadim Kurdyumov, Dominique Langevin, Georgy Lebon, Libero Liggieri, Hartmut Linde, Gustavo Luengo, Dmitry V. Lyubimov, Alcina Mendes-Tatsis, Ken C. Mills, Aleksey I. Mizyov, Francisco Monroy, Rodolfo Monti, Ranga Narayanan, Alex A. Nepomnyashchy, Alex Nikolov, Si Ostrach, John F. Padday (deceased), Jacques Pantaloni, Alberto Passerone, Agnes Patterson, Francois Petrelis, Aleksandr F. Phenichnikov, Len Pismen, Yves Pomeau, Alex Ye. Rednikov, Rudolph N. Rudakov, Yuri S. Ryazantsev, Jesus Salan, Albert Sanfeld, Maria Santiago, G. S. Rama Sarma, Robert S. Schechter, Benoit Scheid, Grigory F. Shaidurov, Mihail T. Sharov, Viktor Ya. Shkadov, Oleg E. Shklyaev, Boris L. Smorodin, Vladimir D. Sobolev, Victor M. Starov, Annie Steinchen-Sanfeld, Johannes Straub, R. Shankar Subramanian, Milton Van Dyke, Stefan Van Vaerenbergh, Michele Vignes-Adler, Antonio Viviani, Darsh T. Wasan, J. Eduardo Wesfreid, Andreas Wierschem, Harvey A. R. Williams, Valentin I. Yakushin, Radyadour Kh. Zeytounian, Efim M. Zhukhovitsky (deceased) and Andrey L. Zuev.

For the preparation of the compuscript, Ms. Iouliia Makarova and Ms. Maria-Jesus Martin have been invaluable. Our sincere thanks.

The contents of this book originate to a large extent in our research and in our cooperation over a decade. When the manuscript had been almost completed, we suffered the loss of our admired, inspiring, and great friend, Professor Vladimir Abramovich Briskman. His unexpected death so much affected us that for a year we were unable to resume completion of the book.

Economic support for the collaborative research came from the Spanish Ministerio de Educacion y Cultura (Programa Sabaticos), the Spanish Ministerio de Ciencia y Tecnologia (Grant PB96-0599), the Belgian InterUniversity Poles of Attraction Programme (Federal Office for Scientific, Technical and Cultural Affairs), NATO (Scientific Affairs Office), INTAS, the European Union, the Universidad Complutense de Madrid (Programa Sabaticos) and the European Space Agency (ESA Microgravity Office under the enthusiastic leadership of Dr. Guenther Seibert and his collaborators, Dr. Hannes U. Walter and Dr. Olivier Minster). To all these institutions we express our gratitude.

Contents

1
Introduction

In the Preface we have referred to the work of pioneers in the field of interfacial phenomena, for example, Marangoni and colleagues [1–5], Plateau [6], and others [7–26]; Refs. 11 and 12 in particular are noteworthy, and Myshkis et al. [26] offer documented historical notes. We have also commented on technological processes where interfacial phenomena play a significant role and provide further references of interest from recent basic studies and application-oriented work [27–37].

In the present chapter we recall the basic equations to be used in this and in subsequent chapters. We also discuss scales involved in interfacial phenomena under various circumstances, providing heuristic arguments and dimensionless groups to help find significant (yet simple enough for easy theoretical study) limiting cases where one or another aspect of interfacial phenomena appear transparently.

I. BASIC CONCEPTS AND SCALES

A. Surfaces and Interfaces, and Surface Stresses

Following Laplace and Gibbs, the real inhomogeneous interfacial region of finite width is replaced by an idealized, much simpler system. The adhering bulk phases are just matched by a sharp boundary, e.g., a surface, and the overall action of the pressure and stress tensors is accounted for by ad hoc forces acting on the surface, e.g., surface tension, which resists curvature of the interface. The surface boundary is endowed with material properties mimicking features of the transition region, e.g., surface viscosities, etc. However, treating cases that involve complex interfacial rheology or anisotropic phenomena is beyond the scope of this book [38]. The reader interested in the adequate treatment of the interfacial region and the determination of its appropriate constitutive relations for simple or multicomponent deforming fluid interfaces and measurement of the surface rheological and transport properties may consult the basic texts by Aris [39], Slattery [40], Edwards et al. [41], or Gatignol and Prud'homme [42].

Interfacial transition regions exhibit properties very much like membranes and are therefore prone to deformability and vibratory and elasticlike (stretch–squash) motions. They are also capable of exhibiting patterns (surface relief) that may be a manifestation of underlying convective motions. Fluid interfaces may be at equilibrium if all stresses balance to zero.

Capillary-gravity waves are transverse vibratory motions of an interface associated with deformability. However, an interface may also exhibit compression–expansion, dilational, longitudinal waves even for incompressible liquids. The latter can be visualized easily if we consider the possible analogy between a monolayer covered surface, and a stretched–squashed elastic membrane. Coverage with a surfactant monolayer gives elasticity to a surface so that it tends to resist the surface expansion and compression that appears as wave motion when the Marangoni effect operates. This is only a particular case given here for the purposes of metaphoric illustration. Similar phenomena do occur in real interfaces that are eventually clean but subject, for example, to heat transfer. We shall return to this subject below.

Transverse vibrations of an interface or an open surface are associated with normal stresses and may appear in inviscid liquids. When the vibratory motions are driven by the balance between the inertia of the liquid and the force of gravity, they are called gravity waves $[(\lambda/2\pi) \gg (\sigma/\rho g)^{1/2}]$; λ, σ, ρ and g denote wavelength, surface tension, density and gravity acceleration, respectively. If rather than gravity there is an electric field, the phenomena share common features. If the restoring force to bring the surface level is surface tension, the transverse surface vibrations are called capillary waves or ripples $[(\lambda/2\pi) \ll (\sigma/\rho g)^{1/2}]$. As a result of viscosity, both waves are confined to a distance d from the surface of about a wavelength $[d \leq (\nu/\omega)^{1/2}]$, ν and ω denote kinematic viscosity and (angular) frequency, respectively; this is particularly true with high frequency ("long" wavelength) motions. Thus the potential flow function penetrates to much greater depth below the surface than the vorticity function, and hence the bulk liquid flow is mainly potential and viscosity may be disregarded. Gravity and capillary waves exhibit dispersion, i.e., their velocity depends on wavelength.

Surface longitudinal, compression–expansion motions are related to tangential stresses; hence they are necessarily dissipative flow motions. These dilational waves do not exhibit dispersion.

A film, that is, a thin liquid layer with two outer open surfaces, may show flexural/antisymmetric wave motions or squeezing/symmetric vibrations.

The forces that give rise to the abovementioned phenomena appear because of the alteration in stresses at the interface between two immiscible fluid phases. For a curved interface there is a difference in pressure between the two fluids given by the Young–Laplace equation (18,19,41). This pressure difference is termed the capillary pressure, and because the normal stress component at the interface must be continuous, then that pressure added to the hydrostatic pressure must balance. A balance can always be achieved under static conditions. In addition, the tangential stress must also be continuous at the interface. However, if there is a gradient in surface tension in the tangential direction, this leads to an additional tangential stress that adds an unbalanced force that will set the fluid into motion (Marangoni effect); that is, the interface cannot be balanced when the fluids are motionless. On the other hand, the conditions of continuity of tangential velocity and no normal velocity across the immiscible interface are unaltered by surface tension.

Suppose that as a result of, say, a spontaneous flow disturbance an interface full of adsorbed surfactant is cleared and the surfactant molecules are displaced from a given location. Then the surfactant concentration is lowered at that location while increasing around. This flow disturbance is counteracted by the surface tension gradient thus created because of the excess surface concentration and hence tends to suppress the abovementioned spontaneous flow. This is very much like an elastic response of the surface and hence the Marangoni effect yields an elasticlike, albeit dissipative, force to the interface. This argument has been used to explain why pollutants may help to stop a drop moving in a temperature gradient due to accumulation of surfactants at its rear role, which creates counterflow and hence drag. Clearly, what has been said also applies to a nonisothermal interface.

The normal stress, orthogonal to an interface, consists of a contribution from pressure and another from viscous stress. For a liquid surface open to air the pressure from the air layer is the uniform ambient value. However, the viscous stress from the air is negligible compared to that from the liquid side because the dynamic (shear) viscosity of air is, generally, negligible compared with that of the liquid ($\eta_{\text{liquid}} \gg \eta_{\text{air}}$; $\rho_{\text{liquid}} \gg \rho_{\text{air}}$; $\eta_{\text{air}}/\eta_{\text{water}} \approx 10^{-2}$; $\nu_{\text{air}}/\nu_{\text{water}} \approx 10$; $\kappa_{\text{air}} \ll \kappa_{\text{water}}$; $\chi_{\text{air}}/\chi_{\text{water}} \approx 10^{2}$). η, κ and χ denote shear viscosity, heat conductivity and heat deffusivity, respectively. The liquid surface is then said to be free. If the liquid surface-to-volume ratio is small, a rigorous formulation of the normal stress balance at the corresponding interface is, generally, unnecessary.

In most cases to be discussed in this book we shall consider liquid layers where surface deformation at interfaces is small or negligible. Deformability would only be accounted for in contributions where externally applied fields, gravity, vibrations, or electrical fields cause deformation to play a significant role.

Let us examine the role of a surface tension gradient as a pressurelike agent (Marangoni effect). We shall take verbatim an argument from Edwards et al. [41, pp. 31–32]: “By analogy with classical pressure gradient-driven Poiseuille flow in

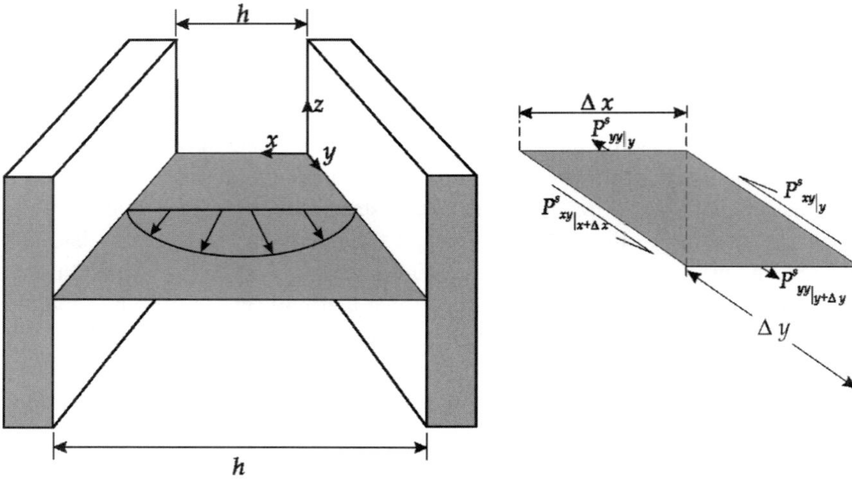

FIG. 1 A surface tension gradient or "surface pressure" is imposed across a fluid–fluid interface between two parallel plates, generating a two-dimensional "Poiseuille flow" in the plane of the interface. (From Ref. 41.)

a rectilinear channel, imagine a time-independent, fully-developed 'surface pressure' gradient acting within the plane of an interface between two parallel plates (Fig. 1).

The balance of interfacial forces acting upon the boundaries of the surface element shown in Fig. 1 may be expressed as

$$(P^S_{xy}|_{x+\Delta x} - P^S_{xy}|_x)\Delta y + (P^S_{yy}|_{y+\Delta y} - P^S_{yy}|_y)\Delta x = 0 \tag{1}$$

where P^S_{xy} and P^S_{yy} are surface stress components.

Division by the areal element $\Delta x \Delta y$ while allowing this element to become vanishingly small yields

$$\frac{\partial P^S_{xy}}{\partial x} + \frac{\partial P^S_{yy}}{\partial y} = 0 \tag{2}$$

expressing the balance of interfacial linear momentum for the surface pressure gradient-driven flow. The constitutive expression for the surface stress component P^S_{yy} is simply

$$P^S_{yy} = \sigma \tag{3}$$

where σ is the interfacial tension.

As the interfacial tension is highly sensitive to surfactant adsorption, any disturbance within the interface that breaks the homogeneity of the surfactant surface concentration thereby raises or lowers the interfacial tension, thus creating a dynamic interfacial tension gradient at the location of the inhomogeneity. Equation (2) in combination with Eq. (3) indicates that the existence of an interfacial-tension gradient $\partial\sigma/\partial y$ creates an interfacial stress P^S_{xy}. In turn, this results by interfacial motion in the system attempting to re-establish an equilibrium state of uniform surfactant concentration.

Together, Eqs (2), (3), and Newton's viscosity law combine to yield

$$\frac{\mathrm{d}\sigma}{\mathrm{d}y} + \eta^S \frac{\mathrm{d}^2 v_y^o}{\mathrm{d}x^2} = 0 \tag{4}$$

where the interfacial tension gradient is a constant of the interfacial flow owing to the fully developed nature of the interfacial motion. η^S and v_y^o denote shear viscosity and velocity, respectively. Surfactant concentration inhomogeneities are not the only source of such gradients. Temperature gradients constitute another source owing to the sensitive dependence of interfacial tension upon temperature.

Boundary conditions imposed upon Eq. (4) again correspond to no-slip

$$v_y^o = 0 \text{ at } x = 0 \tag{5}$$

and

$$v_y^o = 0 \text{ at } x = h \tag{6}$$

at the plate surfaces in Fig. 1, giving the quadratic field

$$v_y^o = \frac{h^2}{2\eta^S}\left(\frac{\mathrm{d}\sigma}{\mathrm{d}y}\right)\left[\frac{x}{h} - \left(\frac{x}{h}\right)^2\right] \tag{7}$$

corresponding to the two-dimensional Poiseuille flow shown in Fig. 1."

B. Surface Tension and Its Variation

Tables 1 to 5 provide values of physical properties (please note that whereas η is, generally, much smaller for gases than for liquids, the opposite is true for ν), heats of solution, and of surface tensions and their variation with temperature for various liquids. Fully miscible liquids should not exhibit surface tension (see, however, Refs. 43–49; a strong concentration, and density, gradient leads to stresses that may act like the Marangoni effect).

Water has a surface tension $\sigma = 72.8$ mN/m at 20°C. Adding inorganic salts increases the surface tension a little. For sea water (at salinity of 35‰) the value is approximately 78 mN/m at the same temperature. The crude oil–brine interface in an oil field has a surface tension of $\sigma = 30$ mN/m. By adding surfactants this value can be reduced to $1-10^{-1}$ mN/m.

TABLE 1 Approximate Values (Obtained from Various Sources) of the Surface Tension of Various Liquids (Interface Liquid-Air; Interface Liquid–Vapor)

Liquid	σ (mN/m = dyne/cm)	T (20°C unless otherwise specified)
Acetic acid	27.5	
Acetone	23.7	
Benzene	28.9	
Carbon monoxide (CO)	9	−189°C
Carbon tetrachloride	26.9	
Chloroform	27.1	
Ethanol/Ethyl alcohol	22.8	
Ethylene glycol	47	
Fluorinerts	15–18	
Helium	0.24	−270°C
Hydrogen	2.3	−255°C
Iron	1850	Melting temperature
Isopropanol	21.7	
Lead	430	Melting temperature
Mercury	470	
Mercury	498	Melting temperature
Methane	17.3	−171°C
Methanol/Methyl alcohol	22.5	
n-Hexadecane	27.6	
n-Hexane	18.4	
Nitrobenzene	43.3	
Nitrogen	10.5	−203°C
Nitrogen	6.6	−183°C
n-Octane	21.8	
Oxygen	13.2	−180°C
Pentane	16.5	
Silicon	10^{-3}	Melting temperature
Silicone oils	10–20	
Sulfur	61	120°C
Toluene	28.5	
Water	72.8	

The quantity $d\sigma/dT$ is negative for nearly all substances that are relatively easily obtained in an almost pure state (order of magnitude is 10^{-1}–10^{-2}). It is significant that it is negative also for water between 0°C and 4°C, i.e., in the range in which the density of the liquid increases with temperature; however, in this region $-d\sigma/dT$ is smaller (0.144 mN/mK) than at 20–25°C. When purification of the sample is very difficult, the temperature coefficient of σ may be negative within a certain temperature interval; in some instances, σ as a function of T appears to have both a maximum and a minimum. Engineers have on occasion

TABLE 2 Approximate Values (Obtained from Various Sources) of the Interfacial Tension Between Various Fluid-Pairs (Liquid–Liquid Interface)

Fluid-pair	σ (mN/m)	T (20°C)
Aniline–water	5.9	
Benzene–water	35	
Butyl alcohol–water	1.6	
Carbon disulfide–water	48	
Carbon tetrachloride–water	45	
Castor oil/Ricin oil–silicone oil	3.5	
Ethyl acetate–water	6.8	
Ethyl ether–water	10.7	
Fluorinert–silicone oil	4–8	
Hydrocarbons–water	50	
n-Heptane–water	51.2	Solubility 10 mg/kg
n-Hexane–water	50.8	Solubility 3 mg/kg
Nitrobenzene–water	26	
n-Octane–methanol	4	
n-Octane–water	51.7	Solubility 0.7 mg/kg
n-Pentane–water	50.2	Solubility 38 mg/kg
Vegetable oil (olive oil)–silicone oil	1.2	

considered these features as anomalies caused by impurities, as often occurs in liquid metals. Indeed, the irregular change of σ with T observed in many samples of mercury may be caused by dissolved gases (up to 0.2 cm^3 gas in one cubic centimeter of liquid mercury after a vacuum distillation). Having observed positive values for $d\sigma/dT$, for instance, for copper and germanium, careful purification of the metals changed the sign of the coefficient. The $d\sigma/dT$ for pure liquid tin is, generally, negative but small additions of tellurium produce an increase of σ with

TABLE 3 Illustrative Examples (Obtained from Various Sources) of Variation of Surface Tension with Temperature

Fluid-pair	$\sigma_1 = \left\lvert \frac{d\sigma}{dT} \right\rvert$ (mN/mK)
Castor oil/Ricin oil–silicone oil	10^{-2}
Fluorinert–air	10^{-1}–10^{-2}
Fluorinert–silicone oil	10^{-2}
n-Hexadecane–air	10^{-1}
n-Octane–methanol	10^{-5}
Silicon–vapor (melting point)	10^{-1}
Silicone oil–air	10^{-1}–10^{-2}
Water–air	10^{-1}

TABLE 4 Approximate Values (Obtained from Various Sources) of the Bulk Physical Properties of Various Fluids

Liquid	ρ (kg/m^3)	η (N s/m^2) $\times 10^3$	ν (m^2/s) $\times 10^6$	κ (W/m K)	χ (m^2/s) $\times 10^6$	α (K^{-1}) $\times 10^3$	*Pr*
Air	1	1.5×10^{-4}	15.6	2×10^{-2}	2.2	0.7	≤ 1
Ethanol/Ethyl alcohol	$\leq 10^3$	1.2	1	0.15		1	
Fluorinerts	$\geq 10^3$	$\geq 10^5$	10	$\geq 10^{-1}$	10^{-2}	1	$\geq 10^2$
Glycerin/Glycerol	$\geq 10^3$	10^{-3}–1	10^3–10^6		10^{-3}	0.5	10^2–10^3
Helium				0.15		3.7	
Mercury and molten metals	10^4	1	10^{-1}	10	10^{-1}–10^{-2}	0.2	$\geq 10^{-2}$
Methanol/Methyl alcohol	$\leq 10^3$	1	$\geq 10^{-1}$	$\geq 10^{-1}$	$\geq 10^{-1}$	1	≤ 10
n-Octane	$\leq 10^3$	1	$\geq 10^{-1}$	$\geq 10^{-1}$	$\geq 10^{-1}$	1	≤ 10
Olive oil	10^3	10	10			0.7	
Silicon	2×10^3 (melting point)	10^{-7} N s/cm^2 (melting point)	3×10^{-3} (melting point)	10^2 (solid at room temperature)			
Silicone oils	$\leq 10^3$	1–10^3	1–10^3	10^{-1}	$\geq 10^{-2}$	1	10–10^3
Water (room temperature)	10^3	1	0.9	$\geq 10^{-1}$	10^{-3}	0.2	≤ 10

Pa (Pascal) = N/m^2 = kg/ms^2; Poise = 10^{-1} kg/ms; Stokes = 10^{-4} m^2/s.

TABLE 5 Approximate Values of Integral Heats of Solution

Solvent	Solute	T (°C)	Q (kJ/gmol solute)
Acetyl acetone	Water	25	−4.27
Aniline	Hexane	25	−4.90
Ethyl acetate	Water	25	−4.82
Hexane	Aniline	19	−5.53
Isobutanol	Water	25	−1.236

Source: A. Mendes-Tatsis.

temperature. Some alloys of silver and aluminum had a surface tension which, within a certain range, increased with T when the temperature was gradually raised, but during the subsequent cooling, σ monotonically increased on lowering T and, at every T, remained higher than the heating half-cycle. Presumably, the surfactant impurities were destroyed at the high temperatures and the sample was purer during the cooling period than during the preceding heating period. However, there are substances that, within a temperature interval, are liquid crystals that may have a maximum in that region and a minimum near the transition into the normal liquid state. Let us simply acknowledge that, generally, water–alcohol solutions, dodecylammonium (DAC), (molten) binary alloys, etc., exhibit a minimum in their surface tension as temperature is increased.

In a binary liquid system an initially equilibrated interface may have a change in interfacial tension if the desorption rate from the interface of the two components is different. Similar and related phenomena occur in ternary systems. Also thermal effects resulting from heat of solution may provide interfacial motions as a manifestation of the Marangoni effect. By producing a temperature change they alter locally the value of the interfacial tension. Furthermore, indirectly, they may affect the interfacial tension as a result of solubility variation with temperature. These effects are, generally, small in comparison with direct concentration-induced fluctuations of interfacial tension. However, they could be significant when the latter are absent or small, or when buoyancy is also affecting the system. Such may be the case for the interfaces ethyl acetoacetate/water (weak interfacial motions observed) or acetyl acetone/water (strong interfacial motions observed; interfacial tension about 6.2 mN/m, thus showing a sizeable mutual solubility) when a third chemical (like ethanol) transfers from one phase to another.

C. Scales and Dimensionless Groups

When a motionless horizontal liquid layer (depth, d) is heated from below two mechanisms may destabilize the conductive steady state at mechanical equilibrium. One is buoyancy due to the liquid thermal expansion, α, which for most liquids is positive thus leading to a heavy top stratification bound to be unstable (a notable exception is water at around 4°C as at this temperature $\alpha = 0$, becoming negative for lower temperature values). For a reference density ρ_0, buoyancy contributes with a force proportional to $\alpha\rho_0 g d^2$ in the presence of gravity (acceleration, g).

If the liquid layer is open to ambient air (or to another fluid) held at a (mean) temperature different than that at the layer's bottom, then surface tension would be altered according to local fluctuations of the temperature at the interface or free open surface. Local inhomogeneities of surface tension yield spontaneous flows (Marangoni effect) and hence surface tension gradients, $\partial\sigma/\partial x$, contribute with a force proportional to $|\mathrm{d}\sigma/\mathrm{d}T|$.

The ratio between buoyancy and surface-tension-gradient forces is the *dynamic* Bond number (also called *modified* Bond number)*

$$Bd \equiv \frac{\alpha\rho_0 g d^2}{-\mathrm{d}\sigma/\mathrm{d}T} \tag{8}$$

*The quantity g appearing in Eq. (8) and in subsequent equations refers to its actual (effective) value in a given environment. The term microgravity is commonly (albeit improperly) used for the free-fall conditions aboard a spacecraft like the International Space Station (ISS, now under construction) or in the various Spacelabs launched by NASA, ESA, and the former Soviet Union (Mir Station). Up there the gravitational acceleration is g save the difference due to the orbit being about 300 km from the Earth's surface. Then g is to be added to inertia and for a mass, M, we have $Mg - Ma = 0$ (friction forces are considered negligible and assured to vanish in a rarefied environment). Consequently, the "effective" gravity or effective weight vanishes. Indeed, this is why we see astronauts floating up in the Spacelab cabin. Free-fall conditions do not really occur undisturbed. Let us consider, for illustration, the platforms offered by the various space agencies around the world: (i) Drop towers (Madrid-INTA, Bremen-ZARM; the drop tower seems to have first been used in modern times in Bristol. In 1782 a patent was awarded to plumber William Watts, for "making small shot solid throughout, perfectly globular in form and without the dimples, scratches and other imperfections which other shot heretofore manufactured usually had on their surface"), drop tubes, and sounding rockets. The velocity vector is generally oriented along the gravity vector. Accordingly, vertical motion upwards occurs with decreasing velocity (deceleration) while downwards motion is with increasing velocity, save a free-fall interval; (ii) aircraft (KC135/Boeing 707, Caravelle, Airbus 330) in parabolic flight: normal (n) and along (s) the trajectory we have $g_n - V^2/R = 0$ and $g_s - \mathrm{d}V/\mathrm{d}t = 0$, the latter depending on the pilot's ability; (iii) platforms placed in circular orbits (taken as a point at distance R from the Earth's center) $g - V^2/R = 0$. In general, however, the right hand side is never strictly zero (astronauts move, there is some aerodynamic drag, motors do not operate smoothly, etc.). This means that in practice $g - a = g_R + g(t) \neq 0$, where g_R denotes a steady (or quasisteady) component miscalled residual $-g$ and $g(t)$ is a time-dependent (practically random) part called g-jitter. For the specific case of the ISS the following predictions have been made: for g_R, it is expected an aerodynamic drag about $10^{-7}g_0$, a radiation pressure about $10^{-8}g_0$ (g_0 is the g-value at sea level on Earth). For points distant from the ISS center of mass, and due to rotation an acceleration gradient is expected of about $10^{-7}g_0/m$. Pulse-like (single or compensating) accelerations, due to thruster firings ($10^{-4}\,g_0$), crew activities (10^{-2}–$10^{-3}\,g_0$), docking/berthing ($10^{-4}\,g_0$), micrometeorite impacts ($10^{-9}\,g_0$). Periodic, high frequency accelerations, due to on-board machineries and natural frequencies excited by internal or external actions ($10^{-6} < g/g_0 < 10^{-2}$; $0.1\,\mathrm{Hz} < f < 300\,\mathrm{Hz}$). The predicted value of g_R (for the ISS Assembly complete) is about $0.5 \times 10^{-6}\,g_0$ ($0.5\,\mu g$) for the USA Laboratory, 1–2 μg, for the ESA-COF microgravity facilities, and 2 μg for the Japanese module. The expected g-jitter is still under evaluation [36].

and hence one expects that the Marangoni effect is going to be the dominant agent for instability of the mechanical equilibrium when

$$\sigma_1 \equiv -\left(\frac{\mathrm{d}\sigma}{\mathrm{d}T}\right) > \alpha\rho_0 g d^2, \qquad -\rho_0\alpha\Delta T = \Delta\rho \tag{9}$$

which is a condition satisfied by thin enough liquid layers (d, small). For $\alpha \sim 10^{-3}/K$, $\rho \sim 1\ \mathrm{g/cm^3}$, $g \sim 10^3\ \mathrm{cm/s^2}$, and $\sigma_1 \sim 10^{-1}$ dyne/cmK (mN/mK), we have $Bd \sim 1$ for $d \sim 10^{-1}$ cm. Note that Eq. (8) may be significant even if the surface remains flat. Needless to say, viscosity tends to dampen the flow disturbances. On the other hand, heat diffusivity also does that, indirectly, by homogenizing the temperature at the surface. This damping will affect approximately equally both buoyancy and surface tension gradients. The dynamic Bond number as well as most of the dimensionless groups that follow below can be interpreted as ratios between forces, velocities, or time scales occurring in a problem [47,48]. We shall make use of one or another interpretation according to circumstances.

The Prandtl number, Pr, is given by

$$Pr = \frac{\nu}{\chi} \tag{10}$$

when heat is involved or

$$Pr_{\mathrm{D}} \equiv Sc = \frac{\nu}{D} \tag{11}$$

when mass diffusion occurs [when it is called the Schmidt number (Sc)]. Oils tend to have high Prandtl numbers (viscous effects dominate the flow), whereas liquid metals and helium have rather low Prandtl numbers (heat diffusivity is high and inertial effects tend to dominate the flow). If heat and mass transfer occur simultaneously, as in drying processes or other double-diffusive phenomena, the Lewis number, $Le = D/\chi$ (or its inverse), is of interest. Liquids tend to have Lewis numbers much smaller than unity because, generally, $D \ll \chi$.

The Galileo number, Ga, is given by

$$Ga = \frac{gd^3}{\nu\chi} \tag{12}$$

or with χ^2 in the denominator, or even $\widetilde{Ga} = gd^3/\nu^2$, if heat is not involved; hence $Ga = \widetilde{Ga}\,Pr$. Ga is sometimes called after Archimedes. Note that Ga is a dimensionless measure of gravity. When Ga has high values the response time of gravity is so short that deformational perturbations are quickly damped and hence the liquid surface can be considered effectively flat or better level with the corresponding equipotential of gravity. If, however, Ga has lower values, the viscous and/or heat diffusion times are very short, temperature is quickly made uniform and gravity is relatively so weak that deformational perturbations have time to grow. We shall discuss further its role after the Rayleigh and capillary (or crispation) numbers are defined below.

If the horizontal, lateral extent of the liquid layer is L, then

$$V_M \equiv \frac{-\left(\frac{d\sigma}{dT}\right)\Delta T\left(\frac{d}{L}\right)}{\eta} \tag{13}$$

is the velocity scale due to the Marangoni effect induced by the temperature difference, ΔT, across the layer depth, d. For simplicity, the subscript M is used rather than Ma. The quantity d/L (or its inverse) defines the aspect ratio of the system, called slenderness for liquid bridges. In most cases studied in this book the horizontal extent, L, of the fluid layer is much larger than its transverse depth, d, and hence the aspect ratio would be $d/L \ll 1$.

Two flow-related dimensionless Reynolds numbers can be formed. One is the usual Reynolds number

$$Re = \frac{V_M L}{\nu} \tag{14}$$

which measures the relative importance of inertial effects when compared with frictional, viscous effects in the flow. Generally, every change of scale (space and time) that conserves the Reynolds number carries incompressible flows that satisfy the Navier–Stokes equations into solutions of the same equations. If $Re(d/L)^2 \ll 1$ the flow will be so viscous-dominated that inertia will be negligible. Accordingly, the effects of the surface tension gradient will penetrate into the bulk liquid by viscosity and the liquid layer depth, d, should be the proper length scale along the orthogonal axis to the surface. If, however, $Re(d/L)^2 \gg 1 (d/L \gg Re^{-1/2})$ a boundary or skin layer flow occurs with viscous and inertia terms being of the same order therein. Then $\delta = (L/d)Re$ would be the appropriate length scale, much smaller than the liquid depth. The other group is the thermal Reynolds number or Peclet number.

$$Re_T = Pe = \frac{V_M L}{\chi} \tag{15}$$

which is also the (modified or specific) Marangoni number in the problem. Note that $Pe = Re\,Pr$. For liquids with Prandtl numbers different from unity, the velocity and temperature boundary/skin layer effects will be unequal.

If $\Delta T = \beta d$ and L is taken infinitely large, the Marangoni number is defined as

$$Ma = -\left(\frac{\mathrm{d}\sigma}{\mathrm{d}T}\right)\frac{\beta d^2}{\eta\chi} \tag{16}$$

which is the ratio of the capillary force due to the surface tension gradient to the viscous drag in the flow. It is also the ratio of the time scale for heat diffusion to the time scale for surface flow. Dealing with solutal surfactant problems the corresponding Marangoni number is also termed elasticity number, and is generally defined accounting for diffusion and adsorption processes. Levich [8] (see also Lucassen [14]) noted that if there are adsorbed surfactants, the quantity $(\partial\sigma/\partial C)C_0$ plays the part of a coefficient of elasticity in compression of the surfactant monolayer, where C denotes the excess surfactant surface concentration (C_0 is a reference value). Thus it is the explicit form of V_M that changes according to the gradient and transport processes involved in the problem. Definition (16) or an alternative expression is used according to the peculiarity of the problem treated.

Note that when *Re* and *Pe* have very large values, although inertial effects seem to predominate in the flow, dissipation is, in fact, quite large because of the Marangoni effect, which defines the velocity scale.

Other terms that shall be used in this book include the following. The *static* Bond number is

$$Bo \equiv \frac{(\Delta\rho)gd^2}{\sigma} \tag{17}$$

where $\Delta\rho$ denotes the density contrast between two immiscible liquids separated by the interface. *Bo* is sometimes called after Eötvös. For an air–liquid interface $\Delta\rho \sim \rho_{\mathrm{liquid}}$. The static Bond number is a measure of the hydrostatic head to capillary pressure, and is therefore an estimate of the relative importance of the gravitational and surface tension forces at rest. When its value is large, the capillary pressure effects are negligible in a liquid at equilibrium. Equation (17) defines the capillary length, l, when we set $B_0 = 1$,

$$l = \left(\frac{\sigma}{\Delta\rho g}\right)^{1/2} \tag{18}$$

Accordingly, the capillary length diverges to infinity when $\Delta\rho \to 0$ or $g \to 0$ (save the anomaly $\sigma \to \infty$). The former case was exploited by Plateau to suspend relatively big olive oil drops in a suitable mixture of alcohol and water having approximately the same density as the oil in an arrangement called the Plateau tank. Current experiments are carried out using ricin/castor oil in silicone oil. Thus it seems clear why rain drops are, generally, of millimeter size (in CGS units $\sigma \sim 10^2$, $\Delta\rho \sim 1$, $g \sim 10^3$). As earlier noted transverse surface waves are gravity waves when the wavelength is longer than the capillary length, and are ripples otherwise.

If the scale of stresses in the presence of surface tension is the viscous one the normal stress balance leads to a dimensionless group, the capillary or crispation number, denoted by

$$Ca = \frac{\sigma}{\eta V} \tag{19a}$$

which is the ratio of the surface tension to the viscous force. The capillary number is sometimes defined as Ca^{-1}. Our choice of Eq. (19a) has been dictated by the interest in treating *Ca* and *Ga* in the same way in the stress balances at boundaries. The capillary number determines the relative importance of the normal stress imbalance in deforming a fluid interface. If for V the value V_{M} is used, we have

$$Ca = \frac{-\sigma}{(\mathrm{d}\sigma/\mathrm{d}T)\beta d} \tag{19b}$$

and hence $Bd = Bo\,Ca$, where $Bd(Bd = \rho g d^2/\eta V_{\mathrm{M}})$, or its inverse, is sometimes known as the Stokes number as it measures the ratio of gravitational to viscous forces, a quantity of interest in the study of (Stokes) waves (recall the above given definition of the Galileo number). If the velocity scale is not V_{M} but $V = \chi/d$ then

$$Ca = \frac{\sigma d}{\eta \chi} \tag{19c}$$

and hence *Ca* appears as a dimensionless measure of the surface tension. At low values of the capillary number viscous forces fail to generate pressure capable of deforming the surface at equilibrium and if convection is initiated it does not alter its initial level equilibrium shape. For standard liquids $Ca \underset{\sim}{\sim} 10^5$ or higher. Note that with the new velocity scale we have $Pr/Ca = Bo/\widetilde{Ga}$ and $Ga = Bo\ Ca$ (earlier we saw this ratio equal to the *dynamic* Bond number). When $Ga \gg 1$ at fixed *Bo*, *Ca* is therefore high and σ is large. Alternatively, if we fix the value of *Ca* then *Bo* is large and this means that the effective value of g is large. On the other hand if $Ma/Ca = |(\mathrm{d}\sigma/\mathrm{d}T)|\beta d/\sigma \ll 1$ for a fixed value of the thermal constraint, the relative change in surface tension is small.

When inertial effects dominate over viscous effects, stress should be nondimensionalized with the inertial scale. Then the group that determines the relative importance of deformability is the Weber number (other authors use its inverse)

$$We = \rho V^2 d/\sigma \tag{20}$$

which is a measure ($We = Re/Ca$) of the relative influence of inertia and hydrodynamic pressure to the capillary pressure. It is particularly relevant for drop or bubble migration with significant surface deformation.

The Rayleigh number

$$Ra = \alpha g \beta d^4 / \nu \chi \tag{21}$$

gives the ratio of the buoyancy force to the viscous drag in thermal (also termed natural or free) convection. If the aspect ratio (d/L) significantly enters the problem the expression of Ra may be changed accordingly. It seems of interest to note that if using Eq. (13) we take $V_M = |(\mathrm{d}\sigma/\mathrm{d}t)| \Delta T / \eta \; (d = L)$ and consider the velocity scale produced by buoyancy, $V_R = \rho_0 \alpha \Delta T g d^2 / \eta$, the ratio V_R / V_M is independent of both η (or ν) and χ. Then for the typical values given earlier, from Eq. (9) $V_R / V_M \sim 10 d^2$ whose value is unity for $d \sim 10^{-1}$ cm. Accordingly, we expect that the flow is dominated by the Marangoni effect for (very) thin liquid layers ($d < 10^{-1}$ cm) and by buoyancy otherwise. Note that $Bd = Ra/Ma$. Then although Ra and Ma compare buoyancy with surface tension gradient effects, the group Bd does so whatever the value of the thermal gradient. Furthermore, as $Ra/Ma \sim d^2$ it appears that surface tension gradient effects are dominant for d small; otherwise buoyancy drives the motion.

In a stably stratified layer the square root of Ra provides the Brunt–Väisälä frequency for oscillatory motions (internal waves). It can be interpreted as the ratio of the three time scales involved in natural convection, the time scale associated with the buoyancy-driven flow, and the scales for the decay of temperature or momentum nonuniformity.

Another quantity used in heat transfer problems is the Grashof number

$$Gr = \alpha g \beta d^4 / \nu^2 \tag{22}$$

which is a parameter indicating the relative importance of inertial and viscous forces in thermal convection, but it is not a true ratio of two forces. Some authors use $\hat{G}r = Gr(L/d)$. Note that $Ra = Pr\, Gr$ and thus the Rayleigh number places heat diffusion and momentum (vorticity) diffusion on an equal footing.

The energy flux delivered to or from a liquid to a solid (or to another liquid or gas) has been observed to be proportional to the temperature difference between the two (termed Newton's law of cooling: a body cools down at a rate proportional to the difference in temperature between the body and the surroundings, provided that the temperature of the body is higher than the temperature of the surroundings). It is a phenomenon which occurs at the interface and, in that sense, is a boundary condition. Fourier's law applies at both sides of a solid–liquid interface, but using Newton's law at the liquid side of the interface eliminates the need for solving Fourier's equation in that phase. A detailed discussion of this problem can be found in Refs 49–51. In summary, the Biot number is the useful (ad hoc) group that accounts for the boundary condition

$$Bi = \frac{\kappa_{\mathrm{wall}} d}{\kappa_{\mathrm{liquid}} d_{\mathrm{wall}}}, \tag{23}$$

or $Bi = qd/\kappa$, where q denotes the rate of change with temperature of the heat flux from the liquid and κ its heat conductivity. Definition (23) corresponds to the case of a thin enough solid support. Otherwise the solution of Fourier's heat diffusion in the solid wall needs to be taken into account. Equation (23) expresses the ratio of heat transfer from the body (solid or liquid) to the fluid to heat transfer within the body (solid or liquid). The value $Bi \to \infty$ defines the perfect heat conducting boundary where temperature is held fixed. Copper belongs to this class and glass tends to being a good rather than a bad heat conductor as the ratio $\kappa_{\text{glass}}/\kappa_{\text{oil}}$ is about 10 (51). At an insulating boundary (adiabatic for heat flux disturbances) the imposed heat flux is held fixed, and $Bi = 0$ for heat flux disturbances. Intermediate values of the Biot number define a mixed (also called Robin) condition. For a silicone oil–air interface the value of the Biot number can be taken to be vanishing when the air layer is thick enough relative to a small enough liquid layer depth.[†]

The Nusselt number will also be used

$$Nu = \frac{Hd}{\kappa \Delta T} \tag{24}$$

which is a dimensionless measure of the overall heat transfer including convective and diffusive components relative to the diffusive transfer. H denotes the overall heat transfer per unit surface and κ is the thermal conductivity in the liquid. In the absence of flow the Nusselt number value is unity.

Finally, let us mention the Mach number, which is the ratio of flow velocity to the velocity of sound in the medium. We shall be considering relatively slow convective phenomena in incompressible liquids and hence we shall be working in the realm of vanishing Mach numbers.

II. GOVERNING EQUATIONS AND BOUNDARY CONDITIONS

We shall consider flows obeying the standard thermohydrodynamic equations. For basic material and introductory texts we refer again to the monographs by Aris [39], de Groot and Mazur [57], and by Edwards et al. [41]. Particularly relevant are the monograph by Birkhoff [58], and the careful study by Ostrach [21], and refs 59–61 are also of interest. Pismen [62] and Pomeau [63] have recently provided a breakthrough in our understanding of the fluid–solid boundary conditions (contact angle and related matters) with due consideration of microscopic (van der Waals, Derjaguin) forces, but we shall not consider them here.

[†]In experiments conducted by Koschmieder and Swinney and their collaborators [54–56] the following parameter values were used (in W/mK units): $\kappa_{\text{silicon crystal}} = 154$, $\kappa_{\text{sapphire}} = 36.7$, $\kappa_{\text{copper}} = 393$, $\kappa_{\text{glass}} = 1.74$, $\kappa_{\text{silicone oil}} = 0.16$, and $\kappa_{\text{air}} = 0.025$. Note: 1 cal = 4.1868 J, 1 J = 10^7 erg, W = J/s.

The continuity equation for the density, ρ, is

$$\frac{\partial \rho}{\partial t} = -\text{div}\ \rho \mathbf{v} \tag{25}$$

taken in the incompressible approximation ($\mathrm{d}\rho/\mathrm{d}t = 0$)

$$\text{div}\ \mathbf{v} = 0, \qquad (\text{or } \nabla \cdot \mathbf{v} = 0) \tag{26}$$

and the velocity field is solenoidal. Later we shall allow ρ to vary in the contribution of a body force like buoyancy (or inertia) only if it plays a key role in the problem.

For isotropic Newtonian liquids, the standard case and the only case to be considered in this book, the viscous stress tensor, Π, has the following components

$$\left\{ \Pi_{xx} = 2\eta \frac{\partial \mathbf{v}_x}{\partial x}, \qquad \Pi_{yy} = 2\eta \frac{\partial \mathbf{v}_y}{\partial y}, \right.$$

$$\Pi_{zz} = 2\eta \frac{\partial \mathbf{v}_z}{\partial z}, \qquad \Pi_{xy} = \Pi_{yx} = \eta\left(\frac{\partial \mathbf{v}_x}{\partial y} + \frac{\partial \mathbf{v}_y}{\partial x}\right),$$

$$\left. \Pi_{yz} = \Pi_{zy} = \eta\left(\frac{\partial \mathbf{v}_y}{\partial z} + \frac{\partial \mathbf{v}_z}{\partial y}\right), \qquad \Pi_{zx} = \Pi_{xz} = \eta\left(\frac{\partial \mathbf{v}_z}{\partial x} + \frac{\partial \mathbf{v}_x}{\partial z}\right)\right\}.$$

Note that Π, $\boldsymbol{T}$ or $\boldsymbol{\tau}$ would be used to denote the stress tensor, although $\boldsymbol{\tau}$ may denote a unit tangent vector.

The Navier–Stokes equations are a consequence of Newton's law of motion for a continuum,

$$\frac{\partial \mathbf{v}}{\partial t} + \mathbf{v} \cdot \nabla \mathbf{v} = -\frac{1}{\rho} \nabla p + \nu \nabla^2 \mathbf{v} + \boldsymbol{f}_{\text{ext}} \tag{27a}$$

where $\boldsymbol{f}_{\text{ext}}$ accounts for an external (density) body force like gravity, where, for example, $\boldsymbol{f}_{\text{ext}} = \boldsymbol{g}$. According to circumstances we shall use ∇^2 or $\Delta = \nabla^2$. In most cases considered in this book, buoyancy would be disregarded while maintaining the effect of gravity at an interface (hydrostatic contribution in boundary conditions). The role of the Reynolds number is now clear as the term $\mathbf{v} \cdot \nabla \mathbf{v}$ scales as v^2/d whereas $\nu \nabla^2 \mathbf{v}$ scales as $\nu \mathrm{v}/d^2$ and hence their ratio is just Re. Accordingly, low Reynolds number flows are dominated by viscosity and although they are kinematically reversible, thermodynamically they are not so. At high Reynolds numbers flows are mostly inertial.

The kinematic viscosity, ν, is best suited to account for diffusion of vorticity. Indeed, the degree of circulation in a region of a fluid can be expressed by $\boldsymbol{\omega} = \text{rot}\, \mathbf{v} = \text{curl}\, \mathbf{v} = \nabla \times \mathbf{v}$ (vector product). Flows with zero vorticity everywhere are irrotational and then the velocity potential can be expressed as the gradient of a scalar (potential flow). Using the Navier–Stokes equations become

$$\frac{\partial \boldsymbol{\omega}}{\partial t} + \mathbf{v} \cdot \nabla \boldsymbol{\omega} = \nu \nabla^2 \mathbf{v} + \boldsymbol{\omega} \cdot \nabla \mathbf{v} \tag{27b}$$

Later we shall not use $\boldsymbol{\omega}$ for the vorticity as ω will be denoting frequency. The quantity $\boldsymbol{\omega} \cdot \nabla \mathbf{v}$ in Eq. (27b) is responsible for the stretching of vortex lines with consequent production of vorticity. It is a source term in Eq. (27b) as the body force ($\boldsymbol{f}_{\text{ext}}$) is a source of momentum in Eq. (27a). For steady flows at zero Reynolds number (called after Stokes or creeping flows) we have

$$\nabla^2 \boldsymbol{\omega} = 0 \tag{27c}$$

which shows that the flow is due solely to steady diffusion of vorticity. Solid boundaries are a source of vorticity because of the no-slip condition obeyed by the velocity field (d'Alembert–Prandtl). Note that whereas the dynamic (shear) viscosity, η, is much smaller for gases than for liquids, the opposite is true for the kinematic viscosity, ν.

The heat (Fourier) diffusion equation is

$$\frac{\partial T}{\partial t} + \mathbf{v} \cdot \nabla T = \chi \nabla^2 T$$
$$+ \text{ energy source or sink} \quad \text{(heat due to flow viscosity)} \tag{28}$$

which shows the role of the Peclet number ($Pe = Re_T$) as the term $\mathbf{v} \cdot \nabla T$ scales as $\Delta T/d$ and $\chi \nabla^2 T$ as $\chi T/d^2$ and hence Pe gives the ratio of the former to the latter. We also see the role of the Prandtl number, Pr. Indeed, by suitably choosing the scales of time (d^2/ν or d^2/χ) and velocity (ν/d or χ/d), etc., for $Pr \to 0$, Eq. (28) can be changed to lose its time derivative and inertial term. However, for $Pr \to \infty$, it is Eq. (27a) that loses them. In the former case velocity slaves the evolution of the "frozen" thermal field while in the latter it is the evolution of the temperature field that determines that of the frozen flow velocity.

When mass transport is involved we instead use the mass (Fick) diffusion equation

$$\frac{\partial C}{\partial t} + \mathbf{v} \cdot \nabla C = D \nabla^2 C \tag{29}$$

where C (or Γ in other situations) denotes mass concentration. If chemical reactions are involved it is necessary to account for the various species involved in the problem.

The above recalled equations must obey boundary conditions that in general terms are expressed as follows. At a solid–liquid boundary if no-slip occurs then

$$\mathbf{v} \cdot \boldsymbol{\tau} = 0 \tag{30}$$

with $\boldsymbol{\tau}$ denoting a unit tangent vector to the surface. The no-penetration condition demands

$$\mathbf{v} \cdot \mathbf{n} = 0 \tag{31}$$

where $\mathbf{n}$ is the unit normal vector to the surface.

For a fluid–fluid interface we shall consider various particular cases. If the interface is defined by the equation

$$\xi(x,t) = 0$$

then the unit normal vector to it is $\mathbf{n} \equiv \nabla\xi/|\nabla\xi|$. If fluids are denoted with subscript i ($i = 1{,}2$), when there is no matter transfer we have

$$\mathbf{V}_1 \cdot \mathbf{n} = \mathbf{V}_2 \cdot \mathbf{n} = -\frac{(\partial\xi/\partial t)}{|\nabla\xi|} \tag{32a}$$

If the location of the interface does not change with time the previous equation reduces to

$$\mathbf{V}_1 \cdot \mathbf{n} = \mathbf{V}_2 \cdot \mathbf{n} = 0 \tag{32b}$$

The tangential velocity components on the two sides of the interface are equal to each other. Hence the vector velocity is continuous across the interface

$$\mathbf{V}_1 = \mathbf{V}_2 \tag{33}$$

For a deformable interface the stresses (together denoted here by Π) must account for surface tension in the form of a jump condition (Laplace) and if this surface tension is not constant then the Marangoni effect must be accounted for in a corresponding jump condition. Considering only pressure and the surface tension forces we obtain

$$\mathbf{n} \cdot (\Pi_1 - \Pi_2) = \sigma\mathbf{n}(\text{div } \mathbf{n}) - \nabla_s\sigma \tag{34a}$$

with $\nabla_s[= \nabla - \mathbf{n}(\mathbf{n} \cdot \nabla)]$ denoting the gradient operator taken along the interface. For a level flat surface the normal stress would be continuous. The tangential stress would be continuous if there is no surface tension inhomogeneity. Note that $\text{div}\mathbf{n} = (1/R_1 + 1/R_2)$ with $R_i(i = 1{,}2)$ denoting the principal radii of curvature and hence Eq. (34a) split into normal and tangential parts yields

$$(\Pi_1 - \Pi_2) : \mathbf{nn}(p_1 - p_2) = -\sigma(1/R_1 + 1/R_2) \tag{34b}$$

$$(\Pi_1 - \Pi_2) : \mathbf{n}\boldsymbol{\tau} = -\nabla_s\boldsymbol{\sigma} \cdot \boldsymbol{\tau} \tag{34c}$$

If the surface tension changes with temperature

$$\nabla_s\sigma = \frac{d\sigma}{dT}\nabla_s T \tag{35}$$

and a similar expression holds for surfactant concentration (see, however, [10,14,41,59,64–67]). In the simplest case of double diffusion or where only surfactants are included in the problem in an isothermal system, all above given quantities can be redefined with the pertinent changes: $\chi \to D$, $(\mathrm{d}\sigma/\mathrm{d}T) \to (\mathrm{d}\sigma/\mathrm{d}\Gamma)$, $\beta \to \beta_\Gamma(\Delta T \to \Delta\Gamma)$ and D, Γ, and β_Γ denoting, respectively, mass diffusivity, surfactant concentration and its corresponding gradient.

Equation (34c) clearly expresses that when temperature is nonuniform along an interface, the fluids adjoining it must move because the viscous stress is only zero in fluids at rest. It would not be possible to satisfy the jump balance in tangential stress at the interface if the fluids on both sides of the interface are quiescient. On the other hand, the tangential stress balance in Eq. (34c) links the heat (Fourier) or the mass (Fick) diffusion equation to the Navier–Stokes equations. Thus for liquids with high Prandtl numbers ($\nu \gg \chi$) such as the silicone oils, if the Marangoni effect is strong enough there can be strongly dissipative interfacial turbulence, which is at variance with the (mostly inertial) turbulence in the sense of Reynolds and Kolmogorov.

At a solid–fluid interface we may impose uniform temperature

$$T = \text{constant} \tag{36}$$

or constant heat flux

$$\mathbf{n} \cdot \nabla T = \text{constant} \tag{37}$$

or a combination of both using the corresponding Biot number

$$\mathbf{n} \cdot \nabla T = \pm Bi\ T \tag{38}$$

and hence $Bi = 0$ is a von Newmann (so-called "insulating") condition while $Bi \to \infty$ defines a perfect conductor with the Dirichlet condition. The + or − sign depends on the sign of the gradient and $Bi = 0$ refers to the heat flux disturbances, as earlier noted.

Using the approximations underlying the above given thermohydrodynamic equations we shall apply them in the Boussinesq–Oberbeck (BO) approximation [60,68–71]. In short, this approximation amounts to considering all parameters like viscosity, heat diffusivity, etc., as having values not significantly altered by the action of the (not too strong) thermal gradients. This is only considered when acting upon key agents that, like buoyancy, drive the flow. For instance, density changes due to thermal expansion are incorporated as body force in the Navier–Stokes equations [Eqs (27a–c)] in the presence of gravity. However, the incompressibility condition ($\mathrm{d}\rho/\mathrm{d}t = 0$) is maintained. In the heat equation we shall neglect the contribution of frictional viscous heating. As noted earlier, in the main portion of the book the liquid layers will be assumed to be shallow enough, i.e., $(d/L) \ll 1$, and the thermal gradients not too high ($\alpha \sim 10^{-3} - 10^{-4}$,

$\Delta T = 10\,\mathrm{K}$, $\alpha\Delta T \ll 1$, $\rho g \chi_{\mathrm{T}} d \ll 1$, $\alpha g d/C_{\mathrm{p}} \ll 1$, where C_{p} is the specific heat and χ_{T} the isothermal compressibility).

For illustration, let us now particularize the thermohydrodynamic equations and boundary conditions for various illustrative cases to be studied at length in this book.

A. Thermal Convection

Take, for instance, a liquid layer heated by its solid support. Then if the liquid thermal expansion, α, is small

$$\rho = \rho_0(1 - \alpha T) \tag{39}$$

In the BO approximation the Navier–Stokes equations reduce to

$$\frac{\partial \mathbf{v}}{\partial t} + \mathbf{v} \cdot \nabla \mathbf{v} = -\frac{1}{\rho_0} \nabla p + \nu \nabla^2 \mathbf{v} - \alpha T \boldsymbol{g} \tag{40}$$

while the heat equation reduces to Eq. (28) with no heat production terms. The continuity equation is just the incompressibility condition in Eq. (26).

At the solid surface with no slip we have

$$\mathbf{v} = 0 \tag{41}$$

At the open surface of the liquid layer in the linear approximation

$$\mathbf{v} = \frac{\partial \xi}{\partial t} \tag{42}$$

with

$$p + \sigma(1/R_1 + 1/R_2) - 2\eta \frac{\partial v_{\mathrm{n}}}{\partial n} = 0 \tag{43}$$

and

$$\eta\left(\frac{\partial \mathrm{v}_{\mathrm{n}}}{\partial \tau} + \frac{\partial \mathrm{v}_{\tau}}{\partial n}\right) = \frac{\partial \sigma}{\sigma T}\frac{\partial T}{\partial \tau} \tag{44}$$

for the normal and tangential stress balances, respectively, where v_{τ} and v_{n} denote velocity components, respectively, along $\boldsymbol{\tau}$ and $\mathbf{n}$, the tangent and (outward pointing) normal vectors to the surface. The ambient air plays no mechanically active role as, generally, $\eta_{\mathrm{air}} \ll \eta_{\mathrm{liquid}}$.

In terms of the Rayleigh and Galileo numbers, when buoyancy in a liquid layer is acting together with the Marangoni effect at its deformable surface, if $Ra \sim 1$, as the BO approximation demands $\alpha\beta d \ll 1$ (generally, as earlier noted, both β and d would be considered small), then $Ra \ll Ga$. Thus as noted earlier surface deformation is negligible for relatively high values of Ga. For low values of *Ga*, surface deformation may be so important that it becomes an effect beyond the domain of applicability of the BO approximation. Accordingly the problem demands consideration of the full thermohydrodynamic equations, particularly the continuity equation beyond incompressibility, and therefore $\mathrm{d}\rho/\mathrm{d}t \neq 0$ [35].

B. High-Frequency Vibrations

Let us consider the influence of linear high-frequency oscillations (amplitude b, frequency Ω) and let the vibrations be such that in the reference frame attached to the container a liquid layer (depth d) with homogeneous liquid density is motionless (no high-frequency mixing exists). An inhomogeneous mass force capable of setting the liquid in motion is generated simultaneously with the variation in the density of the liquid, for example, as a result of its thermal expansion, $\alpha \neq 0$. Assuming that the vibration period is much smaller than all characteristic hydrodynamic time scales and the sound velocity is practically unattainable, then the BO approximation can be applied [72].

Using the method of averaging over fast time we obtain a closed system of equations for a slow flow component [72,73]. Substituting the superposition of the average and pulsational components into Eqs (26) and (40) and selecting the leading fast periodic terms we obtain

$$\operatorname{div} \mathbf{v}' = 0 \tag{45}$$

$$\frac{\partial \mathbf{v}'}{\partial t} = -\frac{1}{\rho}\nabla p' + \mathbf{n}\bar{T}\alpha b\Omega^2 \cos \Omega t \tag{46}$$

$$\frac{\partial T'}{\partial t} + \mathbf{v}' \cdot \nabla \bar{T} = 0 \tag{47}$$

In Eq (46) the potential part of the inertial force is equilibrated by the pressure gradient, while the solenoidal part generates a pulsational flow, **w**. Thus, the fast velocity component obeys the following equation:

$$\frac{\partial \mathbf{v}'}{\partial t} = \mathbf{w}\alpha b\Omega^2 \cos \Omega t, \tag{48}$$

with

$$\operatorname{div} \mathbf{w} = 0, \qquad \operatorname{rot} \mathbf{w} = \operatorname{rot}(\mathbf{n}\bar{T}). \tag{49}$$

Equation (48) contains a new slow function, which is described by Eq. (49). This equation has been incorporated into equations for the average component of the motion.

Integrating Eqs (47) and (48) over fast time, it follows that

$$\mathbf{v}' = \mathbf{w}\alpha b\Omega \sin \Omega t, \qquad T' = \mathbf{w}\nabla\bar{T}\alpha b \cos \Omega t. \tag{50}$$

Substituting $\mathbf{v}'$ and T' into the complete system of equations and averaging them with respect to the vibrational period, the equations for the averaged fields are obtained, giving

$$\operatorname{div} \mathbf{v} = 0 \tag{51}$$

$$\frac{\partial \mathbf{v}}{\partial t} + (\mathbf{v} \cdot \nabla)\mathbf{v} = -\frac{1}{\rho_0}\nabla p + \nu\Delta\mathbf{v} - \mathbf{g}\alpha T + \frac{1}{2}(\alpha b\Omega)^2(\mathbf{w} \cdot \nabla)(T\mathbf{n} - \mathbf{w}) \tag{52}$$

$$\frac{\partial T}{\partial t} + \mathbf{v} \cdot \nabla T = \chi\Delta T \tag{53}$$

$$\operatorname{div} \mathbf{w} = 0, \qquad \operatorname{rot} \mathbf{w} = \nabla T \times \mathbf{n}. \tag{54}$$

(overbars have been omitted).

The problem now involves a new dimensionless parameter, which characterizes the intensity of high-frequency vibration effect. This parameter is the vibration analog of the earlier introduced (thermal) Rayleigh number

$$Ra_{\mathrm{v}} = \frac{(\alpha b\Omega\Theta d)^2}{2\nu\chi} \tag{55}$$

where the boundary conditions for the averaged velocity and temperature are similar to those used in the earlier treated case of thermal convection.

It should be noted that the boundary condition for the function $\mathbf{w}$, which is the amplitude of the pulsatory velocity component, must be formulated under the constraint that the pulsatory motion is described in the inviscid approximation. Therefore, on the solid boundary it is necessary to impose a condition ensuring that there is no flow across it, hence $\mathbf{w}_{\mathrm{n}} = 0$. The boundary condition on the free surface is defined bearing in mind that equations are derived under the assumption that the isothermal liquid subject to vibrations is motionless with respect to the walls of the container. In a system with a free surface such a requirement can be satisfied if the vibrations are normal to the plane of the free surface. On this surface it may occur that either the surface tension on the boundary is large and the surface is not deformed by vibrations, hence $\mathbf{w}_{\mathrm{n}} = 0$, or the surface can be easily deformed, so the pulsatory pressure at the liquid boundary remains constant and the tangential component of the pulsatory velocity vanishes and then $\mathbf{w}_{\tau} = 0$.

C. Thin Permeable Partitions or Dividers

Let us now consider the influence of a thin permeable grid-type partition (also called a divider) on the flow of a viscous liquid. A distinctive feature of these partitions is that they have a pronounced effect on large-scale liquid motion, leaving the processes of molecular transfer practically unaltered. Permeable partitions placed in the cavity may greatly affect the threshold of convective instability in the system.

Let a permeable partition be modeled as an array of regularly spaced elements inserted in the bulk of a thin layer. In the description of liquid flows over a distance much greater than a grid spacing the boundary conditions for the flow problem can be formulated in terms of the grid integral characteristics [74,75].

Here the hypothesis for velocity continuity of the flow passing through the partition seems to be more plausible since on solid grid elements the velocity becomes zero at both sides while in the free regions it is continuous:

$$\mathbf{v}^{+} = \mathbf{v}^{-} \tag{56}$$

where the symbols '+' and '−' denote values of functions at different sides of partition.

Moreover, for slow flows it seems reasonable to assume that the force of the flow acting on the partition is proportional to the flow rate at the partition:

$$(\Pi_{ik}^{+} - \Pi_{ik}^{-})\mathbf{n}_k = \alpha_{ik}\mathrm{v}_k, \tag{57}$$

where Π_{ik} is the stress tensor, $\mathbf{n}$ is the normal to the boundary, v_k is the velocity projection on the partition, α_{ik} is the partition resistance tensor (note that here we use the symbol α to characterize the partition although, generally, α will designate the thermal expansion coefficient). This relation also takes into account that the flow velocity near the partition is continuous.

In the system of coordinates, $(\mathbf{n}, \boldsymbol{\tau}_{1,2})$, at the boundary the resistance tensor is diagonal and specified by three values: one normal resistance, α_{n}, and two tangential resistances, α_{τ_1} and α_{τ_2}. Generally, we shall consider that the partition is isotropic and hence $\alpha_{\tau_1} = \alpha_{\tau_2} = \alpha_{\tau}$.

With the above given assumptions Eq. (57) becomes

$$p^{-} = p^{+} = \alpha_{\mathrm{n}}\mathrm{v}_{\mathrm{n}} \tag{58}$$

and

$$\left(\frac{\partial \mathrm{v}_\tau}{\partial \mathbf{n}} + \frac{\partial \mathrm{v}_{\mathrm{n}}}{\partial \tau}\right)^{+} - \left(\frac{\partial \mathrm{v}_\tau}{\partial \mathbf{n}} + \frac{\partial \mathrm{v}_{\mathrm{n}}}{\partial \tau}\right)^{-} = \frac{\alpha_\tau \mathrm{v}_\tau}{\eta}. \tag{59}$$

The ratio, η/d, of the dynamic (shear) viscosity, η, to the characteristic cavity dimension, d, can be taken as the unit for resistance values. The dimensionless resistances will be denoted as α_{n} and α_τ. Incidentally, in a porous medium the diffusion of momentum is much smaller than the heat diffusivity and thus the Prandtl number tends to zero (as it occurs in the Hele–Shaw constrained geometry [51]).

REFERENCES

1. Marangoni, C. *Sull'espansione Delle Gocce di un Liquido Gallegianti Sulla Superficie di Altro Liquido*; Tipografia dei fratelli Fusi: Pavia, 1865.
2. Marangoni, C. Sull'espansione delle gocce sulle superficie liquide. Nuovo Cimento (Ser. 2) **1870**, *3*, 105.
3. Marangoni, C. Ueber die Ausbreitung der Tropfen einer Flussigkeit auf der Oberfläche einer anderen. Ann. Phys. Chem. (Poggendorff) **1871**, *143*, 337–354. (The title in German is the translation of the title of the monograph published in Pavia, 1865).
4. Marangoni, C. Stefanelli, P. Monografia sulle bolle liquide. Nuovo Cimento (Ser. 2) **1873**, *7–8*, 301–356; *9*, 236–256.
5. Marangoni, C. Difesa della teoria dell'elasticità superficiale dei liquidi. Plasticità superficiale. Nuovo Cimento (Ser. 3) **1878**, *3*, 50–68, 97–115, 193–211.
6. Plateau, J. *Statique Experimentale et Theorique des Liquides Soumis aux Seules Forces Moleculaires*; Gauthier-Villars: Paris, 1873, Vol. 1.
7. Tanford, Ch., Franklin, B. *Stilled the Waves. An Informal History of Pouring Oil on Water with Reflections on the Ups and Down of Scientific Life in General*; Duke Univ. Press: Durham, N.C., 1989.
8. Levich, V.(B.)G. *Physicochemical Hydrodynamics*; Prentice-Hall: Englewood Cliffs, N.J., 1962.
9. Levich, V.(B.)G., Krylov, V.S. Surface-tension-driven phenomena. Annu. Rev. Fluid Mech. **1969**, *1*, 293–316.
10. Sternling, C.V., Scriven, L.E. Interfacial turbulence: Hydrodynamic instability and the Marangoni effect. AIChE J. **1959**, *5*, 514–523.
11. Scriven, L.E., Sternling, C.V. The Marangoni effects. Nature (London) **1960**, *187*, 186–188.
12. Ross, S., Becher, P. The history of the spreading coefficient. J. Colloid Interface Sci. **1992**, *149*, 575–579.
13. Davies, J.T., Rideal, E.K. *Interfacial Phenomena*; Academic Press: New York 1963.
14. Lucassen, J. Longitudinal capillary waves. Part 1. Theory. Trans Faraday Soc. **1968**, *64*, 2221–2229; Part 2. Experiments. Trans Faraday Soc. **1968**, *64*, 2230–2235.
15. Lucassen-Reynders, E.H., Lucassen, J. Properties of capillary waves. Adv. Colloid Interface Sci. **1969**, *2*, 347–395.
16. Lucassen, J. Effect of surface-active material on the damping of gravity waves: a reappraisal. J. Colloid Interface Sci. **1982**, *85*, 52–58.
17. Linde, H. Marangoni instabilities. In *Convective Transport and Instability Phenomena*; Zierep, J., Oertel Jr., H. Eds.; Braun-Verlag: Karlsruhe, 1982; 265–296.
18. de Gennes, P.G. Wetting: statics and dynamics. Revs. Mod. Phys. **1985**, *57*, 827–863.
19. de Gennes, P.G., Brochard-Wyart, F., Quere, D. *Gouttes, Bulles, Perles et Ondes*; Belin: Paris, 2002.
20. Kenning, D.B.R. Two-phase flow with non-uniform surface tension. Appl. Mech. Revs. **1968**, *21*, 1101–1111.
21. Ostrach, S. Motion induced by capillarity. In *Physicochemical Hydrodynamics*; Spalding, B., Ed.; Advance Publications: London, 1977; Vol. 2, 571–589.

22. Ostrach, S. Low-gravity fluid flows. Annu. Rev. Fluid Mech. **1982**, *14*, 313–345.
23. Ostrach, S. Fluid mechanics in crystal growth. The 1982 Freeman Scholar Lecture. J. Fluids Engng., Trans. ASME **1983**, *105*, 5–20.
24. Berg, J.C., Acrivos, A., Boudart, M. Evaporative convection. Adv. Chem. Engng. **1966**, *6*, 61–123.
25. Davis, S.H. Thermocapillary instabilities. Annu. Rev. Fluid Mech. **1987**, *19*, 403–435.
26. Myshkis, A.D., Babskii, V.G., Kopachevskii, N.D., Slobozhanin, L.A., Tyuptsov, A.D. *Low-Gravity Fluid Mechanics. Mathematical Theory of Capillary Phenomena*; Springer-Verlag, Berlin: 1987.
27. Wozniak, G., Siekmann, J., Srulijes, J. Thermocapillary bubble and drop dynamics under reduced gravity—survey and prospects. Z. Flugwiss. Weltraumforsch. **1988**, *12*, 137–144.
28. Grotberg, J.B. Pulmonary flow and transport phenomena. Annu. Rev. Fluid Mech. **1994**, *26*, 529–571.
29. Straub, J. The role of surface tension for two-phase heat and mass transfer in the absence of gravity. Exp. Therm. Fluid Sci. **1994**, *9*, 253–273.
30. Quere, D. Fluid coating on a fiber. Annu. Rev. Fluid Mech. **1999**, *31*, 347–384.
31. Velarde, M.G., Ed. *Physicochemical Hydrodynamics. Interfacial Phenomena*; Plenum Press: New York, 1988.
32. Hondros, E.D., Mc Lean, M., Mills, K.C., Eds. *Marangoni and Interfacial Phenomena in Materials Processing*; Royal Society: London, 1998.
33. Velarde, M.G., Zheytounian, R.Kh., Eds. *Interfacial Phenomena and the Marangoni Effect*; Springer-Verlag/CISM: Udine, 2002.
34. Zeytounian, R.Kh. The Benard–Marangoni thermocapillary-instability problem. Phys.-Uspekhi **1998**, *41*, 241–267.
35. Velarde, M.G., Nepomnyashchy, A.A., Hennenberg, M. Onset of oscillatory interfacial instability and wave motions in Benard layers. Adv. Appl. Mech. **2000**, *37*, 167–238.
36. Monti, R., Ed. *Physics of Fluids in Microgravity*; Taylor & Francis: London, 2001.
37. Subramanian, R.S., Balasubramanian, R. *The Motion of Bubbles and Drops in Reduced Gravity*; Cambridge University Press: Cambridge, 2001.
38. Rey, A.D. Tension gradients and Marangoni flows in nematic interfaces. Phys. Rev. E **1999**, *60*, 1077–1080.
39. Aris, R. *Vectors, Tensors and the Basic Equations of Fluid Mechanics*; Prentice-Hall: Englewood Cliffs, N.J., 1962.
40. Slattery, J.C. *Interfacial Transport Phenomena*; Springer-Verlag: New York, 1990.
41. Edwards, D.A., Brenner, H., Wasan, D.T. *Interfacial Transport Processes and Rheology*; Butterworth-Heineman: Boston, 1991.
42. Gatignol, R., Prud'homme, R. *Mechanical and Thermodynamical Modeling of Fluid Interfaces*; World Scientific: London, 2001.
43. Korteweg, D.J. Sur la forme que prennent les equations du mouvement des fluides si l'on tient compte des forces capillaires causées par des variations de densité considerables mais connues et sur la theorie de la capillarité dans l'hypothese d'une variation continue de le densité. Arch. Neerland. Sci. Exactes Naturelles **1901**, *6*, 1–24.

44. Zeldovich, Y.B. About the surface tension of the boundary between two mutually soluble liquids. Zhur. Fiz. Khim. **1949**, *23*, 931–935 (in Russian).
45. Smith, P.G., Van de Ven, T.G.M., Mason, S.G. The transient interfacial tension between two miscible fluids. J. Colloid Interface Sci. **1981**, *80*, 302–303.
46. Kojima, M., Hinch, E.J. Acrivos, A. The formation and expansion of a toroidal drop moving in a viscous fluid. Phys. Fluids **1984**, *27*, 19–24.
47. Joseph, D.D. Fluid dynamics of two miscible liquids with diffusion and gradient stresses. Eur. J. Mech. B/Fluids **1990**, *9*, 565–596.
48. Petitjeans, P. Une tension de surface pour les fluides miscibles. C. R. Acad. Sci. Paris **1996**, *322*, 673–679.
49. Volpert, V.A., Pojman, J.A., Texier-Picard, R. Convection induced by composition gradients in miscible systems. C. R. Mecanique **2002**, *330*, 353–358.
50. Lykoudis, P.S. Nondimensional numbers as ratios of characteristic times. Int. J. Heat Mass Transfer **1990**, *33*, 1568–1570.
51. Guyon, E., Hulin, J.P., Petit, L., Mitescu, C. *Physical Hydrodynamics*; Oxford University Press: Oxford, 2001.
52. Normand, C., Pomeau, Y., Velarde, M.G. Convective instability: a physicist's approach. Revs. Mod. Phys. **1977**, *49*, 581–624.
53. Cross, M.C., Hohenberg, P.C. Pattern formation ousdide of equilibrium. Revs. Mod. Phys. **1993**, *65*, 851–1112.
54. VanHook, S.J., Schatz, M.F., Swift, J.B., McCormick, W.D., Swinney, H.L.: Long-wavelength surface-tension-driven Benard convection: experiment and theory. J. Fluid Mech. **1997**, *345*, 45–78.
55. Koschmieder, E.L., Switzer, D.W. The wavenumbers of supercritical surface-tension-driven Benard convection. J. Fluid Mech. **1992**, *240*, 533–548.
56. Koschmieder, E.L. *Benard Cells and Taylor Vortices*; Cambridge University Press: Cambridge, 1993.
57. de Groot, S.R., Mazur, P. *Non-equilibrium Thermodynamics*; North Holland: Amsterdam, 1961.
58. Birkhoff, G. *Hydrodynamics. A Study in Logic, Fact and Similitude*; Princeton Univ. Press: Princeton, N.J., 1960.
59. Miller, C.A., Neogi, P. *Interfacial Phenomena. Equilibrium and Dynamic Effects*. Marcel Dekker: New York, 1985.
60. Gershuni, G.Z., Zhukhovitskii, E.M. *Convective Stability of Incompressible Fluids*. Keter: Jerusalem, 1976.
61. Joseph, D.D. *Stability of Fluid Motions* (2 vols.); Springer-Verlag: Berlin, 1976.
62. Pismen, L. Mesoscopic hydrodynamics of contact line motion. Colloids & Surfaces A **2002**, *206*, 11–30.
63. Pomeau, Y. Recent progress in the moving contact line problem: a review. C. R. Mecanique **2002**, *330*, 207–222.
64. Palmer, H.J., Berg, J.C. Hydrodynamic stability of surfactant solutions heated from below. J. Fluid Mech. **1972**, *51*, 385–402.
65. Stone, H.A. A simple derivation of the time-dependent convective-diffusion equation for surfactant transport along a deforming interface. Phys. Fluids **1990**, *A2*, 111–112.
66. Ravera, F., Liggieri, L., Steinchen, A. Sorption kinetics considered as a renormalized diffusion process. J. Colloid Interface Sci. **1993**, *156*, 109–116.

67. Ravera, F., Liggieri, L., Passerone, A., Steinchen, A. Sorption kinetics at liquid–liquid interfaces with the surface-active component in both phases. J. Colloid Interface Sci. **1994**, *163*, 309–314.
68. Perez-Cordon, R., Velarde, M.G. On the (nonlinear) foundations of Boussinesq approximation applicable to a thin layer of fluid. J. Phys. (Paris) **1975**, *36*, 591–601.
69. Velarde, M.G., Perez-Cordon, R. On the (nonlinear) foundations of Boussinesq approximation applicable to a thin layer of fluid (II). J. Viscous dissipation and large cell gap effects. J. Phys. (Paris) **1976**, *37*, 177–182.
70. de Boer, P.C.T. Thermally-driven motion of strongly heated fluids. Int. J. Heat Mass Transfer **1984**, *27*, 2239–2251.
71. de Boer, P.C.T. Thermally-driven motion of highly viscous fluids. Int. J. Heat Mass Transfer **1986**, *29*, 681–688.
72. Gershuni, G.Z., Lyubimov, D.V. *Thermal Vibrational Convection*; Wiley: New York, 1997.
73. Zen'kovskaya, S.M., Simonenko, I.B. On the influence of high frequency vibration on the convection initiation. Izv. Akad. Nauk SSSR, Mech. Zhidk. Gaza. **1966**, *5*, 51–55.
74. Marchenko, V.A., Khruslov, E.Ya. *Boundary Tasks in Domains with a Fine Grained Border*. Nauk Dumka: Kiev, 1974 (in Russian).
75. Conca, C. Numerical results on the homogenization of Stokes and Navier–Stokes equations modeling a class of problems from fluid mechanics. Comp. Meth. Appl. Mech. Engng. **1985**, *53*, 223–258.

2

A First Discussion of Instability Phenomena Driven by the Marangoni Effect

In this chapter we shall illustrate, with a few cases treated in a simplified and rather cavalier manner, how the onset of either steady, patterned convection, or oscillatory motions, in the form of transverse or longitudinal surface waves, may occur in a shallow liquid layer as a result of the Marangoni effect or an electric field acting as a Marangoni-like agent. For waves we shall provide a description of a harmonic oscillator, thus allowing a clear view of the role played by thermal gradients or an electric field.

I. INSTABILITY AND STEADY PATTERNED CONVECTION (BENARD CELLS)

Let us begin with the paradigmatic case of Benard convection [1–5], and so let us find in relatively simple way the threshold for the surface tension gradient–(Marangoni)-driven instability of a horizontal liquid layer heated at its solid support. Later we shall return to this paradigmatic case and variations on it with a more complete study. Consider a region where the surface has become colder relatively to its surroundings. The local decrease in temperature, denoted here by T (negative), induces a positive fluctuation of the surface tension $\Delta\sigma = (\mathrm{d}\sigma/\mathrm{d}T)T$, the surface tends to contract locally, and it becomes squashed, therefore dragging the fluid. Continuity demands that the nearby liquid sinks is areas locally where the surface is relatively stretched. The coupling between this local thermocapillary flow and the temperature field arises from the heat equation and the tangential boundary condition satisfied by the Navier–Stokes equations. These equations were introduced in Chapter 1 and are not explicitly recalled here. Accordingly, neglecting the mechanical influence of air ($\eta_{\text{air}} \ll \eta_{\text{liquid}}$) we obtain

$$\frac{\partial T}{\partial t} - \beta w = \chi\left(\frac{\partial^2 T}{\partial x^2} + \frac{\partial^2 T}{\partial z^2}\right) \tag{1}$$

$$\frac{\eta \partial_u}{\partial z} = \left(\frac{\mathrm{d}\sigma}{\mathrm{d}T}\right)\left(\frac{\partial T}{\partial x}\right) \tag{2}$$

where, for simplicity, a two-dimensional (x, z) problem is considered for a liquid layer of infinite horizontal extent and neglected surface deformation. Equation (2) accounts for the Marangoni effect and β is the temperature gradient across the depth, d, of the liquid layer (considered positive when heating from the bottom side). We have as unknowns u and w, the horizontal and vertical velocity components of the local spontaneous flow. Neglecting buoyancy in the Navier–Stokes equations and using also the incompressibility condition an equation for w can be obtained,

$$\Delta^2 w = 0 \tag{3}$$

whose solutions can be sought using Fourier normal modes, exp(ikx), or just $w \sim \sin kx$. Then Eq. (3) yields

$$\left(-k^2 + \frac{\partial^2}{\partial z^2}\right)^2 w = 0 \tag{4}$$

whose general solution is

$$w = \sin(kx) \sum_{\pm} (a_{\pm} + b_{\pm} z) \exp(\pm kz). \tag{5}$$

To obtain u the continuity equation and Eq. (5) are used.

There are four unknows, $a_{\pm}$ and $b_{\pm}$, and four boundary conditions (b.c.) to be satisfied: $w = 0$ and the tangential boundary condition at the open surface, and the no-slip condition at the solid bottom ($u = w = 0$).

Assuming a short wave instability ($k \sim \pi/d$) as illustrated by the Benard cells in Fig. 1, it is passible to set

$$u = \frac{b_+}{k}(1 + kz) \exp kz \cos kx \tag{6}$$

$$w = b_+ z \exp kz \sin kx \tag{7}$$

where $b_+/k = (\mathrm{d}\sigma/\mathrm{d}T)T/2\eta$ and T accounts now for the temperature amplitude and hence maximum disturbance at the open surface. Then $w \sim u$ and

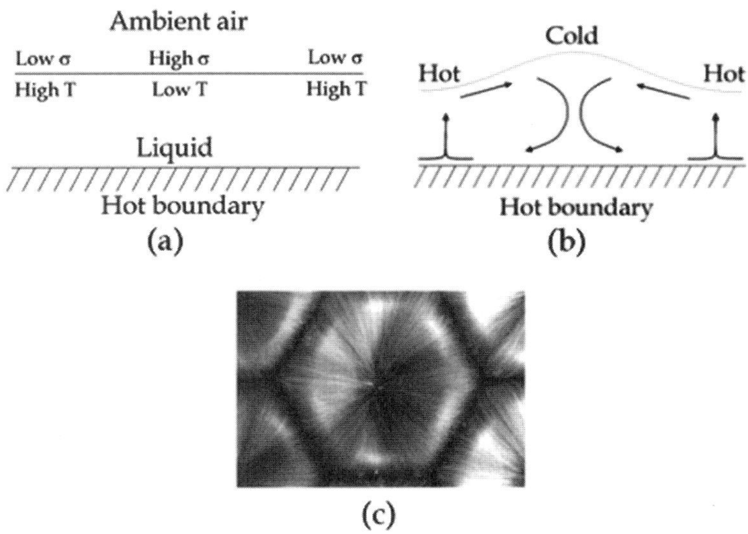

FIG. 1 Instability due to the Marangoni effect and Benard cells. (a) Liquid layer open to air and heated uniformly at a solid support held at constant temperature. At the open surface to ambient air, temperature fluctuations result in local changes of surface tension that yield flow. (b) Liquid flows from relatively hotter to colder regions. (c) Past a critical value of the thermal gradient, steady patterned convection can be observed by direct light illumination of the surface using aluminium flakes added to the liquid. Note that the liquid rises hot beneath surface-depressed regions (cell centers) and it descends under relatively elevated areas (periphery). Exposure time was longer than the time it takes the fluid to travel from the center to the periphery. Silicone oil layer depth was ~ 1 mm and velocity was $\sim 10^{-2}$ cm/s. Benard cells correspond to short wave instability and convection in the liquid layer driven by the surface tension gradient. Similar phenomena occur with surfactants under otherwise isothermal conditions. In this case the localized hot spot is, for example, a localized monolayer accumulation of surfactant molecules, which lower the surface tension and squash the interface and hence flow is initiated moving outwardly from the point where the local disturbance occurred. As surfactant (or solute) molecules exist in the bulk, the possibility of sustaining the initial flow fluctuation arises provided there is a strong enough (surfactant) concentration gradient.

$w \sim (\mathrm{d}\sigma/\mathrm{d}T)T/\eta$. Thus the heat equation (1) can be reduced to

$$\frac{\partial T}{\partial t} = \left[\frac{-\chi\pi^2}{d^2} - \frac{\beta(\mathrm{d}\sigma/\mathrm{d}T)}{\eta}\right]T \tag{8}$$

The quantity in square brackets in Eq. (8) denotes the time scale for decay or growth of temperature fluctuations and, consequently, flow disturbances. There is growth and instability when $-\beta(\mathrm{d}\sigma/\mathrm{d}T)/\eta > \chi\pi^2/d^2$, and so when

$\beta > \beta_c \equiv -\eta\chi\pi^2/(\mathrm{d}\sigma/\mathrm{d}T)d^2$. The quantity $Ma = -(\mathrm{d}\sigma/\mathrm{d}T)\beta d^2/\eta\chi$ is the earlier introduced Marangoni number. Accordingly, instability is expected when Ma reaches a threshold value here estimated as $Ma_c = \pi^2 = 0(1)$. It shows that when the force induced by the surface tension gradient overcomes the frictional viscous damping and acts faster than the natural tendency of heat to diffuse and homogenize temperature we expect instability of the quiescent state and then patterned flow with $k \sim \pi/d$ (Benard cells; Fig. 1). The value $Ma_c = 80$ and others are obtained in more elaborated analyses [6–10] and depend on the actual boundary conditions used for temperature and velocity, as will be described later. Note that when the full dynamics are considered, including surface deformation and its evolution, not just one, but two instability thresholds have been predicted [7] and experimentally observed [10,11]. The short wave instability just described leads to Benard cells with a critical Marangoni number independent of the Galileo number and a long wave instability with threshold value of the Marangoni number proportional to the Galileo number, hence showing the significance of surface deformation.

II. INSTABILITY, TRANSVERSE SURFACE OSCILLATIONS AND CAPILLARY-GRAVITY WAVES

As noted earlier, when the equilibrium surface of a liquid layer open to the ambient air is disturbed, the interface behaves like a membrane; it gets deformed, and then forces tending to return it to the original equilibrium state appear in the liquid. Capillary forces tend to reduce the increased surface and, moreover, if the liquid is in a gravitational field, the disturbance gives rise to gravitational forces that tend to return the interface to its original level shape. Because of inertia, however, the liquid particles may overshoot their original equilibrium position (overstability). As a consequence transverse waves may appear at the air–liquid interface (capillary-gravity waves). Viscosity tends to damp these waves. For a frequency of oscillation Ω and kinematic viscosity ν, the viscous penetration length of the wave is of the order $(\nu/\Omega)^{1/2}$. Generally, for high enough frequencies, $\Omega \gg \nu/d^2$, the (viscous) wave penetration depth is rather small with respect to the layer depth, d, and viscosity can be neglected in the bulk away from the interface.

Let us consider the role of the Marangoni effect upon the abovementioned possible transverse oscillations of the surface. Note that if a spontaneous fluctuation misaligns the surface (deformation) relative to the isotherms, then in an elasticlike manner the Marangoni effect may help bring the surface back to its original level position, or, otherwise, depending on the sign of the thermal gradient, may sustain the disturbance. Accordingly, we shall see how long an instability threshold wave motion can be sustained and we shall provide the oscillation frequency and the critical Marangoni numbers in terms of the Prandtl and the capillary numbers.

As for the case of Benard cells, studied in the preceding section, small, infinitesimal disturbances in the liquid layer obey the continuity, the Navier–Stokes, and the heat (Fourier) diffusion equations introduced in Chapter 1. At the air–liquid interface we have

$$\frac{\partial w}{\partial t} = -\left(\frac{1}{\rho}\right)\left(\frac{\partial p}{\partial z}\right) + \nu\left(\frac{\partial^2 w}{\partial x^2} + \frac{\partial^2 w}{\partial z^2}\right) \tag{9}$$

where again for simplicity we restrict consideration to a two-dimensional (x, z) problem in an infinitely extended horizontal liquid layer. In Eq. (9) t and z denote time and vertical coordinate and w, ρ, and p are vertical velocity, density, and pressure, respectively. As w is the liquid velocity, if we assume the kinematic condition $w = \partial\xi/\partial t$ we have a direct relationship between the velocity of a point at the geometrical interface and that of a material liquid point at every time t. $\xi(x, t)$ denotes the deformation of the interface. Neglecting the mechanical influence of the ambient air, at the interface the balance for the normal and tangential components of the stress tensor yields

$$p - \rho g\xi + \sigma\partial^2\frac{\xi}{\partial x^2} = 2\eta\frac{\partial w}{\partial z} \tag{10}$$

and

$$\eta\left(\frac{\partial u}{\partial z} + \frac{\partial w}{\partial x}\right) = \left(\frac{\partial \sigma}{\partial T}\right)\left(\frac{\partial T}{\partial x} - \beta\frac{\partial \xi}{\partial x}\right) \tag{11}$$

where x is the only considered horizontal coordinate, u is the corresponding horizontal velocity component, g is the gravitational acceleration, σ is the surface tension, T denotes temperature, and β is the temperature gradient induced across the liquid layer (as in the earlier case, β is here also taken positive when the layer is heated from below, i.e., from the liquid side); $\eta = \rho\nu$ is the dynamic (shear) viscosity. Equation (10) shows that we take gravity into account. However, we disregard buoyancy in the bulk.

In the high frequency limit the bottom of the liquid layer can be considered to be at "practical" infinite distance for the interface. Then the liquid depth, d, becomes useless and a suitable scale is the capillary length obtained when the *static* Bond number, $Bo = \rho g l^2/\sigma$, is equal to unity, $Bo = 1$. Then, using Eqs (10) and (11), the heat equation and the kinematic condition, Eq. (9) can be approximated by

$$\frac{\mathrm{d}^2\xi}{\mathrm{d}t^2} + k^2\left[4\nu - \frac{k\beta(\partial\sigma/\partial T)(\chi/2)^{1/2}}{\rho\Omega^{3/2}}\right]\frac{\mathrm{d}\xi}{\mathrm{d}t} + \left[gk + \frac{\sigma k^3}{\rho} + \frac{(\partial\sigma/\partial T)\beta k^3(\chi/2\Omega)^{1/2}}{\rho}\right]\xi = 0 \tag{12}$$

where it is assumed that

$$\xi = -\left(\frac{A}{\Omega}\right)\cos(kx + \Omega t), \tag{13}$$

and similar Fourier normal mode expressions for all the other disturbances, and χ is the thermal diffusivity.

Equation (12) is the harmonic oscillator equation obeyed by a point at the air–liquid interface [12,13]. The damping coefficient may be positive, negative, or vanishing according to the sign and values given to β for given $\partial\sigma/\partial T$.

Using the capillary length as unit, $l = (\sigma/\rho g)^{1/2}$, Eq. (12) takes on a dimensionless, universal form

$$\frac{d^2\zeta}{d\tau^2} + \left[4a^2 + \frac{Ma\,a^3}{(2Pr^3\omega^3)^{1/2}}\right]\frac{d\zeta}{d\tau} + \left[\frac{Ca\,a(Bo+a^2)}{Pr} - \frac{Ma}{a(2Pr^3\omega)^{1/2}}\right]\zeta = 0, \tag{14}$$

where $\zeta = \zeta/l$, $\tau = t\nu/l^2$, $a = lk$, and $\omega = \Omega/\nu k^2$. Three dimensionless groups appear: the Marangoni number, $Ma = -(\partial\sigma/\partial T)\beta l^2/\eta\chi$, the Prandtl number, $Pr = \nu/\chi$, and the capillary number, $Ca = \sigma l/\eta\chi$. The Bond number shows explicitly the relative influence of surface tension and gravity. Further simplification of Eq. (14) can be achieved by insisting on the above invoked high frequency limit.

We see that in order to have vanishing damping we need $Ma = 0(\omega^{3/2})$, with $Pr = 0(1)$. Then $Ca = 0(\omega^2)$, is a reasonable assumption provided we are far from a critical point or if the liquid layer is not too shallow. Then the second term in the spring constant, the coefficient of ζ, can be neglected and Eq. (14) reduces to

$$\frac{d^2\zeta}{d\tau^2} + \left[4a^2 + \frac{Ma\,a^3}{(2Pr^3\omega^3)^{1/2}}\right]\frac{d\zeta}{d\tau} + \left[\frac{Ca\,a(Bo + a^2)}{Pr}\right]\zeta = 0 \tag{15}$$

This equation is the simplest (high frequency) harmonic oscillator approximation to the oscillatory interfacial motion of the open surface in a liquid layer [13]. Yet it contains all the relevant physics. Indeed, the damping coefficient vanishes when we set

$$Ma = \frac{-4(2Pr^3\omega^3)^{1/2}}{a} \tag{16}$$

thus allowing a "free" oscillation of (dimensionless) frequency given by

$$\omega^2 = \frac{Ca\,a(Bo + a^2)}{Pr} = \tilde{G}a\,a\left(1 + \frac{a^2}{Bo}\right), \tag{17a}$$

or else, in dimensional form,

$$\Omega^2 = gk + \frac{\sigma k^3}{\rho} \tag{17b}$$

which is the dispersion relation for gravity-capillary waves in an infinitely deep liquid layer (Laplace–Kelvin law). Note that $Ca/Pr = \tilde{G}a/Bo$, $Pr\,\tilde{G}a = Ga$. The dispersion relation, for a layer of finite depth, is $\omega^2 = \tilde{G}a\,a(1 + a^2/Bo)\tanh a$.

It does appear that for an air–liquid interface the oscillatory instability is to be expected for negative values of the Marangoni number only. For standard liquids ($\partial\sigma/\partial T < 0$) this means that the heating is from the ambient air above or the layer is cooled from inside the liquid below. For liquids that have a minimum or a maximum in their variation of surface tension with temperature, the predicted oscillatory instability can be seen when heating the liquid layer from the solid bottom.

The minimum Marangoni number that sustains the oscillatory motion is

$$Ma_c = -7.93(Ca\,Pr)^{3/4} \tag{18a}$$

with a frequency

$$\omega_c^2 = \frac{6\sqrt{5}Ca}{Pr} \tag{18b}$$

and a wave number

$$a_c = \frac{1}{\sqrt{5}}. \tag{18c}$$

It seems interesting to see how temperature and surface deformation appear in the wave motion. To do this we need the explicit forms of the Fourier normal modes. Besides Eq. (13) we have

$$u = (A \exp kz + B \exp mz) \exp(ikx + \lambda t) \tag{19}$$

$$w = -i[A \exp kz + kB \exp(mz/m)] \exp(ikx + \lambda t) \tag{20}$$

where $m^2 = (k^2 + \Omega/\nu)$ and λ is the complex time constant or growth rate. Note that we could take $\exp \lambda t$ or $\exp(-\lambda t)$, according to convention. Then $\mathrm{Re}\,\lambda > 0$ (alternatively, $\mathrm{Re}\,\lambda < 0$) implies instability, $\mathrm{Re}\,\lambda = 0$ denotes a neutrally stable case, and $\mathrm{Im}\,\lambda \equiv \Omega$ denotes the frequency of overstable, oscillatory motions. On the other hand $\mathrm{Re}\,\lambda < 0$ (alternatively $\mathrm{Re}\,\lambda > 0$) implies (linear or local) stability. Linear stability analyses provide *sufficient* conditions for instability or *necessary* conditions for stability. A system may be linearly stable but unstable to finite amplitude disturbances. Generally, in this book, we shall be concerned with neutral states that at instability thresholds are called critical states. The equation provides the dispersion relation between frequency and corresponding wavenumber (or wavelength).

The complete expression of the mode in Eq. (13) is

$$\xi = -\left(\frac{i}{\lambda}\right)\left(\frac{A + kB}{m}\right)\exp(ikx + \lambda t), \tag{21}$$

according to the above indicated kinematic boundary condition at $z = 0$. Equation (13) provides the real part of Eq. (21).

In the earlier high frequency approximation the boundary condition involving the Marangoni effect in Eq. (11) yields

$$B = \sqrt{2}(3 + i)\frac{A}{\omega^{1/2}} + 0(\omega^{-1/2}) \tag{22}$$

which shows indeed that for transverse oscillations the rotational part is negligible in comparison with the potential one in the Fourier normal mode solutions.

From the velocity field, Eqs (19) and (20), it follows that the temperature disturbance along the interface is given by

$$T(0) - \beta\xi = \left(\frac{A\beta}{\Omega^{3/2}}\right)\chi^{1/2}k \sin\left(kx + \Omega t + \frac{\pi}{4}\right). \tag{23}$$

Comparison of Eqs (23) and (21) shows that temperature and interfacial deformation are clearly out of phase: the extrema of the former disturbance do not sit on the extrema of the latter, i.e., on the crests and troughs of the wave, respectively, for the hottest and coldest spots (Fig. 2). Then the wave propagation can be enhanced or suppressed by the Marangoni flow, according to the sign of the temperature gradient. When for a standard liquid the heating is from the air side (β negative) the Marangoni flow brings liquid from the hot point, located at $x = 5\pi/4$, to the cold point, located at $x = \pi/4$. Such flow favors the natural motion assumed for the wave, right to left. In the opposite case when for the same standard liquid we heat from the liquid side or cool it from the air side (β positive) the hot point is at $x = \pi/4$ whereas the cold point is at $x = 5\pi/4$. Now the Marangoni flow rather pushes the interfacial disturbance to the right, opposes the wave propagation and eventually suppress it.

Thus we have seen that transverse vibrations, and hence capillary-gravity waves, at the open surface of a liquid can be excited, and eventually sustained, by the Marangoni effect. In the description presented above we have used the capillary length as unity and hence the liquid layer was tacitly taken to be of infinite depth but not quite, only in the sense that viscous effects were appreciable only very near to the open surface of the liquid. If the bottom of the layer affects the open surface it is expected to play a dissipative stabilizing role, hence lowering the above given threshold value [14]. Recall that linear theory provides sufficient

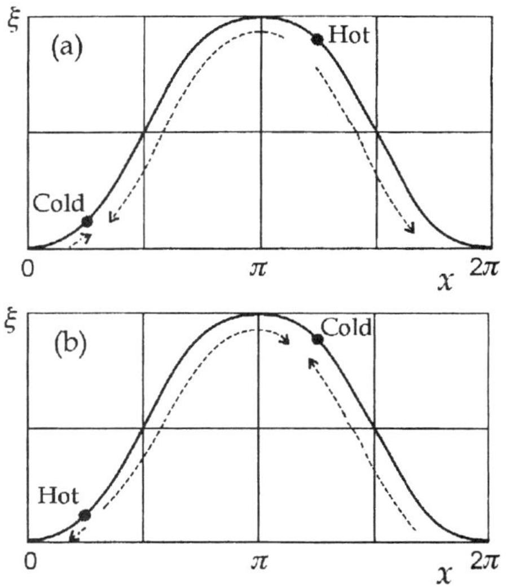

FIG. 2 Transverse waves: a schematic and slightly exaggerated sketch of the phase shift between temperature disturbances (hot and cold spots, only) and surface deformation, ξ, in a capillary-gravity wave moving from right to left. (a) The liquid layer is heated from the ambient air ($\beta < 0$); (b) The heating is done from the liquid ($\beta > 0$). The Marangoni effect acts according to the arrows (flow from hot to cold regions), and hence favors flow disturbance in the first case but tends to suppress it in the latter.

conditions for instability that may not be necessary. Furthermore, an approximate linear theory (as demonstrated here for high frequency motions) may very well be off a realistic value of the threshold by orders of magnitude. On the other hand the asymptotics used are also limited in scope. Thus to have a complete picture of overstability including better quantitative estimates of its characteristic quantities, using a two-layer system with the full thermohydrodynamic equations is needed together with appropriate description of boundaries and boundary layers at the solid bottom and at the open surface [15–25]. In the full two-layer problem, oscillatory instability may appear with the sign of the Marangoni number depending on the fluids considered, as shown by various authors; this will be discussed later in this book. For experimental results see Refs. 26–31.

Finally, let us mention that when heating from the liquid side there is another transverse surface wave, which coexists with monotonic instability, and which the Marangoni effect can excite, although not satisfying the dispersion (Laplace–Kelvin) relation for capillary-gravity waves (Fig. 3). Its interest is only for exotic cases such as (low Prandtl number) liquid helium layers near the lambda line (low values of the capillary number) [32]. This question will be addressed further below.

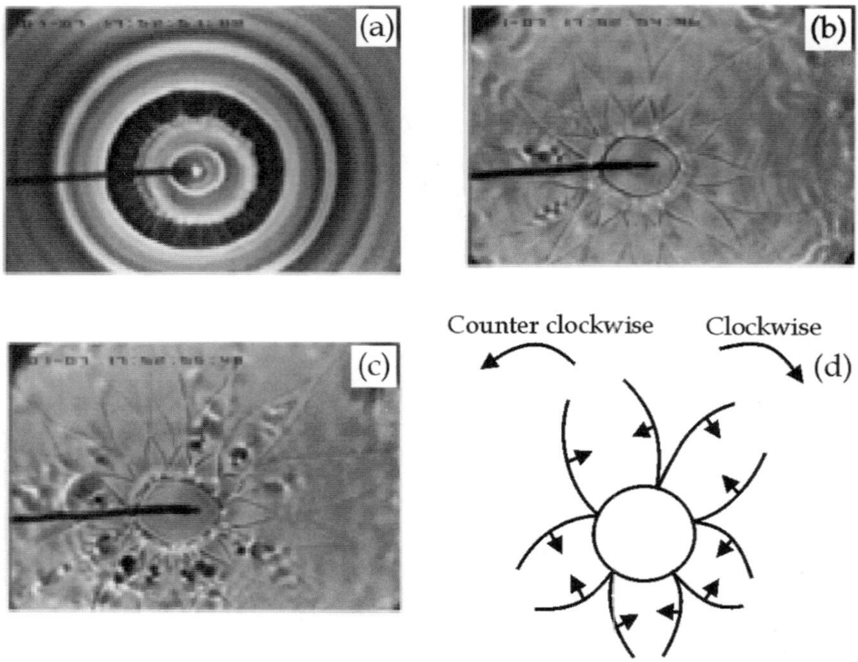

FIG. 3 Exotic capillary-gravity waves excited by the solutocapillary Marangoni effect. Sequence of events in a typical experiment lasting 5 s. (a) Traveling ripples following drop deposition on the water surface with subsequent drop spreading. (b) A petal-like (unsteady) structure appears in the primary film due to surface wave collisions. (c) Interfacial turbulence. (d) Schematic representation of the propagating wave fronts forming daisy-like petals. (From Ref. 31.)

III. INSTABILITY, LONGITUDINAL SURFACE OSCILLATIONS AND DISSIPATIVE COMPRESSION–DILATIONAL WAVES

Let us now illustrate how the Marangoni effect can also excite and, past an instability threshold, sustain waves with no significant deformation of the interface. For mass transfer, the study of this problem was pioneered by Sternling and Scriven [33]. For this reason we shall consider mass rather than heat transfer [18,19,34].

Consider a two-layer system with liquids at rest having an interface between them located at $z = 0$, taken as reference level along the vertical axis; z is the vertical coordinate. Let η, ν, D, and ρ denote the corresponding dynamic (shear) viscosity, kinematic viscosity, mass diffusivity, and density in each liquid. We shall denote with subscript "1" and "2" the lower and upper liquids, respectively. Let us assume that a solute, which is expected to act as a surfactant at the interface, is distributed in each phase with a given bulk gradient and that it may be adsorbed at the interface according to Langmuir's law [35]. Then if we consider a disturbance at the interface that may eventually be amplified thus leading to interfacial instability the expected evolution of such disturbance is, as in the earlier case studied, governed by the continuity, Navier–Stokes, and mass (Fick) diffusion equations valid on each side of the interface and at the interface itself. For the simplest two-dimensional problem, (x, z), again considering an infinitely extended system along the horizontal interface, in dimensionless form we have

$$\operatorname{div} \mathbf{v}_1 = \operatorname{div} \mathbf{v}_2 = 0 \tag{24}$$

$$\frac{\partial \mathbf{v}_1}{\partial t} + \nabla p_1 - \nabla^2 \mathbf{v}_1 = N_\rho \frac{\partial \mathbf{v}_2}{\partial t} + \nabla p_2 - N_\eta \nabla^2 \mathbf{v}_2 = 0 \tag{25}$$

and

$$\frac{\partial C_1}{\partial t} - w_1 - Sc^{-1} \nabla^2 C_1 = \frac{\partial C_2}{\partial t} - w_2 - Sc^{-1} N_D \nabla^2 C_2 = 0 \tag{26}$$

where $\mathbf{v}_i$ $(i = 1,2) = (u_i, w_i)$ with u and w denoting the horizontal and vertical velocity components of the disturbance velocity field and p denoting pressure. Here, in the absence of heat we use $Sc = \nu/D$ (Schmidt number). $N_D = D_2/D_1$, $N_\eta = \eta_2/\eta_1$, and $N_\rho = \rho_2/\rho_1$, x accounts for the horizontal coordinate, and C is the bulk concentration of the surfactant. The solutal or surfactant Marangoni number is defined here as $E = -(\partial\sigma/\partial C_1)\beta_1 l^2/\eta_1 D_1$ to emphasize its earlier mentioned elasticity character. σ is the liquid–liquid interfacial tension, while β denotes bulk concentration gradient of the surfactant. The parameter l is a characteristic length that may very well be the capillary length but need not be so. In this case, as for the earlier case with transverse waves, it is of no interest to use the liquid depth, d, as viscous motions are really significant only in a boundary/skin layer near the interface. The capillary length here is given by $l = [\sigma_0/(\rho_1 - \rho_2)g]^{1/2}$ where σ_0 is a reference value. $Bo = (\rho_1 - \rho_2)gl^2/\sigma_0$ is the corresponding Bond number. When surface deformation and surface accumulation of the surfactant can be neglected [35] the disturbance evolution Eqs. (24)–(26) for longitudinal motions obey the following boundary conditions (b.c.) at $z = 0$

$$w_1 = w_2 = 0 \tag{27}$$

$$\frac{\partial w_1}{\partial z} = \frac{\partial w_2}{\partial z} = 0 \tag{28}$$

$$\left(\frac{E}{Sc}\right) a^2 C_1 - N_\eta \frac{\partial^2 w_2}{\partial z^2} + \frac{\partial^2 w_1}{\partial t^2} = 0 \tag{29}$$

$$\Gamma Sc \frac{\partial C_1}{\partial t} = \frac{\partial (C_2 - C_1)}{\partial z} \tag{30}$$

$$C_1 = N C_2 \tag{31}$$

where Γ is the Langmuir adsorption number [20–22,35], i.e., the slope of the Langmuir adsorption law at the interface (per unit length in the two-dimensional problem). N is the ratio of C_1 to C_2 when C is evaluated from each side at the interface.

Again because of the linearity of the problem we may seek solutions using Fourier normal modes

$$w_1 = B_1(-\exp az + \exp m_1 z) \tag{32}$$

$$w_2 = B_2[-\exp az + \exp(-m_2 z)] \tag{33}$$

$$p_1 = -B_1\left(\frac{\lambda}{a}\right) \exp az \tag{34}$$

$$p_2 = B_2 N_\rho \left(\frac{\lambda}{a}\right) \exp(-az) \tag{35}$$

$$C_1 = R_1 \exp q_1 z + \left(\frac{B_1}{\lambda}\right) \exp az - \left[B_1 \frac{Sc}{\lambda(Sc-1)}\right] \exp m_1 z \tag{36}$$

and

$$C_2 = R_2 \exp(-q_2 z) + \left(\frac{B_2}{\lambda}\right) \exp(-az) - \left[\frac{B_2(Sc/N_\mathrm{D})}{\lambda((Sc/N_\mathrm{D}) - (N_\rho/N_\eta))}\right] \exp(-m_2 z) \tag{37}$$

where a denotes the wave number of a Fourier mode and, as earlier indicated, λ is the complex time constant whose imaginary part is the dimensionless frequency of motions. Also, $m_1^2 \equiv a^2 + \lambda^2$, $m_2^2 \equiv a^2 + (N_\rho/N_\eta)\lambda$, $q_1^2 \equiv a^2 + Sc\lambda$ and $q_2^2 \equiv a^2 + (Sc\lambda/N_\mathrm{D})$. The quantities B_1, B_2, R_1, and R_2 are the unknown disturbance amplitudes left undetermined in the linear theory.

The time derivative of Eq. (30) yields

$$\Gamma Sc \frac{\partial^2 C_1}{\partial t^2} = \frac{\partial^2 (C_1 - C_2)}{\partial z \partial t}, \tag{38}$$

and taking the z-derivative in Eq. (26) we can estimate the right-hand side of Eq. (38).

After some lengthy, albeit straightforward, calculus for oscillatory disturbances, where $\mathrm{Im}\,\lambda = \omega$, Eq. (38) reduces to

$$(1 + N_\rho^{1/2} N_\eta^{1/2}) \Gamma Sc^{1/2} \frac{\mathrm{d}^2 C_1}{\mathrm{d}t^2} + \sqrt{2\omega}\left[(1 + N_\rho^{1/2} N_\eta^{1/2})\left(1 + \frac{1}{NN_\mathrm{D}^{1/2}}\right) + \frac{Ea^2}{Sc^2 \omega^2}\Pi_1\right]\frac{\mathrm{d}C_1}{\mathrm{d}t} + a\omega\left[(1 + N_\eta)\left(1 + \frac{1}{NN_\mathrm{D}^{1/2}}\right) + \frac{Ea^2}{Sc^2\omega^2}\Pi_2\right]C_1 = 0 \tag{39}$$

with

$$\Pi_1 = \left(\frac{N_\rho N_D}{N_\eta}\right)^{1/2} - 1 \tag{40}$$

and

$$\Pi_2 = N_D^{1/2} - 1 \tag{41}$$

Note that Π does not refer to surface stresses here.

Equation (39) is the simplest harmonic oscillator description of the surfactant oscillations along the interface [36]. It is an equation for longitudinal interfacial surface oscillations whose damping coefficient can be zero with a suitable choice of the value of the solutal Marangoni number.

When the damping coefficient is set to zero we obtain the following two relationships

$$(1 + \sqrt{N_\rho N_\eta})\left(1 + \frac{1}{NN_\mathrm{D}^{1/2}}\right) + \frac{a^2 E \Pi_1}{Sc^2 \omega^2} = 0 \tag{42a}$$

and

$$a\left[(1 + N_\eta)\left(1 + \frac{1}{NN_\mathrm{D}}\right) + \frac{Ea^2}{Sc^2\omega^2}\Pi_2\right] = \omega(1 + N_\rho^{1/2} N_\eta^{1/2})\Gamma Sc \tag{42b}$$

To be satisfied, Eq. (42a) demands that

$$E\Pi_1 < 0 \tag{43a}$$

for example,

$$\mathrm{sgn}(E) = -\mathrm{sgn}\left(\frac{N_\rho^{1/2} N_\mathrm{D}^{1/2}}{N_\eta^{1/2} - 1}\right) \tag{43b}$$

which, together with $\mathrm{sgn}(E) = \mathrm{sgn}(\beta_1)$, ensures that to have oscillatory behavior we need

$$\frac{D_\mathrm{f}}{D_\mathrm{t}} > \frac{\nu_\mathrm{f}}{\nu_\mathrm{t}} \tag{44}$$

where "f" and "t" denote "from" and "to," a way of indicating how the surfactant is being transported *from* and *to* the volume [33].

We now use Lucassen's complex elasticity modulus ε [37], which is related to the solutal Marangoni number by the relationship

$$E = -\frac{\varepsilon l^2 (k^2 + i\Omega/\nu_1)^{1/2}}{\eta_1 D_1} \approx -\frac{\varepsilon l^2 (i\Omega/\nu_1)^{1/2}}{\eta_1 D_1} \tag{45}$$

where $a = kl$ and $\omega = \Omega l^2/\nu_1$. Then, using these new variables, Eq. (45) reduces to

$$\varepsilon m k^2 + i\eta_1 \Omega m^2 \left(\frac{\nu_1}{D_1}\right) \left\{ \frac{[1 + (\rho_2\eta_2/\rho_1\eta_1)^{1/2}][1 + N(D_1/D_2)^{1/2}]}{(\nu_1 D_1/\nu_2 D_2)^{1/2} - 1} \right\} = 0 \tag{46}$$

where $m^2 = k^2 i\Omega/\nu_1 \approx i\Omega/\nu_1$. Equation (46) is a generalization of the case discussed by Lucassen [35].

Then, using Eqs. (42a) and (42b) we obtain

$$E_\mathrm{c} = -\left[\frac{(N_\mathrm{D}^{1/2} + N_\eta)(N_\rho^{1/2} N_\eta^{-1/2} - 1)}{Sc\,\Gamma\Pi_1\Pi_2}\right]^2 \frac{(1 + N^{-1} N_\mathrm{D}^{-1/2}) Sc^2}{\Pi_1} \times (1 + N_\rho^{1/2} N_\eta^{1/2}) \tag{47}$$

where the dispersion relation is

$$\omega_\mathrm{c} = a_\mathrm{c} \left[\frac{(N_\mathrm{D}^{1/2} + N_\eta)(N_\rho^{1/2} N_\eta^{-1/2} - 1)}{Sc^{1/2}\Gamma\Pi_1\Pi_2}\right]^2 \tag{48}$$

which corresponds to dispersion-free waves. It is worth recalling that (long) waves in liquid layers suffer virtually no dispersion. Practical use of this phenomenon was made in the design of water tables, which exploit the analogy between two-dimensional problems in (compressible) gas dynamics and (kinematically) geometrically similar problems involving nondispersive water waves. Figure 4 illustrates the experimental dispersion relation reported by Wierschem et al. [30]. For heat transfer, in the high frequency limit, the corresponding dispersion relation analogue to Eq. (48) is $\omega^2 = |Ma|\,Pr\,a^2/(1 + \mathrm{Pr}^{1/2})$. Note that in the description given above of longitudinal waves we have considered this mode of disturbance for an infinitely deep layer and the asymptotics used suited this approximation. Therefore, for layers of finite depth, other long longitudinal wave modes may appear with a dispersion relation different from the one given above (see later chapters of this book).

The value E_c is the minimum value of the Marangoni number needed to sustain longitudinal interfacial oscillations of frequency ω_c. Because both ω and a must be positive numbers we deduce from Eq. (48) that the following relationship must be satisfied

$$\nu_f/\nu_t < 1 \tag{49}$$

in order to have oscillations. Thus Eqs. (44) and (49) provide conditions needed long for oscillatory longitudinal motions. Note, however, that these are not the only conditions for instability, oscillatory, or otherwise. Other conditions have been predicted using more complete theories [20,33].

Although Lucassen [37] only considered naturally damped motions he clearly emphasized that these interfacial waves were drastically different from capillary-gravity waves. As noted earlier, the latter are transverse motions originating in the geometric deformation of the interface and Laplace law. Thus they may appear in inviscid liquids. The longitudinal oscillations necessarily demand dissipation in viscous liquids and hence surface tension variation (Marangoni effect) as active agent. With high enough values of the Marangoni number (larger than E_c) these waves can be sustained along the interface even if it is not deformed. Recall the earlier argument about the elasticlike role played by the Marangoni effect, which is valid for both surfactants and heat.

A better understanding of the results given above and improved quantitative results demands consideration of the full thermohydrodynamic problem for a two-layer system. Longitudinal waves can then be put in proper context relative to capillary-gravity waves and even to internal waves if buoyancy is also taken into consideration both in theory and experiment [17,18,20–25,27–30,38].

IV. INSTABILITY INDUCED BY AN ELECTRIC FIELD AND GRAVITO-ELECTRO-CAPILLARY WAVES

Let us see how the action of an electric field applied tangentially or orthogonally to the interface separating two liquids or to the open surface of a liquid is very much like that of the (thermal or solutal) Marangoni effect. We shall see how either surface waves, true oscillatory motions, or a static surface relief (frozen wavelike pattern) can be excited and sustained past a critical value of the field's strength. We shall provide the oscillation frequency and the critical field's strength in terms of the various other parameters of the problem.

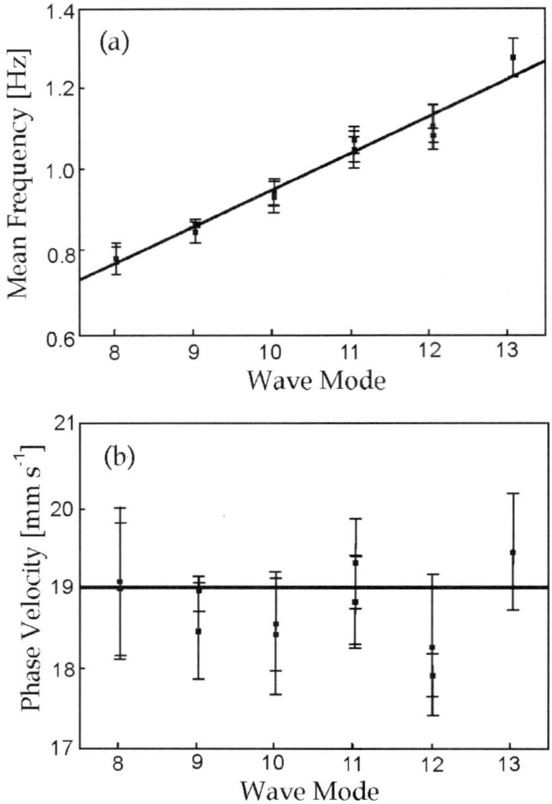

FIG. 4 Longitudinal waves excited by the solutocapillary Marangoni effect. Experimental data obtained when pentane vapor is adsorbed and, subsequently, absorbed by liquid toluene. (a) Dispersionless relation between frequency and wave number; (b) (phase) wave velocity. Error bars indicate maximum and minimum values observed. (From Ref. 30.)

The basic evolution equations are now the continuity and the Navier–Stokes equations, augmented with Maxwell's equations for an electrohydrodynamic medium [39–44]. In the linear approximation we have

$$\mathrm{rot}\,\boldsymbol{E} = 0 \tag{50}$$

$$\mathrm{div}\,\boldsymbol{E} = 0 \tag{51}$$

$$\mathrm{div}\,\boldsymbol{V} = 0 \tag{52}$$

$$\rho\frac{\partial \boldsymbol{V}}{\partial t} = -\nabla p + \nabla\cdot\boldsymbol{T} - \rho g\mathbf{k} \tag{53}$$

where E, $\boldsymbol{V}$, ρ, p, and $\boldsymbol{T}$ denote the electric field intensity, the velocity field, the density, the pressure field, and the overall stress tensor, respectively. Note that because heat is not involved in the case treated here, to avoid confusion with other uses of Π we use $\boldsymbol{T}$ for the electrohydrodynamic tensor [39]:

$$\boldsymbol{T} = \varepsilon \boldsymbol{E}\boldsymbol{E} - \tfrac{1}{2}\varepsilon E^2 \boldsymbol{I} + \eta\{\nabla \boldsymbol{V} + (\nabla \boldsymbol{V})^T\}, \tag{54}$$

where ε is the dielectric constant, η is the dynamic viscosity, and $(\nabla \boldsymbol{V})^T$ the transpose of $(\nabla \boldsymbol{V})$. Recall that $\boldsymbol{D} = \varepsilon \boldsymbol{E}$, with $\boldsymbol{D}$ denoting electric induction or flux density $\mathbf{k} = (0, 0, 1)$. Using Eqs. (50) and (51) we have

$$\nabla \cdot \boldsymbol{T} = \eta \nabla^2 \boldsymbol{V}. \tag{55}$$

Infinitesimal disturbances upon the originally quiescent state are now defined by

$$\boldsymbol{E} = \boldsymbol{E}_0 + \boldsymbol{e} \tag{56}$$

$$\boldsymbol{V} = \mathbf{v}\,\text{(the initial state is a state of rest)} \tag{57}$$

$$P = P_0 - \rho g z + p \tag{58}$$

$$\boldsymbol{T} = \boldsymbol{T}_0 + \tau \tag{59}$$

These disturbances $\{\boldsymbol{e},\mathbf{v},\rho,\tau\}$ obey the following evolution equations:

$$\nabla^2 \boldsymbol{e} = 0(\operatorname{div} \boldsymbol{e} = \operatorname{rot} \boldsymbol{e} = 0) \tag{60}$$

$$\operatorname{div} \mathbf{v} = 0 \tag{61}$$

$$\rho \frac{\partial \mathbf{v}}{\partial t} = -\nabla p + \eta \nabla^2 \mathbf{v} \tag{62}$$

For simplicity in the following we shall restrict consideration to two-dimensional geometry (x,z). The solutions of Eqs. (60) and (62) can be sought using Fourier normal modes with $\exp(\lambda t + iax)$, as

$$w_1 = A_1 \exp az + B_1 \exp m_1 z, \qquad w_2 = A_2 \exp(-az) + B_2 \exp(-m_2 z),$$

$$p_1 = \frac{\rho_1 A_1 \lambda}{a} \exp az, \qquad p_2 = \frac{\rho_2 A_2 \lambda}{a} \exp(-az), \qquad e_{z_1} = F_1 \exp az,$$

$$e_{z_2} = F_2 \exp(-az), \qquad e_{x_1} = iF_1 \exp az, \qquad e_{x_2} = -iF_2 \exp(-az) \tag{63}$$

where $m_i^2 = a^2 - \lambda/\nu_i$, $i = 1,2$; a denotes the wave number of a Fourier mode and ν is the kinematic viscosity; $\eta = \rho\nu$ is the dynamic (shear) viscosity. As in earlier cases, λ is a complex quantity and $\mathrm{Im}\,\lambda = \omega$ gives the frequency of expected motions. A_i,B_i, and $F_i(i = 1,2)$ are amplitude coefficients left undetermined by the linear theory. (x,z) and (u,w) denote the coordinate and velocity components considered and the subscripts x and z also denote x- and z-components, respectively. These solutions must satisfy the appropriate boundary conditions:

$$\mathbf{n} \times [\boldsymbol{E}] = 0 \tag{64}$$

$$\mathbf{n} \cdot [\sigma \boldsymbol{E}] + \nabla_\Sigma \cdot (Q_\Sigma \mathbf{v}) + \frac{\partial Q_\Sigma}{\partial t} = 0 \tag{65}$$

$$[\mathbf{v}] = 0 \tag{66}$$

$$\mathbf{n}[P] = \mathbf{n} \cdot [\boldsymbol{T}] + \mathbf{n}S\frac{\partial^2 \xi}{\partial t^2} \tag{67}$$

$$\frac{\partial \xi}{\partial t} = w_\Sigma \tag{68}$$

$$\mathbf{n} = (- \partial\xi/\partial x,1) \tag{69}$$

$\xi = \xi(x)$ denotes the interface deformation, ∇_Σ is the gradient along the interface, Σ, $Q_\Sigma = \mathbf{n} \cdot [\varepsilon \boldsymbol{E}]$ (Q_Σ is the surface charge density), and

$$\begin{aligned}\mathbf{n} \cdot [\boldsymbol{T}] = \mathbf{i}\Bigg\{ & -\frac{\partial \xi}{\partial x}[\varepsilon E_x^2] + [\varepsilon E_z E_x] + \frac{1}{2}\frac{\partial \xi}{\partial x}[\varepsilon E^2] \\ & + \left[\eta\left((\mathbf{n} \cdot \nabla)\mathbf{v} + \frac{\partial(\mathbf{n} \cdot \mathbf{v})}{\partial x}\right)\right]\Bigg\} \\ & + \mathbf{k}\Bigg\{ -\frac{\partial \xi}{\partial x}[\varepsilon E_x E_z] + [\varepsilon E_z^2] - \tfrac{1}{2}[\varepsilon E^2] \\ & + \left[\eta\left((\mathbf{n} \cdot \nabla)w + \frac{\partial(\mathbf{n} \cdot \mathbf{v})}{\partial z}\right)\right]\Bigg\}\end{aligned} \tag{70}$$

In this case note that σ denotes the electrical conductivity and use is made of S to account for the interfacial tension. The bracket [ ] denotes the jump of the enclosed quantity across the interface, $[A] = A_2 - A_1$ ("1" and "2" refer to liquid below and above, respectively); $\mathbf{i} = (1, 0)$ and $\mathbf{k} = (0, 1)$. To the lowest order approximation, Eq. (70) gives $[P_0] = [\varepsilon E_{0z}^2] - \frac{1}{2}[\varepsilon E_0^2]$, which is the Korteweg–Helmholtz pressure, which can initially raise or decrease the surface level against the gravity field [39].

A. Tangentially Applied Electric Field

When the electric field is applied tangential to the liquid–liquid or air–liquid interface, $E_{0x} = E_0$, $E_{0z} = 0$, disturbances must obey the following boundary conditions

$$[e_x] = 0 \tag{71}$$

$$[w] = \left[\frac{\partial w}{\partial z}\right] = 0 \tag{72}$$

$$\frac{\partial \xi}{\partial t} = w \tag{73}$$

$$-\frac{\partial \xi}{\partial x} E_0[\sigma + \varepsilon\lambda] + [(\sigma + \varepsilon\lambda)e_z] = 0 \tag{74}$$

$$-\frac{\partial \xi}{\partial x}[\varepsilon]E_0^2 + [\varepsilon e_x]E_0 + \left[\eta\left(\frac{\partial u}{\partial z} + \frac{\partial w}{\partial x}\right)\right] = 0 \tag{75}$$

$$[p] = -[\varepsilon e_x]E_0 + 2\left[\eta \frac{\partial w}{\partial z}\right] + S\frac{\partial^2 \xi}{\partial t^2} + g[\rho]\xi \tag{76}$$

Note that

$$\mathbf{n}\cdot[T] = \mathbf{i}\left\{-\frac{1}{2}\frac{\partial \xi}{\partial x}[\varepsilon]E_0^2 + [\varepsilon e_z]E_0 + \left[\eta\left(\frac{\partial u}{\partial z} + \frac{\partial w}{\partial x}\right)\right]\right\} + \mathbf{k}\left\{-[\varepsilon e_z]E_0 + 2\left[\eta\left(\frac{\partial w}{\partial z}\right)\right]\right\} \tag{77}$$

Imposing the boundary conditions (71)–(76) upon the general solution Eq. (63) from Eqs. (60)–(62) after some elementary albeit lengthy calculation we obtain

$$\left(\sum \rho\right)\frac{\mathrm{d}^2\xi}{\mathrm{d}t^2} + \left\{-ag[\rho] + a^2E_0^2\left([\varepsilon]\Lambda + (\pi_1 + \pi_2)\left([\varepsilon] - \Lambda\sum \varepsilon\right)\right) + a^3S\right\}\xi = \left\{4a^2[\eta]\pi_1 + 2a^2[\eta]\pi_2 - \frac{4a\lambda\rho_1\rho_2}{\rho_2(m_1 - a) + \rho_1(m_2 - a)}\right\}\frac{\mathrm{d}\xi}{\mathrm{d}t} \tag{78}$$

which describes interfacial motions with a harmonic oscillator. Note that we have written the dissipative part in the right-hand side (r.h.s.) of Eq. (78). Note that here

$\sum$ denotes summation over $i = 1, 2$. Also

$$\Lambda = \frac{[\sigma + \lambda\varepsilon]}{\sum(\sigma + \lambda\varepsilon)} \tag{79}$$

$$\pi_1 = \frac{\rho_1(m_2 - a) - \rho_2(m_1 - a)}{\rho_1(m_2 - a) + \rho_2(m_1 - a)} \tag{80}$$

and

$$\pi_2 = \frac{2a(m_1 - a)(m_2 - a)[\eta]}{\lambda\{\rho_1(m_2 - a) + \rho_2(m_1 - a)\}} \tag{81}$$

Equation (29) can be decomposed into real and imaginary parts

$$\Lambda = \Lambda_R + i\Lambda_I \tag{82}$$

with

$$\Lambda_R = \frac{(\sigma_2^2 - \sigma_1^2) + \omega^2(\varepsilon_2^2 - \varepsilon_1^2)}{(\sigma_1 + \sigma_2)^2 + \omega^2(\varepsilon_1 + \varepsilon_2)^2} \tag{83a}$$

and

$$\Lambda_I = \frac{2\omega(\sigma_2\varepsilon_1 - \sigma_1\sigma_2)}{(\sigma_1 + \sigma_2)^2 + \omega^2(\varepsilon_1 + \varepsilon_2)^2} \tag{83b}$$

Equation (78) can be simplified if, as is earlier done, we recall that the possible gravity-electro-capillary waves would penetrate to a depth of about $(\nu/\omega)^{1/2}$ for a frequency $\omega = \omega(a)$. Then, in the high-frequency approximation, $\omega/\nu \gg a^2$, after elementary albeit again lengthy calculation, Eq. (78) becomes

$$\left(\sum\rho\right)\frac{d^2\xi}{dt^2} + \left\{ -ag[\rho] + \frac{a^2E_0^2}{\sum\rho\sqrt{\nu}}\left\{2\left[\varepsilon\rho\sqrt{\nu}\frac{[\sigma^2] + \omega^2[\varepsilon^2]}{(\sum\sigma)^2 + \omega^2(\sum\varepsilon)^2} - [\varepsilon][\rho\sqrt{\nu}]\right]\right\} + a^3S\right\}\xi$$

$$= \frac{1}{\sum\rho\sqrt{\nu}}\left\{ -2a\rho_1\rho_2\sqrt{2\omega\nu_1\nu_2} - 4a^2[\eta][\rho\sqrt{\nu}] - 4a^2E_0^2\frac{[\varepsilon\rho\sqrt{\nu}](\varepsilon_2\sigma_1 - \varepsilon_1\sigma_2)}{(\sum\sigma)^2 + \omega^2(\sum\varepsilon)^2}\right\}\frac{d\xi}{dt} \tag{84}$$

which is the approximate harmonic oscillator description of gravity-electro-capillary waves excited by the electric field tangential to the liquid–liquid or air–liquid interface [45]. The spring constant, the coefficient ξ, gives the dispersion relation $\omega = \omega(a)$. It is noteworthy that the electric field contributes to the dispersion relation with a^2E^2 (or a^2ED), which indicates waves in the intermediate wave number range between gravity waves and ripples. As the damping coefficient incorporates the term with E_0^2 and the terms containing jumps across the interface in viscosity and density it appears that a suitable electric field intensity can indeed sustain the oscillations. Let us consider a few particular cases.

1. $\sigma_1 = \sigma_2 = 0$. This corresponds to two dielectric liquids. Then Eq. (84) reduces to

$$\left(\sum \rho\right)\frac{\mathrm{d}^2\xi}{\mathrm{d}t^2} + \left\{-ag[\rho] + \frac{a^2E_0^2[\varepsilon]}{\sum \rho\sqrt{\nu}}\left\{2\frac{[\varepsilon\rho\sqrt{\nu}]}{\sum \varepsilon} - [\rho\sqrt{\nu}]\right\} + a^3S\right\}\xi$$
$$= \frac{1}{\sum \rho\sqrt{\nu}}\{-2a\rho_1\rho_2\sqrt{2\omega\nu_1\nu_2}\}\frac{\mathrm{d}\xi}{\mathrm{d}t} \tag{85}$$

which is the equation for a strictly damped oscillator.

We see that no wave can be sustained; a disturbance spontaneously arising at the interface decays due to the viscosity of the liquids. In a strict dielectric there is no conductivity and thus there is no mechanism to transfer energy from the imposed electric field to the velocity field. The vanishing value of the spring constant defines a static (frozen wavelike) pattern, hence a mere change in surface relief.

2. $\rho_2 = 0, \rho_1 = \rho, \sigma_2 = 0, \sigma_1 = \sigma, \varepsilon_2 = \varepsilon_0, \varepsilon_1 = \varepsilon_0\varepsilon_r$. This is the case of an air–liquid interface. Equation (84) becomes

$$\rho\frac{\mathrm{d}^2\xi}{\mathrm{d}t^2} + \left\{ag\rho + a^2E_0^2\left(-2\varepsilon_0\varepsilon_r\frac{-\sigma^2 + \omega^2\varepsilon_0^2(1-\varepsilon_r^2)}{\sigma^2 + \omega^2\varepsilon_0^2(1+\varepsilon_r)^2} + \varepsilon_0(1-\varepsilon_r)\right) + a^3S\right\}\xi$$
$$= 4a^2\left\{-\eta + \frac{\varepsilon_0^2\varepsilon_r\sigma_0^2}{\sigma^2 + \omega^2\varepsilon_0^2(1+\varepsilon)^2}\right\}\frac{\mathrm{d}\xi}{\mathrm{d}t} \tag{86}$$

where we see that an oscillation can be sustained if the electric field strength is above the value

$$E_0^2 = \left(\frac{\eta\sigma}{\varepsilon_0^2\varepsilon_r}\right)^{1/2} \tag{87}$$

3. For arbitrary values of the various material parameters, we can conclude, from inspection of Eq. (84), that in order to have sustained oscillations a

necessary condition is

$$\frac{\rho_1\sqrt{\nu_1}}{\rho_2\sqrt{\nu_2}} > \frac{\varepsilon_2}{\varepsilon_1} > \frac{\sigma_2}{\sigma_1} \quad \text{or} \quad \frac{\rho_1\sqrt{\nu_1}}{\rho_2\sqrt{\nu_2}} < \frac{\varepsilon_2}{\varepsilon_1} < \frac{\sigma_2}{\sigma_1} \tag{88}$$

which for $\rho_1 = \rho_2$, $\nu_1 = \nu_2$ reduces to

$$1 > \frac{\varepsilon_2}{\varepsilon_1} > \frac{\sigma_2}{\sigma_1} \quad \text{or} \quad 1 < \frac{\varepsilon_2}{\varepsilon_1} < \frac{\sigma_2}{\sigma_1} \tag{89}$$

B. Orthogonally Applied Electric Field

When the applied field is orthogonal to the liquid–liquid or to an air–liquid interface, $E_{0x} = 0$, $E_{0z} = E_0$, the following boundary conditions for disturbances must be satisfied:

$$[e_x] + \frac{\partial \xi}{\partial x}[E_0] = 0 \tag{90}$$

$$[w] = \left[\frac{\partial w}{\partial z}\right] = 0 \tag{91}$$

$$\frac{\partial \xi}{\partial t} = w \tag{92}$$

$$[(\sigma + \varepsilon\lambda)e_z] + [\varepsilon E_0]\frac{\partial u}{\partial x} = 0 \tag{93}$$

$$\frac{\partial \xi}{\partial x}[\varepsilon E_0^2] + [\varepsilon e_x E_0] + \left[\eta\left(\frac{\partial u}{\partial z} + \frac{\partial w}{\partial x}\right)\right] = 0 \tag{94}$$

$$[p] = [\varepsilon e_z E_0] + 2\left[\eta\frac{\partial w}{\partial z}\right] + S\frac{\partial^2 \xi}{\partial t^2} + g[\rho]\xi \tag{95}$$

Let us denote the current intensity by J. Then $J_0 = \sigma_1 E_{01} = \sigma_2 E_{02}$ (when $\sigma = 0$, $D_0 = \varepsilon_1 E_{01} = \varepsilon_2 E_{02}$). In this case the evolution equation for the surface deformation is

$$\left(\sum \rho\right)\frac{d^2\xi}{dt^2} + \left\{-ag[\rho] - a^2[E_0]\frac{\varepsilon_2 E_{02}(\sigma_1 + \lambda\varepsilon_1) - \varepsilon_1 E_{01}(\sigma_2 + \lambda\varepsilon_2)}{\sum(\sigma + \lambda\varepsilon)}\right.$$

$$\left. + a^2\Gamma\frac{[\varepsilon E_0]\sum E_0(\sigma + \lambda\varepsilon)}{\sum(\sigma + \lambda\varepsilon)} + a^3 S\right\}\xi$$

$$= \left\{-a\sum \rho\nu(m + a) + \frac{\Gamma}{[\rho\nu(m + a)]}\left(a - \frac{2a^2[\eta]}{[\rho\nu(m + a)]}\right)\right\}\frac{d\xi}{dt} \tag{96}$$

which is the counterpart of Eq. (78) for the perpendicular case [45]. The factor Γ is given as

$$\Gamma = -\frac{2a[\eta] - [\rho\nu(m + a)] + (a[\varepsilon^2 E_0^2]/\sum(\sigma + \lambda\varepsilon))}{\sum \rho\nu(m + a) + (a[\varepsilon E_0]^2/\sum(\sigma + \lambda\varepsilon))} \tag{97}$$

As for the tangential field case let us consider a few particular cases.

1. $\sigma_1 = \sigma_2 = 0$. Again for two dielectrics, using the argument provided for the tangential case, one expects no sustained wave. However, a static instability is possible. In the limit $\omega/\nu \ll a^2$ we have

$$\left(\sum \rho\right)\frac{d^2\xi}{dt^2} + \left\{-ag[\rho] - a^2\frac{[\varepsilon]^2}{\sum \varepsilon}E_{01}E_{02} + a^3 S\right\}\xi = -2a^2\sum \eta\frac{d\xi}{dt} \tag{98}$$

which again is an equation describing damped oscillations.

The spring constant provides a new static (frozen wavelike) profile different from the flat level position.

2. $\rho_1 = \rho$, $\rho_2 = 0$, $\sigma_1 \to \infty$, $\sigma_2 = 0$, $E_{01} \sim 0$, and $\varepsilon_1 \sim \varepsilon_2 \sim \varepsilon_0$. This is the case of a liquid metal (e.g., mercury)–air interface. No sustained waves are expected. We look for static instability by taking again $\omega/\nu \ll a^2$. Equation (96) becomes

$$\rho\frac{d^2\xi}{dt^2} + (a\rho g - a^2\varepsilon_0 E_{02}^2 + a^3 S)\xi = -2a^2\frac{d\xi}{dt} \tag{99}$$

which, once more, describes a damped oscillator. As in the preceding case, when the spring constant vanishes we can have a new equilibrium position provided the field intensity reaches the threshold value $E_{02_c}^2 = 2(\rho g S)^{1/2}/\varepsilon_0$. Thus the level position of, say, the mercury–air interface when the value E_{02_c} is reached undergoes a transition to a static wavy profile. We simply have a change of surface relief from flat to patterned.

3. $\sigma_1, \sigma_2 \gg \omega\varepsilon$. The values $\varepsilon \sim 10^{-10}$, $\sigma \sim 10^{-6} - 10^{-8}$, $\omega \sim 1$–10 fit well for deionized water. Field values of 1–10 kV per centimeter are typical of

experiments. If $\eta \sim 10^{-3}$ (all units in CGS system) then $\varepsilon E^2 \ll \omega\eta$. Using now the high-frequency approximation, $\omega/\nu \gg a^2$, the harmonic oscillator equation can be reduced to

$$\left(\sum \rho\right)\frac{d^2\xi}{dt^2} + \left\{ -ag[\rho] - a^2[E_0]\frac{\varepsilon_2 E_{02}\sigma_1 - \varepsilon_1 E_{01}\sigma_2}{\sum \sigma} - 2a^2 J_0 \frac{[\varepsilon E_0]}{\sum \sigma}\frac{[\rho\sqrt{\nu}]}{\sum(\rho\sqrt{\nu})} + a^3 S \right\} = \frac{1}{\sum(\rho\sqrt{\nu})}\left\{ -2a\rho_1\rho_2\sqrt{2\omega\nu_1\nu_2} + \frac{2\sqrt{2}a^3[\eta]J_0[\varepsilon E_0]}{\omega^{3/2}\sum\sigma}\right\}\frac{d\xi}{dt} \tag{100}$$

with $J_0 = \sigma_i E_{0i}$ (continuity of the current). Thus in terms of electric current we see that sustained oscillations are possible provided $J_0 > J_{0_c}$ with

$$J_{0_c}^2 = \frac{2(\sum\sigma)\sqrt{-[\rho]gS}}{\frac{[\eta][\varepsilon/\sigma]\sum\rho}{\rho_1\rho_2\sqrt{\nu_1\nu_2}} + 2\frac{[\rho\sqrt{\nu}]}{\sum(\rho\sqrt{\nu})}[\varepsilon/\sigma] - [\sigma][\varepsilon/\sigma^2]} \tag{101}$$

For a two-layer system with large but finite liquid depths, h_1 and h_2, Eq. (101) amounts to an easily obtained critical voltage (ΔV) ($\sigma_1 E_{01} = \sigma_2 E_{02} = J_0$ and $\Delta V = E_{01}h_1 + E_{02}h_2$, with $h_1, h_2 \ll a_c^{-1}$).

Thus we have shown that an electric field applied to the interface between two liquids brings consequences very much like those found with the Marangoni effect in interfacial heat and mass diffusion processes.

To close this section let us refer the reader to Refs. 46–55 where the reader may find more complete results than those provided here (for illustration) and material for further development.

REFERENCES

1. Benard, H. Les tourbillons cellulaires dans une nappe liquide. Premiere partie: Description generale des phenomenes. Rev. Gen. Sci. Pures Appl. (Paris) **1900**, *11*, 1261–1271; Deuxieme partie: Procèdes mecaniques et optiques d'examen. Lois numeriques des phenomenes. Rev. Gen. Sci. Pures Appl. (Paris) **1900**, *11*, 1309–1328.
2. Benard, H. Les tourbillons cellulaires dans une nappe liquide transportant de la chaleur par convection en regime permanent. Ann. Chim. Phys. (Paris) **1901**, *23*, 62–143.
3. Hershey, A.W. Ridges in a liquid surface due to the temperature dependence of surface tension. Phys. Rev. **1939**, *56*, 204.

4. Block, M.J. Surface tension as the cause of Benard cells and surface deformation in a liquid film. Nature (London) **1956**, *178*, 650–651.
5. Koschmieder, E.L. *Benard Cells and Taylor Vortices*; Cambridge University Press: Cambridge, 1993.
6. Pearson, J.R.A. On convection cells induced by surface tension. J. Fluid Mech. **1958**, *4*, 489–500.
7. Smith, K.A. On convective instability induced by surface-tension gradients. J. Fluid Mech. **1966**, *24*, 401–414.
8. Manneville, P. *Dissipative Structures and Weak Turbulence*; Academic Press: San Diego, C.A., 1990.
9. Colinet, P., Legros, J.C., Velarde, M.G., *Nonlinear Dynamics of Surface-Tension-Driven Instabilities*; Wiley-VCH: Berlin, 2001.
10. VanHook, S.J., Schatz, M.F., Swift, J.B., McCormick, W.D., Swinney, H.L. Long-wavelength surface-tension-driven Benard convection: experiment and theory. J. Fluid Mech. **1997**, *345*, 45–78.
11. Schatz, M.F., Neitzel, G.P. Experiments on thermocapillary instabilities. Annu. Rev. Fluid Mech. **2001**, *33*, 93–129.
12. Bezdenezhnykh, N.A., Briskman, V.A., Cherepanov, A.A., Sharov, M.T. Control of the stability of liquid surfaces by means of variable fields. Fluid Mech. Sov. Res. **1986**, *15*, 11–32.
13. Garcia-Ybarra, P.L., Velarde, M.G. Sustained gravity-capillary (Laplace) waves at interfaces and Marangoni–Benard instability. In *Physicochemical Hydrodynamics. Interfacial Phenomena*; Velarde, M.G., Ed.; Plenum: New York, 1988; 179–181.
14. Velarde, M.G., Nepomnyashchy, A.A., Hennenberg, M. Onset of oscillatory interfacial instability and wave motions in Benard layers. Adv. App. Mech. **2000**, *37*, 167–238.
15. Velarde, M.G., Chu, X.L. The harmonic oscillator approach to sustained gravity-capillary (Laplace) waves at liquid interfaces. Phys. Lett. A **1988**, *131*, 430–432.
16. Garcia-Ybarra, P.L., Velarde, M.G. Oscillatory Marangoni–Benard interfacial instability and capillary-gravity waves in single- and two-component liquid layers with or without Soret thermal diffusion. Phys. Fluids *30*, 1649–1655.
17. Chu, X.L., Velarde, M.G. Sustained transverse and longitudinal waves at the open surface of a liquid. Physicochem. Hydrodyn. **1988**, *10*, 727–737.
18. Levchenko, E.B., Chernyakov, A.L. Instability of surface waves in a nonuniformly heated liquid. Soviet Phs. JETP **1981**, *54*, 102–105.
19. Takashima, M. Surface tension driven instability in a horizontal liquid layer with a deformable free surface. I. Stationary convection, J. Phys. Soc. Japan **1981** *50*, 2745–2750; II. Overstability. J. Phys. Soc. Japan **1981**, *50*, 2751–2756.
20. Chu, X.L., Velarde, M.G. Transverse and longitudinal waves induced and sustained by surfactant gradients at liquid–liquid interfaces. J. Colloid Interface Sci. **1989**, *131*, 471–484.
21. Hennenberg, M., Chu, X.-L., Sanfeld, A., Velarde, M.G. Transverse and longitudinal waves at the air–liquid interface in the presence of an adsorption barrier. J. Colloid Interface Sci. **1992**, *150*, 7–21.
22. Rednikov, A.Ye., Colinet, P., Velarde, M.G., Legros, J.C. Oscillatory instability and high-frequency wave modes in a Marangoni–Benard layer with deformable free surface. J. Non-Equilib. Thermodyn. **2000**, *25*, 381–405.

23. Rednikov, A.Ye., Colinet, P., Velarde, M.G., Legros, J.C. Two-layer Benard–Marangoni instability and the limit of transverse and longitudinal waves. Phys. Rev. E **1998**, *57*, 2872–2884.
24. Chu, X.-L., Velarde, M.G. Dissipative hydrodynamic oscillators. V. Interfacial oscillations in finite containers. Il Nuovo Cimento **1989**, *11* D, 1615–1629.
25. Rednikov, A.Ye., Colinet, P., Velarde, M.G., Legros, J.C. Rayleigh–Marangoni oscillatory instability in a horizontal liquid layer heated from above: coupling and mode mixing of internal and surface dilational waves. J. Fluid Mech. **2000**, *405*, 57–77.
26. Linde, H. Marangoni instabilities. In *Convective Transport and Instability Phenomena*; Zierep, J., Oertel Jr., H., Eds.; Braun-Verlag: Karlsruhe, 1982; 265–296.
27. Linde, H., Velarde, M.G., Wierschem, A., Waldhelm, W., Loeschcke, K., Rednikov, A.Ye. Interfacial wave motions due to Marangoni instability. I. Traveling periodic wave trains in square and annular containers. J. Colloid Interface Sci. **1997**, *188*, 16–26.
28. Wierschem, A., Velarde, M.G., Linde, H., Waldhelm, W. Interfacial wave motions due to Marangoni instability. II. Three-dimensional characteristics of surface waves in annular containers. J. Colloid Interface Sci. **1999**, *212*, 365–383.
29. Linde, H., Velarde, M.G., Waldhelm, W., Wierschem, A. Interfacial wave motions due to Marangoni instability. III. Solitary waves and (periodic) wave trains and their collisions and reflections leading to dynamic network (cellular) patterns in large containers. J. Colloid Interface Sci. **2001**, *236*, 214–224.
30. Wierschem, A., Linde, H., Velarde, M.G. Properties of surface wave trains excited by mass transfer through a liquid surface. Phys. Rev. E **2001**, *64*, 022601-1-4.
31. Santiago-Rosanne, M., Vignes-Adler, M., Velarde, M.G. Dissolution of a drop on a liquid surface leading to surface waves and interfacial turbulence. J. Colloid Interface Sci. **1997**, *191*, 65–80.
32. Chu, X.-L., Velarde, M.G. Dissipative hydrodynamic oscillators. VI. Transverse interfacial oscillations not obeying the Laplace–Kelvin law. Il Nuovo Cimento **1989**, *11* D, 1631–1641.
33. Sternling, C.V., Scriven, L.E. Interfacial turbulence: Hydrodynamic instability and the Marangoni effect. AIChE J. **1959**, *5*, 514–523.
34. Reichenbach, J., Linde, H. Linear perturbation analysis of surface-tension-driven convection at a plane interface (Marangoni instability) J. Colloid Interface Sci. **1981**, *84*, 433–443.
35. Davies, J.T., Rideal, E.K. *Interfacial Phenomena*; Academic Press: New York, 1963.
36. Velarde, M.G., Chu, X.L. Dissipative hydrodynamic oscillators. I. Marangoni effect and sustained longitudinal waves at the interface of two liquids. Il Nuovo Cimento **1989**, *D* 11, 709–716.
37. Lucassen, J. Longitudinal capillary waves. Part 1. Theory. Trans Faraday Soc. **1968**, *64*, 2221–2229; Part 2. Experiments. Trans Faraday Soc. **1968**, *64*, 2230–2235.
38. Wierschem, A., Linde, H., Velarde, M.G. Internal waves excited by the Marangoni effect. Phys. Rev. E **2000**, *62*, 6522–6530.
39. Melcher, J.R. *Continuum Electromechanics*; MIT Press: Cambridge, M.A., 1981.
40. Melcher, J.R. *Field-Coupled Surface Waves*. MIT Press: Cambridge, M.A., 1963.

41. Melcher, J.R., Taylor, G.I. Electrohydrodynamics: a review of the role of interfacial shear stresses. Annu. Rev. Fluid Mech. **1969**, *1*, 111–146.
42. Ostroumov, G.A. *Interaction of Electrical and Hydrodynamic Fields*; Nauka: Moscow, 1979 (in Russian).
43. Castellanos, A. Coulomb-driven convection in electrohydrodynamics. IEEE Trans Elect. Insul. **1991**, *26*, 1201–1215.
44. Castellanos, A. Electrohydrodynamics: basic equations and dimensionless numbers. In *Fluid Physics*; Velarde, M.G., Christov, C.I., Eds.; World Scientific: Singapore, 1995; 14–32.
45. Chu, X.L., Velarde, M.G., Castellanos, A. Dissipative hydrodynamic oscillators. III. Electrohydrodynamic interfacial waves. Il Nuovo Cimento **1989**, *D* 11, 727–737.
46. Malkus, W.V.R., Veronis, G. Surface electroconvection. Phys. Fluids **1961**, *4*, 13–23.
47. Melcher, J.R., Schwarz, W.J. Interfacial relaxation overstability in a tangential electric field. Phys. Fluids **1968**, *11*, 2604–2616.
48. Melcher, J.R., Smith, Jr., C.V. Electrohydrodynamic charge relaxation and interfacial perpendicular-field instability. Phys. Fluids **1969**, *12*, 778–790.
49. Saranin, V.A. On convective stability of a low-conductivity liquid in an electrical field. Izv. Akad. Nauk SSSR, Mech. Zhidk. Gaza **1976**, *5*, 16–23.
50. Zhakin, A.I. Stability of a horizontal free surface of a low-conductivity liquid in a tangential alternating electrical field. Magnitnaya Gidrodinamika (Magnetohydrodynamics) **1981**, *3*, 7480–7490.
51. Briskman, V.A., Saranin, V.A. About the opportunity of control of hydrodynamic processes with the help of an electrical field. In *The Hydrodynamics and Heat-Mass-Transfer in Microgravity Condition*; Nauka: Moscow, 1982; 147–154.
52. Zhakin, A.I. Linear and nonlinear waves in liquid non-conductor. Izv. Akad. Nauk SSSR, Mech. Zhidk. Gaza **1983**, *1*, 98–106.
53. Castellanos, A., González, A. Interfacial electrohydrodynamic instability: The Kath and Hoburg model revisited. Phys. Fluids A **1992**, *4*, 1307–1309.
54. Saranin, V.A., Zharov, A.N., Belonozhko, D.F. Oscillatory instability of a interface of conductivity liquids in normal electrical field. Pisma JTF (Letters JTPh) **1997**, *23*, 41–44.
55. Birikh, R.V., Lyushnin, A.V. Influence of Marangoni convection on the injection mechanism of instability. Zh. Tec. Fiz. (J. Tech. Phys.) **2000**, *70*, 19–23.

3

Excitation of Isothermal Liquid Surface Instability by Variable (Vibrational and Electric) Force Fields

In this chapter we shall consider the parametric excitation of oscillations in an isothermal incompressible liquid. The investigation of this phenomenon was pioneered by Faraday [1], who observed wave generation (now called Faraday ripples) on the free surface of a liquid in a container executing vertical oscillations. An explanation for this effect was provided by Lord Rayleigh [2], who attributed it to parametric resonance, describing the wave excitation by the Mathieu equation [3]. Indeed, as in earlier treated cases, the liquid surface acts like a membrane and its natural vibrating frequencies are surface waves that depend on the gravitational acceleration. If a container filled with a liquid executes vertical oscillations, as a whole, the gravitational acceleration and natural frequency are periodic variables. We have here the basic ingredients for the onset of parametric resonance. This effect is also perfectly illustrated by Kapitza's pendulum, that is, a pendulum with a vertically oscillating suspension point [4,5].

Parametric wave excitation under the action of different agents is still today the subject of considerable theoretical and experimental interest [6–16]. Our understanding of the problem came from the development of parametric wave theory for an ideal, inviscid liquid. Taking viscosity into account leaves only the calculation of the excitation thresholds where energy pumping compensates for viscous dissipation. Although stationary surface waves with half the driving frequency are observed in liquids of different kinematic viscosities it was recently shown that the mechanism that produces the instability is very different for a highly viscous fluid compared to the nearly inviscid case [15]. Furthermore, much effort has been spent on the nonlinear aspects of wave evolution, leading to our understanding of supercritical wave structures, onset, and interaction of patterns, defects, and their evolution, etc. ([16] and Refs. therein). In this chapter we shall only study two cases. First we shall investigate the parametric generation of liquid oscillations in communicating containers or vessels and then the onset of gravito-electro-capillary surface waves. Later in the book we shall consider the influence of a surface tension gradient, and so the Marangoni or thermocapillary effect on Faraday-ripple excitation.

I. PARAMETRIC LIQUID PENDULUM

Let us consider a U-tube partially filled with a liquid as illustrated in Fig. 1. In the gravity field on Earth a liquid equilibrium state may exist and any natural, spontaneous oscillatory disturbances of the equilibrium are damped by viscosity. It has been shown that the natural frequency depends on gravity (inertial) acceleration [18,19]. Consequently, the U-tube is a good candidate for observing a parametric resonance as in Faraday experiments, albeit in a simpler one-dimensional geometry.

The threshold of excitation of the main parametric resonance was predicted and measured [19]. Subsequently, the problem was studied with nonlinear damping added ad hoc to the standard Mathieu equation, thus mimicking the role of viscous friction [20]. Numerical results obtained with a fourth-order Runge–Kutta method showed peculiar nonlinear behaviors in the system. However, the damping added to simulate fluid friction is far from realistic. In Refs. 21 and 22 the interest focused on the experimental determination of instability boundaries of the equilibrium states with respect to both small random disturbances (self-excitation) and driven finite-amplitude disturbances. The analysis of nonlinear excitation

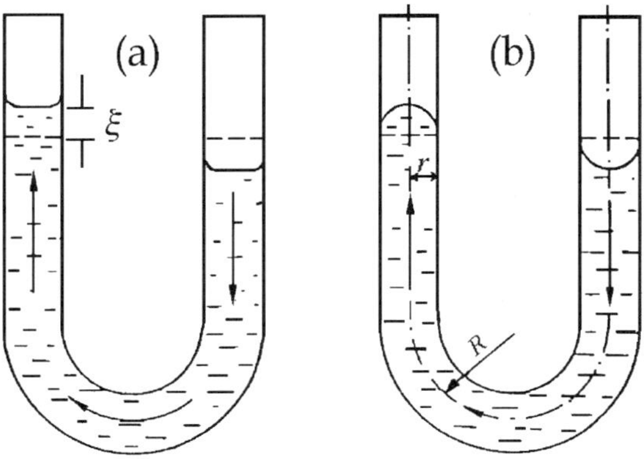

FIG. 1 Liquid pendulum showing two types of oscillation in a U-tube: (a) piston oscillations, and (b) meniscus oscillations. (From Ref. 22.)

regimes permitted the identification of regions of existence of stationary oscillations.

A. Formulation of the Problem: Equations for Disturbances

We shall restrict consideration to small-amplitude oscillations of liquid in the U-tube having inner diameter d (Fig. 1). The total length of the liquid column inside the tube is $2L$ with $L > R > d$ where R is the radius of curvature of the tube.

Disregarding surface waves (ripples) in each elbow of the U-tube let us focus our attention on the oscillations observed in both sides as the liquid moves from one tube elbow to the other. Two possibilities exist at the gas–liquid–tube wall interface according to experiment [21,22]: (1) the edges of the menisci move freely along the wall of the tube, so that when oscillating each meniscus preserves its equilibrium shape and hence capillary forces are negligible ("*a*"-type or "piston" oscillations), and (2) the edges of the menisci adhere to the tube wall, and, at small amplitude oscillations, the surface of each meniscus bends as a membrane with fixed anchoring ("*b*"-type or "meniscus" oscillations).

Piston-type oscillation is possible only if the meniscus edges are detached from the wall, i.e., there exists an initial shear stress. This oscillation was experimentally observed when the wall was fully wetted by the liquid. Adhesion of the meniscus was observed with incomplete wetting and practically always occurred when no special care was taken to clean the contact surfaces. The initial shear stress is caused by the adhesion between the liquid and the solid surface, i.e., for the same reason that contact angle hysteresis exists in flow along tubes.

B. Natural Oscillations

The oscillatory solutions (natural oscillations) satisfying the vertical projection of the Euler equations can be easily obtained in terms of one-dimensional flow approximation and ordinary boundary conditions on the free surface. For the frequency Ω of natural oscillations of a- and b-type we have the following expressions:

$$\Omega_a^2 = \frac{g_0}{L}, \qquad \Omega_b^2 = \frac{g_0}{L} + \left(\frac{4\sigma}{\rho L d^2}\right) j_{mn}^2 \tag{1}$$

where g_o is the reference value of the gravitational acceleration, σ is the surface tension, ρ is the liquid density, and j_{mn} are roots of the zeroes of Bessel's n-order function of the first kind; $m = 1, 2, \ldots, n = 0, 1, 2, \ldots$. The theoretical values agree well with experimental data (Fig. 2).

The damping factor, δ, for the oscillation frequency, γ, can also be estimated. At high enough frequencies we may introduce a dimensionless

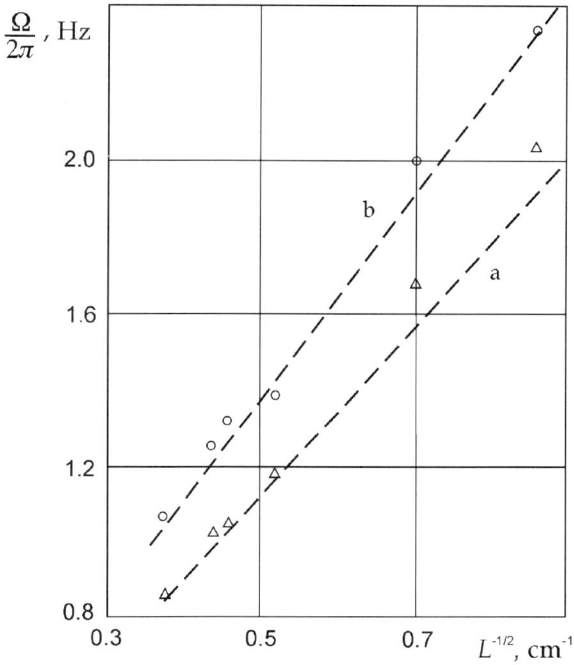

FIG. 2 Parametrically excited liquid pendulum showing dependence of the natural frequencies (a) Ω_a and (b) Ω_b, on total liquid column length. (From Ref. 22.)

dissipation parameter, ε, such that

$$\varepsilon = \left(\frac{2\nu}{\gamma d^2}\right)^{1/2} \ll 1 \tag{2}$$

Thus, a wall skin layer, with thickness εd, is formed in the liquid because of its kinematic viscosity, ν. There the oscillating flow is damped. Assuming that energy dissipation occurs mainly in such a skin layer and representing δ in terms of dissipation we obtain

$$\delta = \delta_a = \delta_b = \varepsilon\gamma \tag{3}$$

which agrees with the calculated value of δ_a for piston-type oscillations [19] and the experimental data on naturally damped oscillations [17,18] (Fig. 3).

C. Parametric Oscillations

Now let the container with liquid, as a whole, experience vertical oscillations or vibrations with amplitude a and frequency ω. Then in the noninertial reference

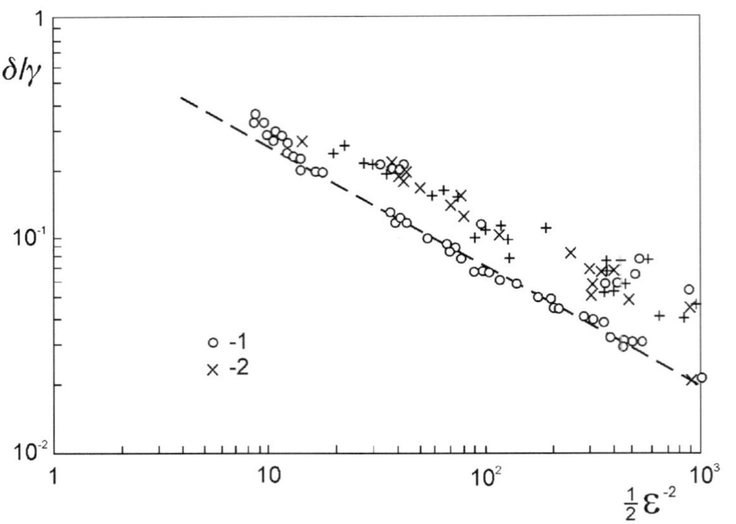

FIG. 3 Parametrically excited liquid pendulum showing dependence of the damping factor, δ/γ, on the dissipation parameter, ε. Results taken from Ref. 17 (○) and from Ref. 18 (× and +); the dashed line corresponds to Eq. (3). (From Ref. 22.)

system attached to the tube the *effective* gravity is a modulated function of time

$$g = g_0 - a\omega^2 \cos \omega t. \tag{4}$$

Modulation does not disturb the equilibrium conditions in the liquid and merely causes pressure oscillations. However the *effective* natural frequencies appear to be varying parameters, which might be responsible for excitation of parametric resonance. In this case, the evolution of small surface displacements, ξ, is described by the Mathieu equation including damping [3,19],

$$\frac{d^2\xi}{dt^2} + 2\delta\frac{d\xi}{dt} + (\Omega^2 + q \cos \omega t)\xi = 0 \tag{5}$$

with Ω and δ defined by Eq. (1) and Eq. (3), respectively. The modulation frequency is equal to the vibrational frequency, ω, and the modulation amplitude, q, is equal to $a\omega/L$. As noted earlier, surface waves are not being considered; therefore we disregard the possible Faraday-ripple parametric excitation. In experiments this was easily achieved because the frequency of surface capillary-gravity waves with wave number k is $[(\sigma k^3/\rho) + gk]^{1/2}$, and hence largely different from either value in Eq. (1).

To define the main region of parametric resonance, where the frequency of liquid oscillations, γ, is about half the frequency of the drive, ω, and is close to the natural frequency, $\Omega(\gamma = \omega/2 \approx \Omega)$, we need to expand the solution to Eq. (5) in a power series of the small modulation amplitude [2,23]. A linear approximation with respect to q gives

$$\left[\left(\frac{a}{L}\right)^2 - \varepsilon^2\right]^{1/2} = \pm 2\left(\frac{\Omega^2}{\omega^2} - \frac{1}{4}\right) \tag{6}$$

From this it follows that the threshold value of the dimensionless vibrational amplitude is $(a/L)^* = \varepsilon$.

D. Experimental Set-Up and Experimental Procedure

The set-up for the experiments to investigate excitation conditions and stationary parametric oscillations in a liquid consisted of a U-tube partially filled with liquid, a vibration machine to create vertical vibrations in the tube, and a device to record the liquid oscillations [21,22]. Nearly harmonic vertical motions of the vibration machine and the U-tube fixed to the platform were generated by an electric motor, reductor, and eccentric. The amplitude and frequency of vibrations could be varied from 3.0 to 30 mm (± 0.5) and from 0.3 to 5 Hz, (± 0.02) respectively. The vibration machine was equipped with a measuring system including a capacity-type displacement gauge, an amplifier, and an oscilloscope. These devices were used to measure the amplitude (absolute error, ± 0.05 mm) and the frequency of liquid oscillations relative to the wall of the tube.

To gain a better understanding of the oscillatory behavior several runs were made with different boundary conditions, using tubes of different inner diameter values ($d = 1.85$, 1.25, and 0.76 cm), various liquids (water or 96% ethanol), and changing the natural oscillation frequency by varying the length of the liquid column ($2L = 137 \div 689$, ± 1 mm), and the vibration amplitude and frequency.

The design of the experimental set-up allowed all measurements to be carried out by quasi-statically changing the frequency at fixed vibration amplitude. Liquid oscillations in the tube were observed in two (a, ω) regions, one having $\omega/2 \approx \Omega$ (first or main resonance region) and the other $\omega \approx \Omega$ (second resonance region). We also investigated stability with respect to finite-amplitude disturbances. The latter were excited without applying the initial velocity, simply by disturbing the level of the liquid surface in both U-tube elbows from their equilibrium positions.

E. Parametric Oscillations in the Absence of Initial Shear Stress

With complete wetting one can observe self-induced piston-type oscillations. Fig. 4 shows the neutral curve [Eq. (6)] in the dimensionless coordinates $\beta =$

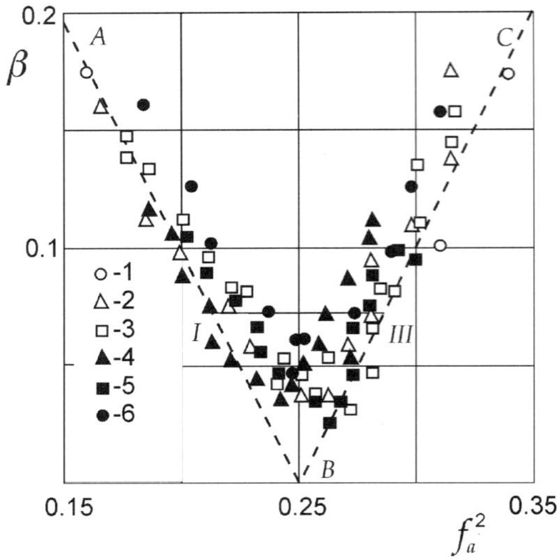

FIG. 4 Parametrically excited liquid pendulum, showing region of self-excited piston oscillations in the absence of initial shear stress. Dashed lines *BA* and *BC*, define the neutral curve Eq. (6) in the dimensionless coordinates $\beta \equiv [(a/L)^2 - \varepsilon^2]^{1/2}$ and $f_a^2 = \Omega_a^2/\omega^2$. Experimental data correspond to the main instability region for water columns of various lengths (○, △, □, ▲, ■; $2L = 13.7$, 28.4, 37.0, 45.6, and 53.5 cm, respectively) and to ethanol (●; $2L = 37.2$ cm). (From Ref. 21.)

$[(a/L)^2 - \varepsilon^2]^{1/2}$ and $f_a^2 = \Omega_a^2/\omega^2$ (dashed line) and experimental data corresponding to the main instability region for water columns of various lengths $2L = 13.7$, 28.4, 37.0, 45.6, and 53.5 cm (○, △, □, ▲, ■) and for an ethanol column of length $2L = 37.2$ (●) in a pipe with $d = 1.85$ cm. For intermediate values of β the experimental results agree well with theory.

The dimensionless amplitude, $A = A'/L$, of the stationary liquid oscillations plotted as a function of f_a for the section *I–III* (Fig. 4) at a fixed value of $a/L = 0.081$ (in the experiment with ethanol, $2L = 37.2$ cm and $d = 1.85$ cm) is shown in Fig. 5. Symbols △ and ▽ refer to experimental data obtained with increasing and decreasing frequency, respectively.

Figure 6 presents the equilibrium region, *ABC*, unstable to small random disturbances and the region, *DEFGH*, where the liquid oscillations have finite amplitude (experiment with water, $2L = 53.5$ cm and $d = 1.85$ cm). The arrows indicate the direction of frequency variation and symbols ○, □, and △ correspond to self-excitation, oscillation breakdown, and equilibrium instability to finite-amplitude disturbances, respectively. The experimental curves in Figs. 5 and 6 show that the excitation of piston oscillations occurs in a jumplike manner and exhibit hysteresis.

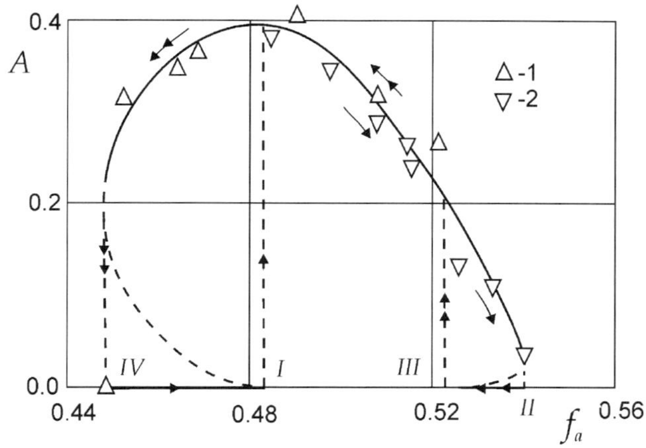

FIG. 5 Parametrically excited liquid pendulum, showing excitation, development, and disappearance of 'piston' oscillations in the absence of initial shear stresses. Symbols △ and ▽ refer to experimental data obtained at increasing and decreasing frequency, respectively. (From Ref. 21.)

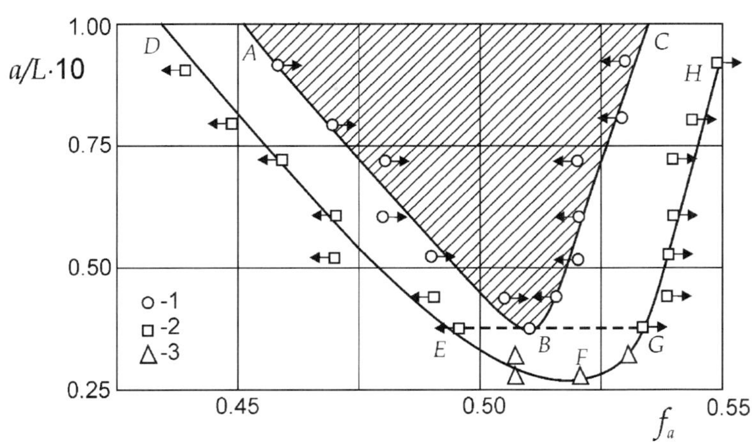

FIG. 6 Parametrically excited liquid pendulum, showing regions of equilibrium instability and of finite amplitude 'piston' oscillations in the absence of initial shear stresses; ○ self-excitation, □ disappearance of oscillations, and △ finite amplitude instability. (From Ref. 21.)

If for a quiescent state the modulation frequency is lowered (this corresponds to shifting from left to right the relevant point on the plot), self-excitation of oscillations occurs at point *I* in Fig. 5 (curve *AB* in Fig. 6). An increase in frequency shifts the self-excitation to point *III* in Fig. 5 (curve *BC* in Fig. 6). Oscillations are induced from the equilibrium in a jumplike manner: the amplitude of the steady-state motion abruptly rises from zero to a value, which, according to our expectation, is by two orders of magnitude far beyond the upper limit of the measuring instrument range. Furthermore, there is hysteresis: the stationary oscillations remain beyond the self-excitation region. Oscillations disappear at point *II* or point *IV* in Fig. 5 (curves *GH* or *DE* in Fig. 6) and are related to a decrease or an increase in the frequency. Straight dashed lines in Fig. 5 represent amplitudes for the transition to stationary oscillation (*I* and *III*) and broken (*II* and *IV*) one. The *ABED* and *CBGH* regions in Fig. 6 are free from self-excitation, although, together with oscillations following from the *ABC* instability region, they may occur with finite-amplitude disturbances. In the latter case the initial displacements of the liquid column must exceed the values qualitatively described in Fig. 5 by curved dashed lines fitting the instability portions of the amplitude curves for parametric oscillations. The subcritical region is bounded by the experimental curve *EFG* (Fig. 6), below which no amplitude disturbance can induce stationary oscillations.

F. Parametric Oscillations in the Presence of Initial Shear Stress

Experiments show that the adhesion of the meniscus to the wall of the tube (dry friction) does not permit piston-type oscillations, at least for weak disturbances. These conditions favor self-excitation of meniscus oscillations only, which in turn might facilitate the subsequent excitation of piston motions. Therefore, in the presence of initial shear stress at the liquid–gas–wall three-phase boundary, there are two ways of exciting piston oscillations: either from the equilibrium state under the action of finite-amplitude disturbances or, as a secondary effect, from instability following stationary meniscus oscillations.

Figure 7 shows the boundaries of the instability regions for weak random disturbances of *a*- and *b*-type (*ABC* and *DEG* broken lines, respectively) determined from Eq. (6) and corresponding to experimental data for water. Since the frequency shift of the regions is defined by the difference between Ω_a and Ω_b, the boundaries of the region for meniscus oscillations, using as coordinates β, and f_a^2 (Fig. 7), would not be the same for different values of Ω_b. Therefore to illustrate the excitation of meniscus oscillations we use data only from one particular case.

First, let us consider piston oscillations excited by finite-amplitude disturbances. Unlike the case described in the previous section, such disturbances can be excited only in the region between the boundaries *AB* and *BC*. Only at these

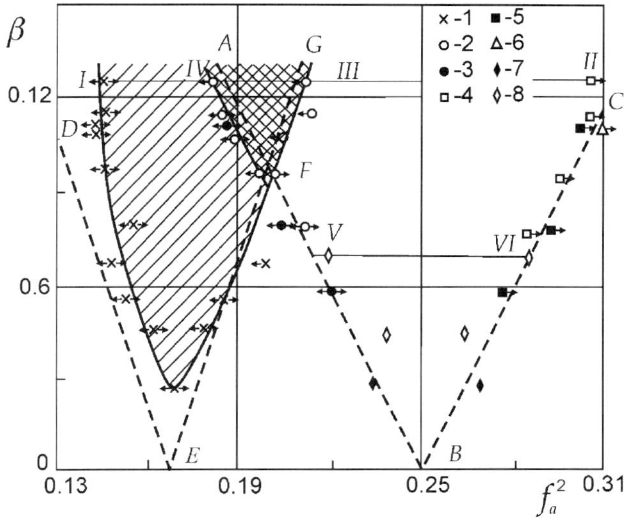

FIG. 7 Parametrically excited liquid pendulum, showing regions of equilibrium instability relative to piston and meniscus oscillations in the presence of initial shear stress determined from Eq. (6) (dashed lines *ABC* and *DEG*, respectively) and experimental data for water: ×, excitation of meniscus oscillations; ○, ● excitation of piston oscillations in the absence of finite-amplitude disturbances; □, ■, △ breakdown of the oscillation; and ◆, ◇, finite-amplitude instability. For symbols ×, ○, and □, $2L = 46.7$ cm and $d = 1.85$ cm; for symbols ●, ■, and ◆, $2L = 51.8$ cm and $d = 1.85$ cm; and for symbols △ and ◇, $2L = 50$ cm and $d = 1.25$ cm. (From Ref. 21.)

boundaries do the stationary oscillations disappear. In Fig. 8, curve 1 describes a jumplike excitation of piston oscillations that occur in section *V–VI* of the resonance region displayed in Fig. 7 (water experiment, $2L = 50$ cm, $d = 1.25$ cm, $a/L = 0.084$; notation in Fig. 8 is similar to that in Fig. 5). We have shown the plot of the parametric oscillation amplitude vs. frequency as a closed curve. The experimental points on the upper part of the curve (solid line) are the amplitudes of stationary oscillations. The experimental points on the lower part of the curve (broken line) correspond to an initial shift of liquid surface from equilibrium level (critical values of finite-amplitude disturbances), enough to excite oscillations. It appears that disturbances of larger initial value grow but disturbances of smaller initial value decay. The broken line illustrates unstable stationary oscillations. Note that at zero initial shear stress the plot of the parametric oscillation amplitude vs. frequency in a subcritical region (for example *EBGFE* in Fig. 6) has the same form (closed curve).

When no finite-amplitude disturbances are imposed, either an increase or decrease in frequency results in the onset of equilibrium instability (shaded region

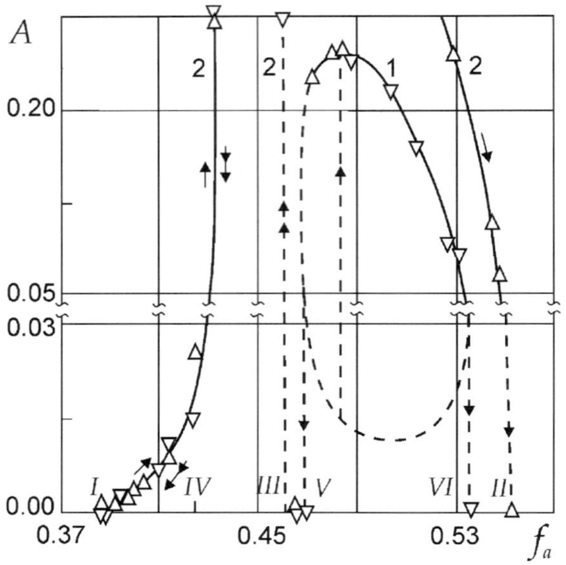

FIG. 8 Parametrically excited liquid pendulum, were curve 1 describes a jumplike excitation mode for piston oscillations that occur in section V–VI of the resonance region displayed in Fig. 7 (water experiment: $2L = 50\,\text{cm}$, $d = 1.25\,\text{cm}$, $a/L = 0.084$). Excitation of meniscus oscillations and their transition to a piston mode are illustrated by curve 2 (water experiment: $2L = 46.7\,\text{cm}$, $d1.85\,\text{cm}$). Notation is similar to that of Fig. 5. (From Ref. 21.)

in Fig. 7) in the vicinity of the *DEG* line. The onset of instability always comes with the appearance of meniscus oscillations, although the stationary motions are dissimilar at different boundary sections. The stationary meniscus oscillations occur at the high-frequency boundary (left boundary in Fig. 7). A similar oscillation mode is observed at the low-frequency boundary for modulation amplitudes lying below point *F*. The *FG* boundary defines the region of stationary piston oscillations occurring as follows. First, self-excited meniscus oscillations appear, which keep growing until at sufficiently large amplitudes they act like finite disturbances, then initiating piston oscillations. At the boundary *AF* the stationary meniscus oscillations most surely play the same role; when decreasing the frequency they change to piston motions. Thus the region where the zones of parametric resonance for piston and meniscus motions coincide (the cross-hatched zone in Fig. 7) is that where piston oscillations can be excited in the absence of any other imposed disturbances.

The excitation of meniscus oscillations and their transition to a piston mode are illustrated by line 2 in Fig. 8 (water, $2L = 46.7$ cm and $d = 1.85$ cm) for the amplitude of modulation $a/L = 0.128$ corresponding to the abcissa *I–IV–III–II* in Fig. 7. The amplitude curve is cut at the top, because at $A > 0.275$ the fluctuations of the liquid surface go beyond the upper limit of the instrument range. The plot shows that excitation of meniscus oscillations at point *I* occurs in a soft manner. Hysteresis has not been detected and the amplitude grows slowly while penetrating deeper into the instability region. This justifies why the calculated and experimental curves in Fig. 7 differ slightly (the amplitude recorded in experiments was $A' > 0.1$ mm). At point *IV* (boundary *AF* in Fig. 7) the amplitude of stationary oscillations rapidly increases and the meniscus oscillations become piston-type. An inverse transition takes place at the same frequencies and hence a subcritical bifurcation occurs at point *IV*.

Excitation of piston oscillations at point *III* (*FG* boundary) occurs abruptly in a jumplike manner. As the modulation frequency decreases, the steady piston oscillations continue until they disappear at point *II* (*BC* boundary). Figure 7 shows that in all cases the experimental points corresponding to the transition from meniscus to piston motions, and vice versa, are located close to the boundary *AB* while the points corresponding to the disappearance of oscillations are close to *BC*. Hence, regardless of how excitation occurs in the presence of initial shear stress, stationary piston oscillations can be stable only in the region of parameter values where self-excitation occurs in the absence of initial stresses.

II. PARAMETRIC EXCITATION OF WAVES WITH A VARIABLE ELECTRIC FIELD

Let us now consider the parametric excitation of gravito-electro-capillary waves on a flat surface very much as described in Chapter 2.

A. Equations for Perturbations and Stability Conditions

Let us consider a flat liquid–gas interface in a homogeneous gravitational field, **g**, orthogonal to the surface. The coordinate origin lies on the interface and the z-axis points down into the liquid. At $g_z > 0$ the quiescent state is assumed stable, where g_z is the projection of **g** into the vertical axis. Capillary-gravity waves, which can be excited on such a surface, naturally, decay because of viscous dissipation. Now let us also impose a homogeneous, albeit time-variable electric field, normal to the surface of the liquid, conductor, or dielectric, $E_z = E_0 \cos \omega t$. If the liquid is a good dielectric or conductor, there is no volumetric electric force and to study the evolution of the liquid under oscillatory driving it is necessary to add, as in the earlier cases, Maxwell's equations and the corresponding electrohydrodynamic stress tensor in the boundary conditions on the liquid surface.

The amplitude, ξ, of small periodic oscillations with wave number, k, of the surface of a viscous liquid then satisfies a Mathieu-type equation with again a dissipative term [Ref. 3, compare with Eq. (5)]:

$$\frac{d^2\xi}{dt^2} + 2\delta\frac{d\xi}{dt} + (\Omega^2 + q\cos 2\omega t)\xi = 0 \tag{7}$$

where here k denotes wave number and

$$\delta = 2\nu k^2, \qquad q = \frac{E_0^2 k^2}{8\pi\rho}, \qquad \Omega^2 = \left(\frac{\sigma}{\rho}\right)k^3 - q + gk \tag{8}$$

Owing to the square-law relationship between the force and the electric field strength, the modulation frequency is equal to twice the frequency of a correspondingly modified field. As seen from the dispersion relation, Eq. (8), for capillary-electro-gravity waves, Ω^2 is a function of the modulation amplitude, q, which is a result of to the abovementioned quadratic relationship between the electric force and the field strength (better *ED* as earlier noted). For this reason a static instability is possible in a variable normal field ($q > 0$) at $\Omega^2 = 0$. This instability is analogous to the Frenkel instability in a constant field. Since we are interested here in modulation-type mechanisms of instability we can assume that they are more dangerous than the static instability only observed at quite low frequencies [24].

The modulation of the parameter ($q \neq 0$) may result in the appearance of oscillations (parametric resonance) in domains of parameters (q, ω) adjacent to the states described by the resonance condition $\Omega = n\omega$ (n is an integer). For liquids of low viscosity the boundaries of instability can be found by a power-series expansion of the amplitude of modulations. For the main resonance boundary ($n = 1$) we obtain

$$q_1 = 4\delta\omega \tag{9}$$

The threshold modulation amplitude, which, owing to the dynamic nature of the instability is a function of the dissipation level, increases monotonically with increasing frequency. The lowest critical value of the electric field intensity obtained from Equation (9) is

$$E_1 = 8(\pi\rho\nu\omega)^{1/2} \tag{10}$$

Because for the unbounded surface of a fluid the spectrum of wave numbers and natural frequencies is continuous, the condition $\Omega = \omega$ is satisfied whatever the value of the driving frequency and determines the wave number of a perturbation leading to instability. The assumption about the smallness of the modulation amplitude, $q \ll \Omega^2$, restricts the validity of Eq. (10) to suitably high frequencies, $\omega \gg 4\nu k^2$.

B. Experimental Results for Interface Instability in a Variable Electric Field

Parametric instability in an electric field has been studied experimentally [8]. The set-up (Fig. 9) consisted of a liquid-filled container, electrodes with clamps to keep them horizontal, a high-voltage source of controllable frequency and amplitude, a Schlieren system with a vertical optical axis, steady and stroboscopic light sources, and a photorecording system. An electrode, with diameter smaller than the transverse dimension of the container, was placed horizontally at a height of about 1 cm above the liquid surface. A variable potential difference with frequency range $20-50\,s^{-1}$ and amplitude up to 15 kV was produced between the electrode and the surface. Since the time-averaged magnitude of the (Korteweg–Helmholtz) electric

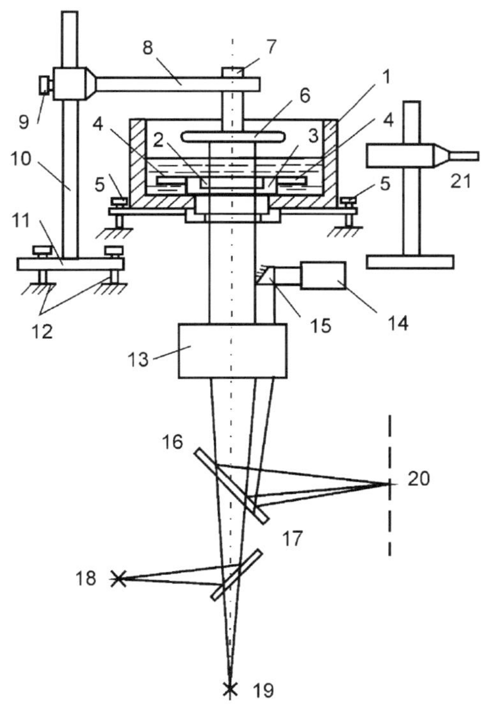

FIG. 9 Parametric excitation of waves by variable electric field. Diagram of set-up to study instabilities in transparent electrically conducting liquids: 1, container; 2, glass with current conducting coating; 3, glass base; 4, protective metal plates; 5, adjusting screws; 6, upper electrode; 7, upper-electrode insulator; 12, parts of the upper electrode suspension system; 13, objective lens of the Schlieren system; 14, illuminated scale; 15, rotating prism; 16 and 17, half-silvered mirrors; 18, stroboscopic light source; 19, constant light source; 20, focal plane of the Schlieren system; and 21, cathetometer. (From Ref. 8.)

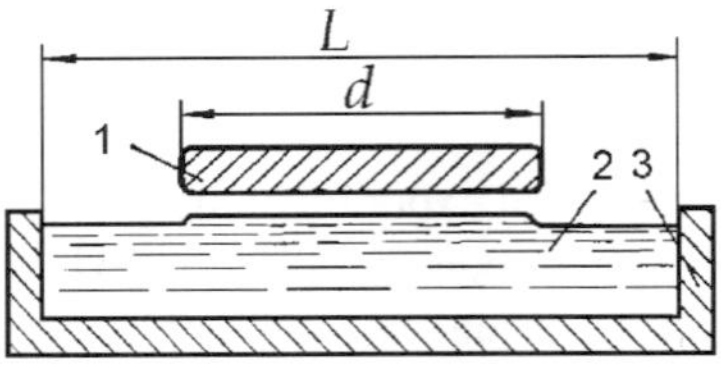

FIG. 10 Parametric excitation of waves by variable electric field. Position of electrode relative to the liquid surface: 1, electrode; 2, liquid; and 3, container. (From Ref. 8.)

force field is not zero, the liquid beneath the electrode was lifted up (Fig. 10). However, its surface was flat under precritical conditions. No induced waves were observed at the applied frequencies, i.e., the amplitude of these waves was below the sensitivity threshold of the Schlieren system (10^{-2} to 10^{-3} mm). When the critical field intensity was attained, axisymmetric standing waves became visible on the surface. The stroboscopic observation showed that the wave frequency coincides with the frequency of the field that is equal to half the frequency of the parameter modulation. This circumstance, and also the fact that there is a threshold for the onset of wave motion, shows that we indeed have a parametrically excited instability. Increasing the field strength, the wave amplitude sharply increases and the wave becomes two-dimensional. Typical Schlieren pictures of such waves on a free water surface are shown in Fig. 11. The photographs are obtained by stroboscopy using light with the frequency of the field.

Threshold field strengths vs. νf, with $f = \omega/2\pi$, are plotted in Fig. 12. The measurements [25] were performed using distilled water ($\nu = 0.01$ cSt;

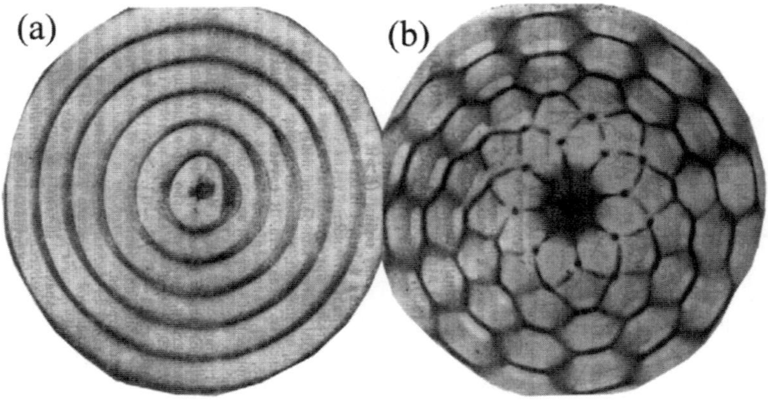

FIG. 11 Schlieren photographs of parametrically excited waves with a variable electric field: (a) axisymmetric standing waves near the excitation threshold, and (b) two-dimensional waves at high values of the electric field intensity. (From Ref. 8.)

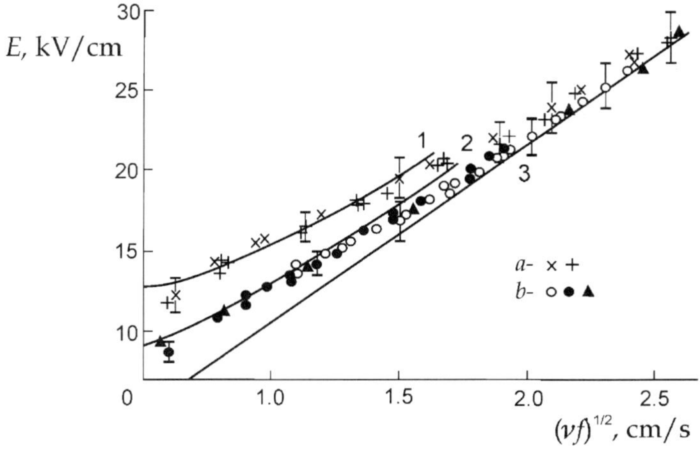

FIG. 12 Parametric excitation of waves with a variable electric field, showing parametric instability of the liquid surface in a variable electric field. Electrode diameter: *a*, 8 cm; *b*, 15 cm. Curves 1 and 2 are plots of Eq. (11) and curve 3 a plot of Eq. (10). (From Ref. 8.)

1 St $= 10^{-4}\,\mathrm{m^2/s}$) in a 38 cm × 48 cm container over a frequency range of 25 to 750 Hz, using two electrodes with diameters $d = 8$ cm (points *a*, Fig. 12) and $d = 15$ cm (points *b*). Taking into account the easy contamination of the water surface by impurities and the need for high accuracy to be able to compare analytical and experimental data, the properties of the liquid surface were measured directly by a suitable technique in the same experiment in which stability was investigated.

As seen from Fig. 12, at high enough frequencies all the experimental points follow satisfactorily line 3 [Eq. (10)], corresponding to the behavior of unbounded liquid surfaces. At lower frequencies, the experimental results show dependence on the electrode diameter. The fact that the measured thresholds, E, are significantly higher than the theoretical prediction indicates the existence of an additional dissipative mechanism such as emission of surface waves from the area under the electrode. An estimate of the emitted energy yields the critical field intensity

$$E_1 = 8(\pi\rho\nu\omega)^{1/2}\left(1 + \frac{1}{d\delta_*}\right)^{1/2} \tag{11}$$

where $\delta_* = \delta/(\partial\Omega/\partial k)$ is a lateral damping factor, expressed in terms of the damping factor, δ. Curves 1 and 2, which are plots of Eq. (11), fit well the experimental results at low frequencies.

REFERENCES

1. Faraday, M. On a peculiar class of acoustical figures; and on certain forms assumed by groups of particles upon vibrating elastic surfaces. Phil. Trans. Royal Soc., London **1831**, *121*, 299–340.
2. Lord Rayleigh. On the crispation of a fluid resting upon a vibrating support. Phil. Mag. **1883**, **16**, 50–58.
3. Jordan, D.W., Smith, P. *Nonlinear Ordinary Differential Equations. An Introduction to Dynamical Systems*; Oxford University Press: Oxford (3rd edn), 1999.
4. Kapitza, P.L. Dynamic stability of a pendulum when its point of suspension vibrates. Soviet Phys. JETP **1951**, *21*, 588–592.
5. Kapitza, P.L. Pendulum with a vibrating suspension. Uspekhi. Fiz. Nauk **1951**, *44*, 7–15.
6. Landau, L.D., Lifshitz, E.M. *Mechanics*. Pergamon Press: Oxford, 1988.
7. Benjamin, T.B., Ursell, F. The stability of the plane free surface of a liquid in vertical periodic motion. Proc. Royal Soc., London A **1954**, *225*, 505–515.
8. Briskman, V.A., Shaidurov, G.F. Parametric instability of the surface in alternating electric field. Sov. Phys.-Dokl. **1968**, *13*, 540–542.
9. Briskman, V.A. Parametric stabilization of fluid interface. Sov. Phys.-Dokl **1976**, *21*, 66–67.
10. Bezdenezhnykh, N.A., Briskman, V.A., Cherepanov, A.A., Sharov, M.T. Control of the stability of liquid surfaces by means of variable fields. In *Hydromechanics and Transfer Processes in Microgravity*. Ural Sci. Center, USSR Acad. Sci.: Sverdlovsk, 1983; 37–56. [Engl. Transl.: *Fluid Mechanics—Soviet Research*. Scripta Technica: Washington, D.C., 1986; 15/1; 11–32].
11. Christiansen, Bo, Alstrom, P., Levinsen, M.T. Ordered capillary-wave states. Phys. Rev. Lett. **1992**, *68*, 2157–2160.
12. Ezersky, A.B., Rabinovich, M.I., Reutov, V.P., Starobinets, I.M. Spatio-temporal chaos in the parametric excitation of a capillary ripple. Sov. Phys. JETP **1986**, *64*, 1228–1336.
13. Nevolin, V.G. Parametric excitation of surface waves. Inzh.-Fiz. Zh. (J. Eng. Phys.) **1985**, *49*, 1482–1494.
14. Miles, J.W., Henderson, D. Parametric forced surface waves. Annu. Rev. Fluid Mech. **1990**, *22*, 143–165.
15. Cerda, E.A., Tirapegui, E.L. Faraday's instability in viscous fluid. J. Fluid Mech. **1998**, *368*, 195–228.
16. Colinet, P., Legros, J.C., Velarde, M.G. *Nonlinear Dynamics of Surface-Tension-Driven Instabilities*. Wiley-VCH: Berlin, 2001.
17. Clarion, C. Mouvement oscillatoire avec viscosité et l'inertie. Pubs. Sci. et Tech. Ministere de l'Air: Paris, No. 303, 1955.
18. Kravchenko, A.E., Smirnov, A.G. Experimental study of natural damping oscillations of liquids in U-pipes. In *Physical Hydrodynamics and Fluid Kinetics*. Rostov-na-Donu State University: Rostov-na-Donu, 1968; pp 3–44 (in Russian).
19. Grigoryan, S.S., Zhigachev, L.I., Kogarko, B.S., Yakimov, Yu.L. Parametrical resonance in communicating vessels with vertical variable accelerative forces. Izv. Akad. Nauk SSSR, Mech. Zhidk. Gaza **1969**, *2*, 148–150 (in Russian).

20. Tanigawa, H., Hirata, K., Yano, H. Nonlinear dynamics of liquid column behaviors. Doshisha University Sci. Engng. Rev. **1970**, *38*, 78–82.
21. Briskman, V.A., Ivanova, A.A., Shaidurov, G.F. Parametric oscillations of liquid in communicating vessels. Izv. Akad. Nauk SSSR, Mech. Zhidk. Gaza **1976**, *2*, 36–42 (in Russian).
22. Briskman, V.A., Ivanova, A.A., Cherepanov, A.A., Shaidurov, G.F. Parametric instability of the equilibrium of a liquid in communicating vessels. In *Hydrodynamics, No. 8* (Uch. Zap. Perm Univ. No. 362). Perm University: Perm, 1976; 97–113.
23. McLachlan, N.W. *Theory and Applications of Mathieu Functions*. Clarendon Press: Oxford, 1951.
24. Briskman, V.A., Shaidurov, G.F. Mechanisms of instability of liquid surfaces in constant and variable electric fields. In *Hydrodynamics, No. 2*. (Uch. Zap. Perm. Univ. No. 216). Perm. Univ: Perm, 1970; 229–240 (in Russian).
25. Sharov, M.T. Stability of the equilibrium of a liquid surface subjected to harmonically fluctuating electric field. In *Transient Processes in Liquids and Solids*. Ural Sci. Center, USSR Acad. Sci.: Sverdlovsk, 1983; 85–98 (in Russian).

4
Stabilizing Influence of High-Frequency Vibrations on the Possible Instability of an Isothermal Liquid Surface

In this chapter we shall discuss mechanisms for stabilizing the surface of a liquid using variable high-frequency force fields and hence including parametric stabilization of a potentially unstable liquid surface. The stability control method suggested here is a feedback method [1]. Its implementation is apparently a hard problem as it requires a priori knowledge of all possible manifestations of instability of the state being controlled. However, the task is simplified by the fact that the various modes of controlling the stability of a surface by means of variable fields share common features. Variable fields may induce surface oscillations and change the initial shape of a liquid surface. However, we shall consider only those situations where, because of symmetry, equilibrium can be maintained.

I. SUPPRESSION OF RAYLEIGH–TAYLOR INSTABILITY AND FORMATION OF MULTILAYER STRUCTURES

Let us consider a flat gas–liquid interface in a uniform field of mass force, **g**, which represents either gravity or the inertial force in an accelerated spacecraft. Let **g** be directed along the outward normal (liquid to gas) to the surface such that instability is expected (called Rayleigh–Taylor instability) [2]. For stability control one may use a homogeneous, orthogonal to the surface, harmonic oscillating motion (vibrations), experienced by the container as a whole. In the noninertial frame of reference attached to it the value of the effective acceleration, g_*, becomes time-dependent and the stability of the system changes accordingly.

A. Small Disturbances and Stability Diagram

Let the origin of coordinates be on the flat liquid–gas interface with the z-axis pointing downwards into the liquid taken unbounded below. For $g_z > 0$, the equilibrium of the liquid is assumed stable. The quantity g_z is the projection of $\mathbf{g}$ onto the z-axis. We assume that the horizontal dimension, L, of the container is much greater than the capillary length, hence the Bond number $Bo > 1 (Bo = \rho g L^2/\sigma)$. Then if $g_z < 0$, the surface tension, σ, will not be able to maintain the interfacial equilibrium and the Rayleigh–Taylor instability will set in.

Stability may be controlled by means of uniform harmonic oscillations of the container with amplitude b and frequency 2ω in the direction of $\mathbf{g}$. In a noninertial frame of reference, attached to the system, the effective acceleration, g_*, changes with time:

$$g_*(t) = g - b\omega^2 \cos \omega t \tag{1}$$

It can be shown that such vibrations do not disturb the equilibrium and in this case only produce pressure fluctuations.

To analyze stability we introduce small perturbations, ζ, of the surface as Fourier normal modes that are periodic with respect to x and y with a continuous spectrum of wave numbers,

$$\zeta = \xi(t) \exp i(k_1 x + k_2 y)$$

and similar expressions for the perturbations of the velocity, $\mathbf{v}$, and the pressure, p, associated with the surface perturbations.

In the weak viscosity approximation, from the Navier–Stokes equations and boundary conditions for the disturbance amplitude, ξ, in a semi-infinite liquid layer we have the following evolution equation

$$\frac{d^2\xi}{dt^2} + 2\delta \frac{d\xi}{dt} + (\Omega^2 - b\omega^2 k \cos \omega t)\xi = 0 \tag{2}$$

where $\Omega^2 = (\sigma/\rho)k^3 - gk$, $\delta = 2\nu k^2$, and $k^2 = k_1^2 + k_2^2$.

Equation (2) allows the existence of an equilibrium state for the plane surface, $\xi = 0$, and its stability depends on the solutions of the disturbance evolution equation.

Then, if there is no vibration ($b = 0$), short-wave disturbances (for which $\Omega^2 > 0$) represent capillary-gravity waves with natural frequency Ω and damping δ. Long-wave Rayleigh–Taylor disturbances (for which $\Omega^2 < 0$) are seen to grow with time (falling along the $\mathbf{g}$ direction) and make the flat surface unstable.

An effective stabilizing action may be expected from high-frequency vibrations, high compared to the inverse growth rate of the most dangerous

Rayleigh–Taylor disturbances, where

$$\omega^2 < |\Omega_m^2|, \qquad \Omega_m^2 = \Omega^2(\lambda_m).$$

Here λ_m is defined by the lateral dimension of the container. In this case the solution can be expressed as a superposition of the slow-changing component, ξ_0, and the small rapidly oscillating part, ξ_1. Averaging Eq. (2) yields

$$\frac{d^2\xi}{dt^2} + 2\delta\frac{d\xi}{dt} + [\Omega^2 + (b\omega)^2k^2/2]\xi = 0 \tag{3}$$

where the subscript '0' is omitted. At vanishing spring constant

$$\Omega^2 + (b\omega)^2k^2/2 = 0$$

we see that in order to suppress the possible slow Rayleigh–Taylor disturbances the vibration rate must exceed a critical value, such that

$$(b\omega) \geq (b\omega)_1 = \left[2\left(gk_m^{-1} - \left(\frac{\sigma}{\rho}\right)k_m\right)\right]^{1/2} \tag{4}$$

Equation (2) also describes parametric resonance. This implies that vibrations, tending to suppress Rayleigh–Taylor instability, may initiate another type of instability, namely, relative short-wave instability to fast disturbances with $\Omega(k_*) = \omega/2$ (the main parametric resonance; Ref. 3). The stability condition, under the assumed approximations, can be obtained by expanding the solution of Eq. (2) in terms of a small modulation amplitude. To maintain equilibrium the vibration rate should not exceed a threshold value, defined by the level of dissipation:

$$(b\omega)_2 = 2\nu k \tag{5}$$

where k_* is defined by $\Omega(k_*) = \omega/2$.

Thus, stabilization of the flat liquid surface requires that the vibration intensity be kept within certain limits:

$$(b\omega)_2 > (b\omega) > (b\omega)_1 \tag{6}$$

These restrictions imply that the vibration rate is high enough to suppress Rayleigh–Taylor instability, but at the same time too low to initiate parametric resonance. The inequalities of Eq. (6) are compatible starting with some frequency ω_*, where ω_* is determined by the condition $(b\omega)_1 = (b\omega)_2$. Stabilization conditions are illustrated in Fig. 1 [4].

Suppression of Rayleigh–Taylor instability in terrestrial conditions, $g = g_0$, requires a high supply of energy. However, the stabilization conditions are made easier if the effective gravity diminishes as in free fall, aboard a spacecraft, because

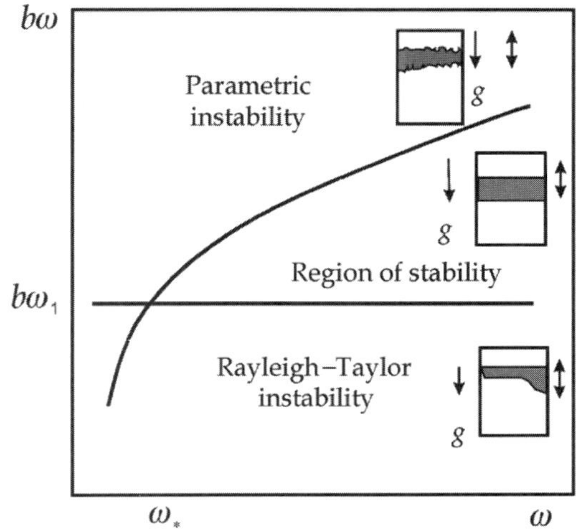

FIG. 1 Suppression of Rayleigh–Taylor instability. Stability diagram of a liquid surface in a vibration field. (From Ref. 4.)

as g/g_0 decreases the minimal stabilization values of vibration rates, $(b\omega)_1$, and frequencies, ω, are significantly lowered. This is illustrated in Fig. 2, where the dimensionless vibration accelerations, $(b\omega^2)_* = (b\omega)_1\,\omega_*$, are plotted vs. g_0/g. For not too large values of g_0/g the ratio $(b\omega^2)_*/g$ becomes less than unity, i.e., control can be provided by forces that are low relative to destabilizing ones.

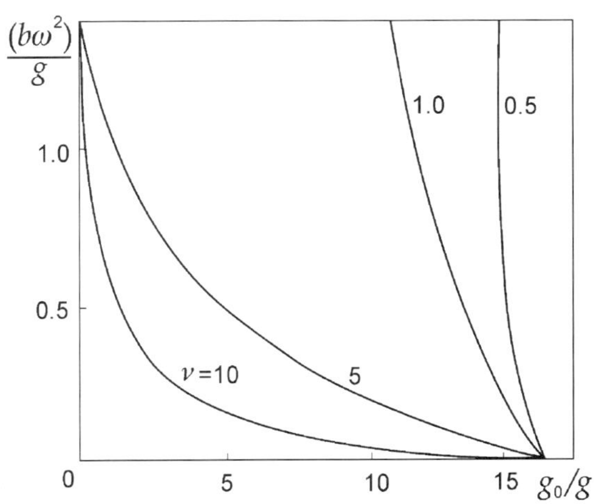

FIG. 2 Suppression of Rayleigh–Taylor instability. Variation of stabilizing vibration force when the effective gravity is reduced. (From Ref. 4.)

B. Experimental Results Illustrating the Suppression of the Rayleigh–Taylor Instability by Means of Vibrations

As noted earlier, parametric excitation of waves—Faraday ripples—on the free surface of a liquid or at the inner interface between two liquids due to homogeneous vibrations has been extensively investigated experimentally [4–6]. The measured stability limits agree with predictions. For this reason we shall only describe experiments aimed at preventing the Rayleigh–Taylor instability when $g_z < 0$. Set-up diagrams and a sketch of the experiment are shown in Fig. 3.

High-frequency stabilization of a statically unstable equilibrium state was first obtained by Wolf [7] who used the following procedure. At the bottom of a cylindrical container (diameter, 2.8 cm) was attached a shaker [Fig. 3(a)]. The vibratory velocity was set above the value of needed for the suppression of the possible Rayleigh–Taylor instability. When the shaker was turned upside down [Fig. 3(b)], the equilibrium of the liquid above the gas remained stable.

In other experiments [8] an inversion of the stable equilibrium state was found as a result of initial excitation of parametric instability of the liquid surface and subsequent development of waves. In the same initial position as in Fig. 3(a)

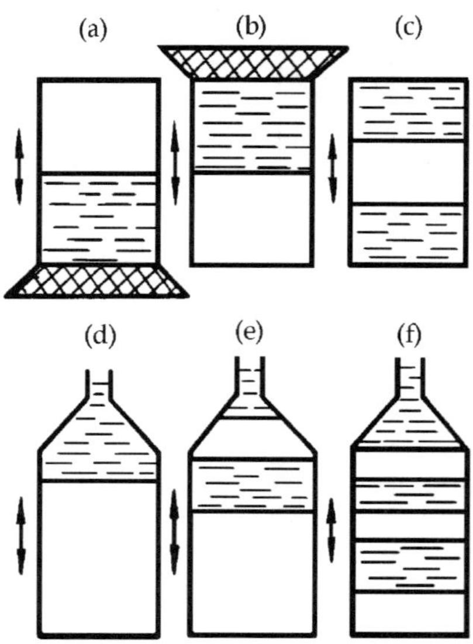

FIG. 3 Diagram of the experiment designed to prevent the onset of Rayleigh–Taylor instability by means of vibration of the container: (a), (b) High-frequency stabilization in Wolf's experiments [6]; (c) Rupture of the liquid layer by parametric instability and then stabilization of the surface by oscillations of smaller amplitude; (d), (f) Step-by-step multilayering of the system. (From Ref. 8.)

the vibratory velocity was raised to values exceeding the threshold for excitation of parametric waves, $(b\omega)_2$, at frequencies $\omega > \omega_*$. As the velocity of the vibrations was increased still further, the waves became unstable and droplets began to be ejected, eventually hitting the cover of the container. Since the modulation amplitude in this case was a priori known to exceed $(b\omega)_1$ a liquid layer gradually accumulated on that cover. Both layers were parametrically unstable and exchanged "spouts" of droplets. When $b\omega$ was then decreased down to a value within the range given by Eq. (6), it was possible to "freeze" both layers, as shown in Fig. 3(c). The photograph for this case was obtained by stabilization of the equilibrium in a 5% solution of water in glycerol (the liquid was colored) in a 2×2-cm container at a frequency of 60 Hz (Fig. 4).

Still another stabilization technique is shown in Fig. 3(d–f). Alternating portions of liquid and gas are slowly fed from the top into a 2-cm diameter cylindrical container placed in a shaker vibrating at 70 Hz with $(b\omega)_2 > (b\omega) > (b\omega)_1$. These techniques permit suspending a liquid layer in air [Fig. 3(e)] and a structure with an arbitrary number of layers [Fig. 3(f)]. Such structures exhibit a stability margin. For example, it is seen in Fig. 5(a) that pushing a wire through a layer does not cause it to break. The layer remains intact

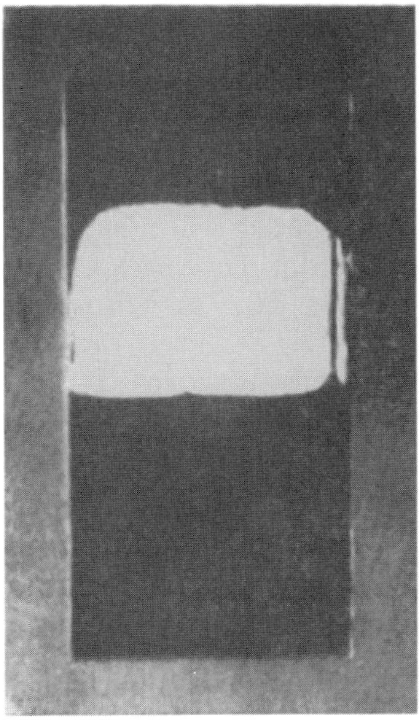

FIG. 4 Suppression of Rayleigh–Taylor instability. Photograph of a liquid–air–liquid stratification that is stabilized by vibration. (From Ref. 4.)

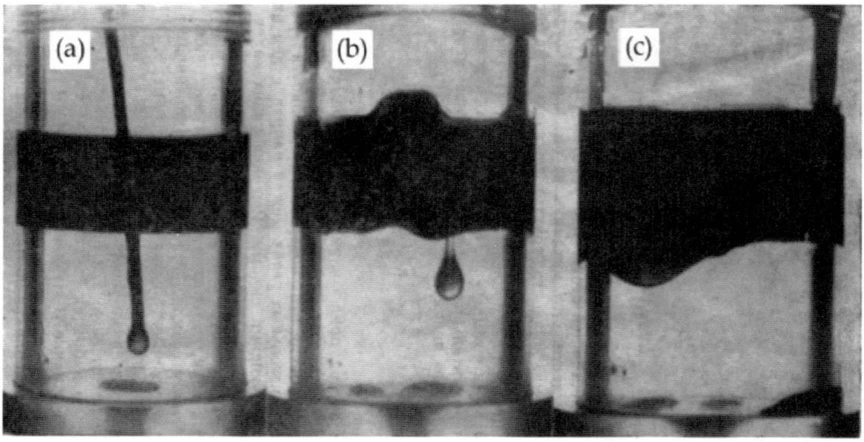

FIG. 5 Suppression of Rayleigh–Taylor instability. Photograph of a liquid layer with two free surfaces in the presence of vibrations. (From Ref. 4.)

also at $(b\omega) > (b\omega)_2$ when both surfaces are parametrically unstable [Fig. 5(b)]. However, if the amplitude is reduced below, Rayleigh–Taylor instability sets in [Fig. 5(c)].

The layered structure described above (with the exception of transient states) consists of plane layers of liquid and gas that are immobile relative to the container walls. The stability or instability of the equilibrium of the layers depends on whether the surface perturbations are damped or grow in time. Vibration can suppress these perturbations, thus preventing break-up of the stratification. Each layer maintained in the suspended state acts as a piston, being held in its position by the difference in gas pressures above and below it. For this reason the thicknesses of the layers and of the spaces between them, as well as their equilibrium positions are independent of the vibrations.

The accelerations used in these stabilization experiments are plotted in Fig. 6 [points ×, ●, and ▲ relate to experiments shown in Fig. 3(c,e,f), respectively]. The experimental points lie above the predicted curve (dashed line) computed from Eq. (4).

Good agreement exists between theory and experiment as illustrated in Fig. 7. The experimental data of Wolf are displayed together with the theoretical prediction [7,9].

From the point of view of applications of the method of parametric stabilization, it is of interest to investigate the possibility of preventing instability of liquid surfaces of different shapes by varying the angle between the direction of vibration and the destabilizing forces. The solution of such problems requires finding new equilibrium surface shapes produced by the averaged vibratory forces

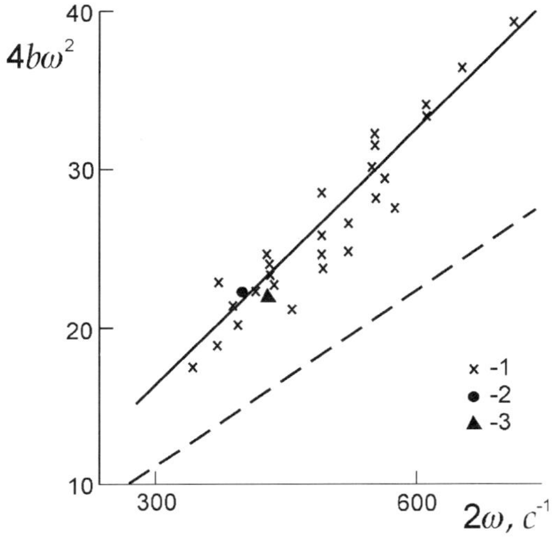

FIG. 6 Suppression of Rayleigh–Taylor instability container, showing limits of Rayleigh–Taylor instability upon application of vibrations. Points ×, ●, ▲ correspond to experiments shown in Fig. 3(c)–(f). The dashed line was computed from Eq. (4). (From Ref. 4.)

and a study of the induced waves, which is outside the scope of our interest here. Note that the possibility of achieving such stabilization rests on two experimental facts. Wolf [7] was able, by applying large enough vibrations, to maintain the equilibrium of the liquid–gas interface irrespective of the deviations of the shaker from the vertical. In experiments [4] vibrations at 70 Hz stabilized vertical liquid zones of different shapes [Fig. 8(a,b)] with solid bases, all 1.5 cm in diameter. Stabilization in this case means an increase in the critical length of the liquid zone with an increase of vibratory amplitudes. Figure 8(c) shows how such a zone disintegrates when vibrations cease.

II. INFLUENCE OF TANGENTIAL VIBRATION ON INTERFACIAL INSTABILITY

Let us now examine the behavior of a liquid–liquid interface with liquids having about the same density and subjected to horizontal high frequency vibrations. In experiments [6,8] a motionless frozen wavelike relief on the interface appears when applying horizontal vibrations. The relief occurs past a threshold value. It is important to note that the wave relief is observed only when the densities of liquids are comparable. The appearance of quasi-equilibrium structures on the fluid

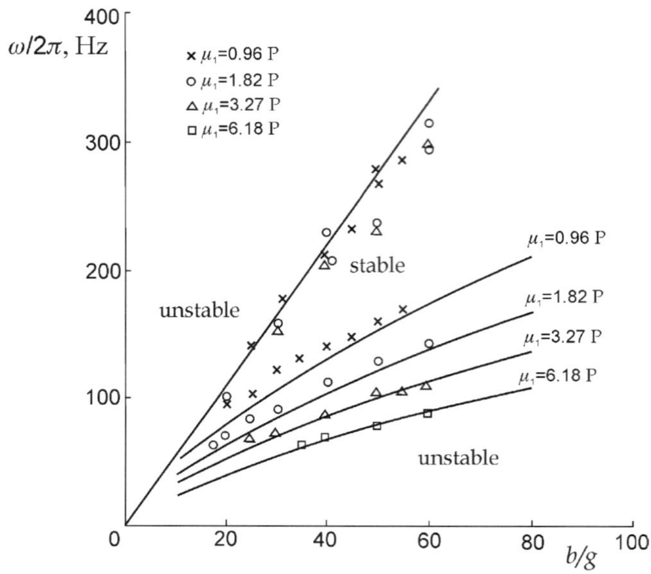

FIG. 7 Suppression of Rayleigh–Taylor instability for an air–oil system. Comparison of the computed stability domain with the experimental results for various values of the dynamic (shear) viscosity of the oil (μ_1). The solid lines have been calculated using $\rho_1 = 0.9\ \text{g/cm}^3$ and $\rho_2 = 0.0013\ \text{g/cm}^3$, $\mu_2 = 0.00018$ P (P $=$ dyne/cm$^2 s = 10^{-2}$ mN/m$^2 s$), and $\sigma = 10$ mN/m. The region of stability is between the upper solid line and one of the lower curves. (From Ref. 9.)

interface permits an analytical approach based on the separation of the motion into mean (averaged) and pulsating components and the application of the averaging method.

A. Equations of Motion and Boundary Conditions in the High-Frequency Approximation

Let us recall the equations describing the average motion of an isothermal system with a liquid interface [11]. Consider two immiscible liquids filling a container subjected to linear high-frequency oscillations according to the law $a\mathbf{k}\sin\omega t$; $\mathbf{k}(0, 0, 1)$. In the noninertial reference system attached to the container the equations of motion can be written in the form

$$\operatorname{div}\mathbf{v} = 0 \tag{7}$$

$$\frac{\partial \mathbf{v}}{\partial t} + (\mathbf{v}\cdot\nabla)\mathbf{v} = -\frac{1}{\rho_m}\nabla p + \nu_m \Delta\mathbf{v} + \mathbf{g} + a\omega^2\mathbf{k}\sin\omega t, \qquad m = 1, 2 \tag{8}$$

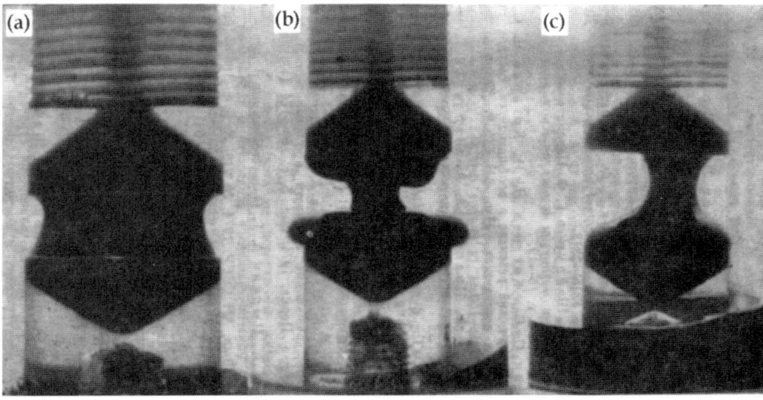

FIG. 8 Suppression of Rayleigh–Taylor instability, showing photograph of a vibration-stabilized liquid zone between two solid bases of diameter 1.5 cm. Stabilization in this case means an increase in the critical length of the liquid zone with the increase of vibratory amplitudes. (c) Shows how such a zone disintegrates when vibrations cease. (From Ref. 4.)

As in earlier cases in this book, **g** is the acceleration of gravity pointing downwards, and the index m relates to the liquids (here 1 is above and 2 is below). On the solid walls the no-slip condition applies and on the interface between the liquids, $\xi(\mathbf{r}, t) = 0$, the following conditions must be satisfied:

stress balance conditions

$$-[\Pi_{ij}]n_i + [p]n_i = \sigma(\nabla\mathbf{n})n_i, \qquad \mathbf{n} = \frac{\nabla\xi}{|\nabla\xi|} \tag{9}$$

continuity of the velocity

$$[\mathbf{v}] = 0 \tag{10}$$

and the kinematic condition

$$\frac{\partial\xi}{\partial t} + \mathbf{v}\cdot\nabla\xi = 0 \tag{11}$$

Here, Π_{ij} denotes the viscous stress tensor, **n** is the unit normal to the interface, and square brackets denotes the jump of the value on the interface: $[f] = f_1 - f_2$.

For high enough frequency of vibration ($\omega \gg \nu/L^2$, where L is now the expected characteristic size of convection), we can consider fast and slow motions. This approximation separating fast (pulsating) and average (mean) parts is possible only for small amplitude of vibrations, $a \ll L$. The amplitude of the velocity of pulsations, $b = a\omega$, is taken to be finite. Then for all variable quantities the pulsating, quickly varying component can be separated from the slow part, whose characteristic time of change is great in comparison with ω^{-1}. Thus there is a hierarchy of times that allows averaging of the equations and using a method of multiple scales. Accordingly, we have a sequence of time scales

$$t_{-1} = \omega t, \qquad t_0 = t, \qquad t_1 = \omega^{-1}, \ldots$$

We assume that all variables in Eqs (7)–(11) depend both on the fast time t_{-1} and on the slow times $t_0, t_1, \ldots$ Time derivatives of any function, $f(t)$, and fields of velocity, pressure, and ξ, which locates a point of the liquid interface, can be expressed as series expansions

$$\frac{\partial f}{\partial t} = \omega \frac{\partial f}{\partial t_{-1}} + \frac{\partial f}{\partial t_0} + \omega^{-1} \frac{\partial f}{\partial t_1} + \cdots, \qquad \mathbf{v} = \mathbf{v}^{(0)} + \omega^{-1}\mathbf{v}^{(1)} + \cdots$$

$$p = \omega p^{(-1)} + p^{(0)} + \omega^{-1} p^{(1)} + \cdots, \qquad \xi = \xi^{(0)} + \omega^{-1}\xi^{(1)} + \cdots \tag{12}$$

Equations (7) and (8) in the leading order in ω^{-1} give

$$\operatorname{div} \mathbf{v}^{(0)} = 0, \tag{13}$$

$$\omega \frac{\partial \mathbf{v}^{(0)}}{\partial t_{-1}} = -\omega \frac{1}{\rho_m} \nabla p^{(-1)} + b\omega \mathbf{k} \sin t_{-1} \tag{14}$$

Integration of these equations with respect to the fast time yields

$$\mathbf{v}_m^{(0)} = b\mathbf{V}_m \cos t_{-1} + \mathbf{u}_m, \qquad p_m^{(-1)} = bP_m \sin t_{-1} \tag{15}$$

where the quantities $\mathbf{V}_m$, $\mathbf{u}_m$, and P_m do not depend on the fast time and satisfy the equations

$$\operatorname{div} \mathbf{V}_m = 0 \tag{16}$$

$$\operatorname{div} \mathbf{u}_m = 0 \tag{17}$$

$$\rho_m(\mathbf{V}_m + \mathbf{k}) = \nabla P_m \tag{18}$$

The amplitudes of the pulsating velocity component are obtained as the solutions of the equations, without viscosity. Therefore they cannot satisfy the non-slip conditions on the solid walls. We can only require the condition

$$V_n|_s = 0 \tag{19}$$

at the boundary S.

The replacement of the no-slip condition by Eq. (19) is acceptable if the thickness of the viscous skin layer for fast motions near the walls is small in comparison with the sizes of the container. In just such a high-frequency approximation, $\omega \gg \nu/L^2$, we obtain the equations for the average components of the motion.

Substituting the fields Eq. (15) in Eqs. (7) and (8) and averaging them over the period of the vibrations, we get the equation for the average velocity, $\mathbf{u}_m$, with the no-slip condition on solid walls

$$\frac{\partial \mathbf{u}}{\partial t} + (\mathbf{u} \cdot \nabla)\mathbf{u} = -\Delta\left(\frac{p}{\rho_m} + b^2 V^2\right) + \nu_m \Delta \mathbf{u} + \mathbf{g}, \qquad \mathbf{u}|_s = 0 \tag{20}$$

From the boundary condition Eq. (11) it follows that the main part ξ does not depend on the fast time scale. Averaging the boundary conditions on the interface Eq. [(9)–(11)] and taking into account Eq. (15), we obtain

$$-[\Pi_{ij}]n_i + [p]n_i + \frac{1}{2}b^2[\rho V_n W_n]n_i = \sigma(\nabla \mathbf{n})n_i \tag{21}$$

$$[\rho W_\tau] = 0, \qquad [W_n] = 0, \qquad \frac{\partial \xi}{\partial t} + \mathbf{u} \cdot \nabla \xi = 0, \qquad \mathbf{W} = \mathbf{V} + \mathbf{k} \tag{22}$$

In Eqs. (20)–(22) the index zero in t_0, in the average pressure $p^{(0)}$ and in the function $\xi^{(0)}$ is omitted; the subscript n denotes the component normal to the interface of the corresponding vectors, while τ denotes tangential components. The tensor Π_{ij} in Eq. (21) is defined using the average velocity field, $\mathbf{u}$.

Equations (16)–(22) completely determine the amplitudes of the pulsation velocity, $\mathbf{V}$, the average velocity fields, $\mathbf{u}$, pressure, p, and the average location of the interface.

B. Shape of the Open Surface of a Liquid in a Container of Finite Size

Let us now consider the case when the liquid only partially fills the container. Assume that the quasi-equilibrium state for which the average velocity, $\mathbf{u}$, vanishes has, a nonzero pulsating component. In this state the open surface of the liquid is not motionless. It makes small fluctuations of the order of amplitude of the displacement of the container. The quasi-equilibrium state can be found from the solution of Eqs. (16)–(22), if we consider that the top liquid is absent and set $\mathbf{u} = 0$. This problem is nonlinear and hence can have a multiplicity of solutions.

For simplicity we investigate the case when the liquid partially fills a container of rectangular cross-section. In the absence of vibrations the open surface is flat and horizontal (level with the gravity equipotential). To obtain the quasi-equilibrium state of the free surface under the influence of vibrations a variational principle can be used [10]. It then follows that in the presence of small vibrations the open surface is not flat but preserves symmetry [Fig. 9(a)] [11]. If the amplitude of the velocity of vibrations increases then a spontaneous symmetry breaking occurs at a critical value of the vibrational parameter [Fig. 9(b)]. At large

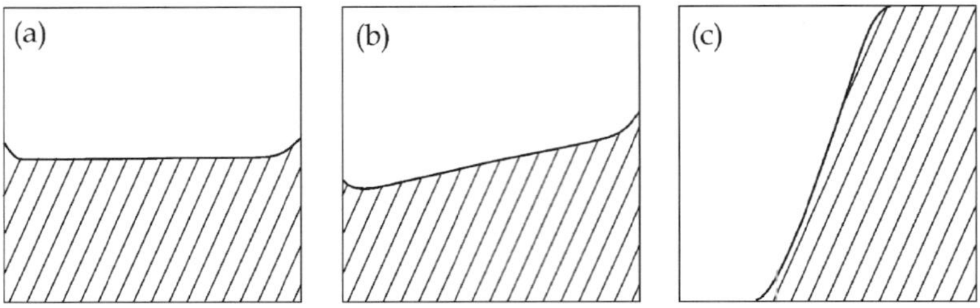

FIG. 9 Influence of tangential vibration on interfacial instability: form of the free surface of a liquid in a vessel of finite size. (From Ref. 12.)

values of the vibrational parameter the static gravity field does significantly influence the form of the free surface. The fluid is placed near one of the vertical boundaries of the container so that the free surface becomes almost perpendicular to the vibration axis [Fig. 9(c)].

C. Instability of a Flat Interface in a Two-Layer System

Another problem of interest is the quasi-equilibrium patterned structures on the interface separating two immiscible liquids of different densities that fill a layer long enough along the horizontal directions and subjected to high-frequency tangential vibrations [11,12]. Equations (16)–(22) also apply to this case.

Let the layer be of infinite horizontal extent in axes x and y, with thickness $2d$, filled with two liquids of density ρ_1 and ρ_2 ($\rho_1 > \rho_2$), respectively. For simplicity we shall examine the case where the liquids occupy equal volumes. Let us choose axes of coordinates such that, with the axis z directed vertically upwards, the bottom and top boundaries of the layer are at $z = -d$ and $z = d$, respectively. Let the vibrations be along axis x.

In the above-described situation a quasi-equilibrium state is possible, i.e., a state such that the average velocity vanishes and all other average characteristics do not depend on time. Let us write the equations and boundary conditions, simplifying Eqs. (16)–(22) for the quasi-equilibrium state. The amplitudes of the pulsating velocity components and the position of the interface are determined from the equations

$$\operatorname{div}\mathbf{W}_1 = 0, \qquad \operatorname{rot}\mathbf{W}_1 = 0, \qquad \operatorname{div}\mathbf{W}_2 = 0, \qquad \operatorname{rot}\mathbf{W}_2 = 0,$$
$$z = -d : W_{1z} = 0, \qquad z = d : W_{2z} = 0 \tag{23}$$

$$z = \zeta(x, y) : W_{1n} = W_{2n}, \qquad \rho_1 W_{1\tau} = \rho_2 W_{2\tau} \tag{24}$$

$$\frac{b^2}{4}[W_{1n}^2(\rho_1 - \rho_2) - (\rho_1 W_{1\tau}^2 - \rho_2 W_{2\tau}^2)] - (\rho_1 - \rho_2)g\zeta + \sigma(\nabla\mathbf{n}) = \text{const.} \tag{25}$$

which possess a quasi-equilibrium solution with a flat interface,

$$\zeta = 0, \qquad W_{1x} = \frac{2\rho_1}{\rho_1 + \rho_2}, \qquad W_{2x} = \frac{2\rho_1}{\rho_1 + \rho_2},$$
$$W_{1y} = W_{1z} = W_{2y} = W_{2z} = 0 \tag{26}$$

satisfying the condition

$$\int_{-d}^{\xi} V_{1x} \mathrm{d}z + \int_{\zeta}^{d} V_{2x} \mathrm{d}z = 0.$$

If, in particular, the top liquid has very low density, then $W_{1x} \ll W_{2x}$. In other words, in this case we have a bottom (heavy) liquid that is practically motionless (in the laboratory reference system). In the case of equal densities ($\rho_1 = \rho_2$) we have $W_{1x} = W_{2x} = 1$, hence $V_{1x} = V_{2x} = 0$, and both liquids are immobile in their respective frame of reference; i.e., they are motionless in the container.

Generally speaking, the above solution is not unique, and hence we shall consider the bifurcation problem from the solution of Eq. (26). Note that this is reminiscent of what an electric field does to a flat interface, as shown in Chapter 2. As the vibrations are directed along the x-axis they have no influence on disturbances polarized along the y-axis; it is enough to consider plane perturbations depending only on x and z. As the vectors $\mathbf{W}_1$ and $\mathbf{W}_2$ are solenoidal in the reference plane, the stream functions, Ψ_1 and Ψ_2, for the perturbations can be introduced satisfying the Laplace equation

$$W_{1x} = \frac{2\rho_2}{\rho_1 + \rho_2} + \frac{\partial \Psi_1}{\partial z}, \qquad W_{1z} = -\frac{\partial \Psi_1}{\partial x} \tag{27}$$

$$W_{2x} = \frac{2\rho_1}{\rho_1 + \rho_2} + \frac{\partial \Psi_2}{\partial z}, \qquad W_{2z} = -\frac{\partial \Psi_2}{\partial x} \tag{28}$$

$$\Delta \Psi_m = 0, \qquad m = 1, 2 \tag{29}$$

As in other cases treated in this book, it is convenient to introduce dimensionless variables by choosing as scale of length the capillary length $[\sigma/(\rho_1 - \rho_2)g]^{1/2}$. In the same scale we shall measure Ψ. The boundary conditions in terms of Ψ_m have the form

$$z = -H : \Psi_1 = 0, \qquad z = H : \Psi_2 = 0 \tag{30}$$

$$z = \zeta(x) : \Psi_1 - \Psi_2 = 2\frac{\rho - 1}{\rho + 1}\zeta, \qquad \rho\left(\frac{\partial \Psi_1}{\partial z} - \frac{\partial \Psi_1}{\partial x}\frac{\partial \zeta}{\partial x}\right) = \frac{\partial \Psi_2}{\partial z} - \frac{\partial \Psi_2}{\partial x}\frac{\partial \zeta}{\partial x}$$

$$B\left[\frac{2\rho}{\rho+1}\frac{\partial \Psi_1}{\partial z} + \frac{2}{\rho+1}\frac{\partial \Psi_2}{\partial z} + \frac{\partial \Psi_1}{\partial z}\frac{\partial \Psi_2}{\partial z} + \frac{\partial \Psi_1}{\partial x}\frac{\partial \Psi_2}{\partial x}\right]$$

$$-\zeta + \frac{\partial^2 \zeta}{\partial x^2}\left(1 + \frac{\partial \zeta^2}{\partial x}\right)^{-3/2} = \text{const.}$$

$$H = d\left[\frac{(\rho_1 - \rho_2)g}{\sigma}\right]^{1/2}, \qquad \rho = \frac{\rho_1}{\rho_2}, \qquad B = \frac{b^2}{4}\left(\frac{\rho_1 - \rho_2}{\sigma g}\right)^{1/2}$$

where H is a dimensionless average depth of the liquid layers and B is a new dimensionless parameter describing the intensity of vibrations. For a bifurcation analysis let us linearize Eqs. (29) and (30) around the quasi-equilibrium state [Eq. (26)]. For perturbations periodic along the x-axis ($\Psi_1, \Psi_2, \zeta \sim \cos kx$) the bifurcation plot using (B, k) as coordinates is

$$B = \frac{(\rho + 1)^3}{8\rho(\rho - 1)}(k + k^{-1})\tanh kH \tag{31}$$

or in dimensional coordinates (b, k)

$$b^2 = \frac{(\rho_1 + \rho_2)^3}{2\rho_1\rho_2(\rho_1 - \rho_2)^2}[\sigma k + (\rho_1 - \rho_2)gk^{-1}]\tanh kd \tag{32}$$

If the amplitude of vibration velocity exceeds the critical value [Eq. (32)], then besides the trivial (unstable) solution [Eq. (26)], the problem also has a periodic solution in x with a quasi-stable surface (frozen) wave relief. Equation (32) also shows that when the heavy liquid is located above ($\rho_2 > \rho_1$), the liquid interface is absolutely unstable. Always there will be a perturbation with large enough long waves resulting in stability loss. Thus, horizontal vibrations do affect the onset of Rayleigh–Taylor instability.

The frozen wavelike relief surface pattern (at $\rho_1 > \rho_2$) is possible only if the interface is between liquids with comparable, albeit unequal, densities. Note that at $\rho_2 \to 0$ the critical value, b^2, diverges to infinity.

The neutral curves for various values of the dimensionless thickness of the layers, H, are shown in Fig. 10. A surface relief with finite wavelength is possible

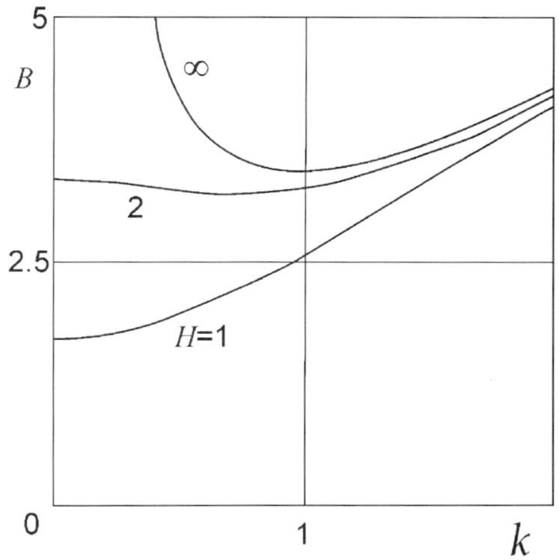

FIG. 10 Influence of tangential vibration on interfacial instability, showing neutral curves for various values of dimensionless thickness of the layers, H. (From Ref. 12.)

only for thick enough layers. For thinner layers ($kH \ll 1$) the neutral curve has the form

$$B = \frac{(\rho + 1)^3 H}{8\rho(\rho - 1)}(k^2 + 1) \tag{33}$$

and hence the most dangerous perturbation has a vanishing wavenumber, $k = 0$. From Eq. (31) it follows that the wave relief is to be expected in layers when $H > \sqrt{3}$, i.e., $d > [3\sigma/(\rho_1 - \rho_2)g]^{1/2}$.

Let us consider what kind of surface wave relief is established at the liquid interface between very thick layers ($kH \gg 1$). The critical amplitude of vibrations is determined in this case by the expression

$$B = \frac{(\rho + 1)^3}{8\rho(\rho - 1)}(k + k^{-1}) \tag{34}$$

whose minimum is at $k = 1$. Thus, for thick liquid layers in a field of high-frequency horizontal vibrations with velocity amplitude exceeding the critical value

$$b^2_{\min} = \frac{(\rho_1 + \rho_2)^3}{\rho_1\rho_2(\rho_1 - \rho_2)}[\sigma(\rho_1 - \rho_2)g]^{1/2} \tag{35}$$

a surface relief with the wavelength

$$\lambda = 2\pi[\sigma/g(\rho_1 - \rho_2)]^{1/2} \tag{36}$$

is expected to appear on the interface.

Results about the behavior of the fluid interface when the values of the parameters are far from the critical ones have been obtained using the finite-difference method [12]. It appears that a quasi-stationary periodic surface relief sets at the interface when the vibrational parameter reaches the critical value B_c. The surface relief, near threshold conditions, is shown in Fig. 11(a) for $\rho = 1.25$ and $B = 11$. With a rise in B the amplitude of the surface wave relief increases and its form is distorted, becoming anharmonic [see Fig. 11(b) for $B = 12$ and Fig. 11(c) for $B = 12.5$].

Further increasing the value of B doubles the spatial period of the surface pattern. At higher supercriticality the preferential orientation of the interface with respect to the vibration axis changes. The layer is split along the horizontal direction into a system of strata, alternating areas of different fluids, separated by almost vertical boundaries.

Experimental results [12] obtained using glycerin and an insulating oil placed into a horizontal cavity of length 160 mm and 16 mm thickness are shown in Fig. 12. The frequency of vibration in this experiment was 50 Hz and the amplitude of the displacement of the container changed in the range 0–2.5 mm. Similar phenomena were found in the direct numerical integration of the system. Note that in experiments with thicker layers a sequence of period doubling bifurcations in space was observed.

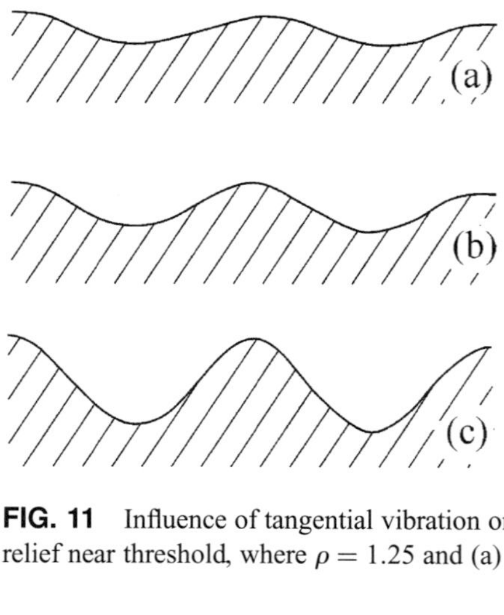

FIG. 11 Influence of tangential vibration on interfacial instability, showing surface wave relief near threshold, where $\rho = 1.25$ and (a) $B = 11$, (b) 12 and (c) 12.5. (From Ref. 12.)

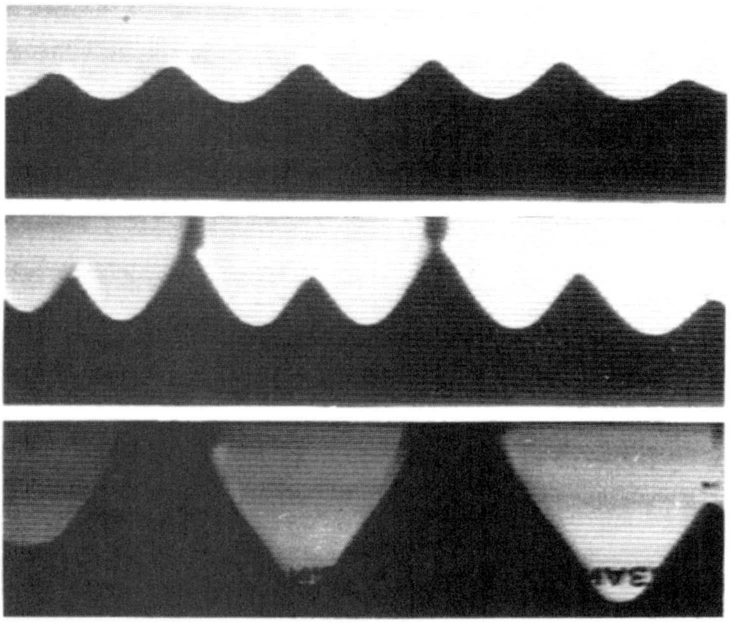

FIG. 12 Influence of tangential vibration on interfacial instability, showing surface wave reliefs experimentally obtained for a glycerin–insulating oil system. (From Ref. 12.)

REFERENCES

1. Ladikov, Yu.P. *Stabilization of Processes in Continuous Media*; Nauka: Moscow, 1978 (in Russian).
2. Drazin, P.G., Reid, W.H. *Hydrodynamic Stability*. Cambridge University Press: Cambridge, 1981.
3. Jordan, D.W., Smith, P. *Nonlinear Ordinary Differential Equations. An Introduction to Dynamical Systems*, 3rd edn; Oxford University Press: Oxford, 1999.
4. Sekerzh-Zen'kovich, S.Ya., Kalininchenko, V.A. Excitation of internal waves in a two-layer liquid by means of vertical vibrations. Dokl. Akad. Nauk SSSR **1979**, *249*, 797–799 (in Russian).
5. Brand, R.P., Nyborg, W.L. Parametrically excited surface waves. J. Acoust. Soc. Amer. **1965**, *37*, 509–515.
6. Hoffmann, F.M., Wolf, G.H. Excitation of parametric instabilities in statically stable and unstable fluid interfaces. J. Appl. Phys. **1974**, *5*, 3859–3863.
7. Wolf, G.N. The dynamic stabilization of the Rayleigh–Taylor instability and the corresponding dynamic equilibrium. Z. Physik B. **1969**, *227*, 291–300.
8. Bezdenezhnykh, N.A., Briskman, V.A., Cherepanov, A.A., Sharov, M.T. Control of the stability of liquid surfaces by means of variable fields. Fluid Mechanics. In

Hydromechanics and Transfer Processes in Microgravity; Sverdlovsk, 1983, 37–56 (in Russian). [Engl. Transl.: Fluid Mechanics—Soviet Research, USA. Scripta Technica: Washington D.C. **1986**, *15*(1), 11–32.

9. Troyon, F., Gruber, R. Theory of the dynamic stabilization of the Rayleigh–Taylor instability. Phys. Fluids **1971** *14*, 2069–2073.
10. Bezdenezhnykh, N.A., Briskman, V.A., Lyubimov, D.V., Cherepanov, A.A., Sharov, M.T. Control of the Stability of Liquid Interface by Means Vibrations, Electrical and Magnetic Fields. 3 Vsecoyus. Seminar po Gidromehanike i Teplomassobmenu v Nevesomosti. Tes. Dokl. (3rd All-Union Seminar on Hydrodynamics and Heat-Mass-Transfer in Microgravity. Abstracts). Chernogolovka, 1984, pp. 18–20 (in Russian).
11. Lyubimov, D.V., Cherepanov, A.A. On the appearance of stationary relief on the fluid interface in the vibrational field. Izv. Akad. Nauk SSSR, Mech. Zhidk. Gaza **1986**, *6*, 8–13 (in Russian).
12. Bezdenezhnykh, N.A., Briskman, V.A., Lapin, A.Yu. Lyubimov, D.V., Lyubimova, T.P., Cherepanov, A.A., Zakharov, I.V. The influence of high frequency tangential vibrations on the stability of the fluid interfaces in microgravity. In *Microgravity Fluid Mechanics*, Rath, H.J., Ed.; IUTAM Symposium, Bremen 1991. Springer-Verlag: Berlin, 1992; 137–144.

5

Thermocapillary Instability of the Free Surface of a Plane Liquid Layer

The dependence of the liquid surface tension on temperature or surfactant concentration breaks mechanical equilibrium as convective motion occurs due to surface forces according to the Marangoni effect. Continuity (mass balance) as well as viscosity help to carry surface motions into the bulk. Along the path, viscosity tends to damp out motions as expressed in the tangential and the normal boundary conditions obeyed by the Navier–Stokes equations.

Surface tension inhomogeneities may be caused by thermal or surfactant concentration gradients placed along or orthogonal to a liquid surface. In the latter case it is known that steady cellular convective motion can be sustained by the Marangoni effect past a threshold, which is the appropriate explanation of Benard cells (Fig. 1, Chapter 2).

We have also remarked that if a temperature gradient is everywhere orthogonal to the free boundary, a liquid layer in weightlessness may remain in mechanical equilibrium. As noted earlier, a spontaneously formed hot spot on the surface gives rise to a surface force, which for standard liquids causes an outward liquid flow from the hot spot. When the temperature gradient is directed downward into the layer, a warmer liquid rising from the bottom to the surface can maintain the spot temperature. An increase of the temperature gradient to a sufficiently high level can help overcoming viscous damping and eventually sustains steady motion through monotonic instability of the initial motionless state, hence breaking the purely diffusive state of mechanical equilibrium. The flow velocity scales with the supercritical temperature difference past threshold. With an outward-directed temperature gradient, an incoming flow will extinguish the temperature spot, hence precluding the onset of monotonic instability. However, this situation favors

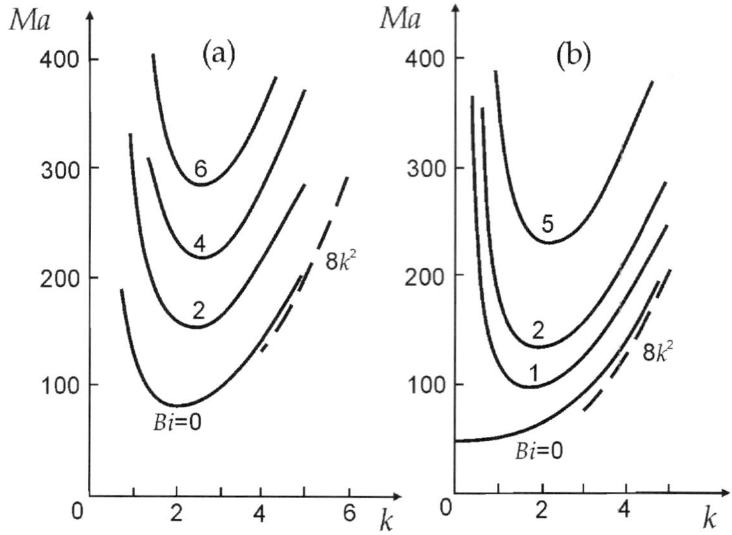

FIG. 1 Surface tension gradient (Marangoni) driven convective instability in a plane layer. Neutral stability curves for several values of the Biot number at the open surface of the liquid: (a) isothermal solid plate; (b) insulating one (heat flux held fix). (From Ref. 1.)

the generation of an elasticlike force, which at sufficiently high motion inertia can lead to overshooting the equilibrium state (overstability) and hence to oscillatory instability as was earlier discussed. In this chapter we shall return to the above stated problem in its variants and hence we shall study how surface tension gradients may alter the equilibrium stability of a plane layer placed on a solid support or a thin film with two free surfaces under various circumstances. For completeness we shall recall a few aspects of the instability resulting from the combined action of the Marangoni effect and buoyancy (Marangoni–Rayleigh convective instability problem). Note that a thermal gradient tangential to the free surface immediately initiates flow in balance with the viscous stress.

I. MONOTONIC SURFACE TENSION GRADIENT (MARANGONI) DRIVEN INSTABILITY IN A PLANE LIQUID LAYER WITH A FLAT FREE SURFACE

For completeness and because of the use that we shall make of Pearson's stability results [1] let us consider again how instability of a plane liquid layer bounded at one side by a solid plate and with an open surface to ambient air occurs as a result of the Marangoni effect. To begin with a state of rest, and hence mechanical equilibrium of the layer, we shall consider that the applied temperature gradient is normal to the layer boundaries. To a first linear approximation the disturbances of transverse velocity and temperature can be considered small and periodic in the layer plane and hence, using Fourier normal modes of wave number, $\mathbf{k}(k_x, k_y)$,

$$\{\mathrm{v}_z, T\} = \{\mathrm{v}(z), \theta(z)\} \exp[i(k_x x + k_y y) - \lambda t],$$

with the amplitudes satisfying the following system of equations:

$$-\lambda\left(\frac{\mathrm{d}^2\mathrm{v}}{\mathrm{d}z^2} - k^2\mathrm{v}\right) = \frac{\mathrm{d}^4\mathrm{v}}{\mathrm{d}z^4} - 2k^2\frac{\mathrm{d}^2\mathrm{v}}{\mathrm{d}z^2} + k^4\mathrm{v} \tag{1}$$

$$-\lambda Pr\theta = \frac{\mathrm{d}^2\theta}{\mathrm{d}z^2} - k^2\theta + Pr\mathrm{v} \tag{2}$$

that follow from the continuity, the Navier–Stokes, and the heat (Fourier) diffusion equations (Chapter 1).

All quantities are dimensionless because we have used as units of length, time, velocity, and temperature, the layer thickness d, d^2/ν, ν/d, and Θ (temperature difference across the layer), respectively. As noted earlier, the complex quantity, λ, defines (linear) stability: neutrally stable (Re $\lambda = 0$), stable (here Re $\lambda > 0$), or unstable (Re $\lambda < 0$). In the latter case we shall have a sufficient condition for monotonic (Im $\lambda = 0$) or oscillatory (Im $\lambda \equiv \omega \neq 0$) disturbances.

The requirement that there is no slip at the solid boundary and hence that all velocity components on the solid boundary of the layer be zero leads to

$$z = 0:\ \mathrm{v} = \frac{\mathrm{dv}}{\mathrm{d}z} = 0 \tag{3}$$

For temperature we consider the two cases defined earlier:

1. The solid surface has perfect conductivity; hence temperature is held fixed there and temperature disturbances are vanishing;
2. A constant heat flux passes through the solid surface, and hence disturbances in the heat flux are vanishing (so called "insulating" or very poor conducting boundary). Thus we have

$$z = 0:\ 1.\ \theta = 0; \qquad 2.\ \frac{\mathrm{d}\theta}{\mathrm{d}z} = 0 \tag{4}$$

For a flat free surface we require that viscous forces be equal to the tangential force provided by the inhomogeneity of the surface tension. Recall that this tangential stress balance condition can be written as

$$\Pi_{ik} n_k = \frac{\partial\sigma}{\partial x_i} \tag{5}$$

where Π_{ik} is the viscous stress tensor on the layer boundary, **n** is the unit vector going away from the liquid, and σ is the surface tension.

With the assumption that v_z is zero everywhere on the free surface, projecting Eq. (5) on axes x and y we obtain

$$\eta \frac{\partial v_x}{\partial z} = \frac{\partial \sigma}{\partial x}, \qquad \eta \frac{\partial v_y}{\partial z} = \frac{\partial \sigma}{\partial y}$$

Differentiating the first equation with x and the second equation with y, adding them and taking into account the continuity equation gives

$$\eta \frac{\partial^2 v_z}{\partial z^2} = -\left(\frac{\partial^2 \sigma}{\partial x^2} + \frac{\partial^2 \sigma}{\partial y^2}\right) \tag{6}$$

For most liquids the surface tension, σ, is a linearly decreasing function of temperature,

$$\sigma = \sigma_0 - \sigma_1 T \tag{7}$$

and, as noted earlier, σ_1 in the majority of cases is positive (normal thermocapillary effect for standard liquids). Using Eq. (7) the boundary condition [Eq. (6)] becomes

$$\eta \frac{\partial^2 v_z}{\partial z^2} = \sigma_1 \left(\frac{\partial^2 T}{\partial x^2} + \frac{\partial^2 T}{\partial y^2}\right) \tag{8}$$

For dimensionless amplitudes of the normal disturbances we set

$$z = 1: \ \frac{d^2 v}{dz^2} = -k^2 Ma\, Pr^{-1} \theta \tag{9}$$

where $Ma = \sigma_1 \Theta d / \eta \chi$ is the Marangoni number and Ma/Pr plays the role of the thermal Reynolds (Re_T) or Peclet number, accounting for the ratio between the thermocapillary and viscous forces.

To formulate the thermal boundary condition on the free surface, using the dimensionless amplitude of the temperature disturbance, we set

$$z = 1: \ \frac{d\theta}{dz} = -Bi\theta \tag{10}$$

where the Biot number is taken as $Bi = qd/\kappa$, q is a heat transfer coefficient, and, as noted earlier $Bi = 0$ defines heat flux held fixed while $Bi \to \infty$ corresponds to the perfect heat conductor (temperature held fixed).

Then, λ is the eigenvalue of the following boundary-value problem:

$$-\lambda\left(\frac{d^2v}{dz^2} - k^2v\right) = \frac{d^4v}{dz^4} - 2k^2\frac{d^2v}{dz^2} + k^4v \tag{11}$$

$$-\lambda Pr\theta = \frac{d^2\theta}{dz^2} - k^2\theta + Prv, \tag{12}$$

$$z = 0\colon\ v = \frac{dv}{dz} = 0, \qquad \text{(i)}\ \theta = 0; \qquad \text{(ii)}\ \frac{d\theta}{dz} = 0,$$

$$z = 1\colon\ v = 0, \qquad \frac{d^2v}{dz^2} = -Ma\,Pr^{-1}k^2\theta, \qquad \frac{d\theta}{dz} = -Bi\theta$$

For monotonic (Im $\lambda = 0$), neutral (Re $\lambda = 0$) disturbances the problem is reduced to

$$\frac{d^4v}{dz^4} - 2k^2\frac{d^2v}{dz^2} + k^4v = 0 \tag{13}$$

$$\frac{d^2\theta}{d^2z} - k^2\theta + Prv = 0 \tag{14}$$

$$z = 0\colon\ v = \frac{dv}{dz} = 0, \qquad \text{(i)}\ \theta = 0; \qquad \text{(ii)}\ \frac{d\theta}{dz} = 0$$

$$z = 1\colon\ v = 0, \qquad \frac{d^2v}{dz^2} = -Ma\,Pr^{-1}k^2\theta, \qquad \frac{d\theta}{dz} = -Bi\theta$$

whose solution is known [1]. The critical values of the Marangoni number as a function of the wavenumber and those of the Biot number for the above given two types of boundary conditions on the solid surface are

$$Ma = \frac{8k(k\cosh k + Bi\sinh k)(\sinh k \cdot \cosh k - k)}{\sinh^3 k - k^2\cosh k} \tag{15a}$$

$$Ma = \frac{8k(k\sinh k + Bi\cosh k)(\sinh k \cdot \cosh k - k)}{\sinh^2 k \cdot \cosh k - 2\sinh k + k^2\cosh k - k^3\sinh k} \tag{15b}$$

The critical numbers Ma defined by these formulas are positive for all values of k and Bi. Neutral stability curves for several values of the Biot number are given in Fig. 1.

The Marangoni number curves pass through a minimum at a certain critical value, k_*, of the wavenumber. For a solid isothermal boundary and small Biot number $k_* = 2$ and $Ma_* = 80$. In the case where the lower boundary is thermally "insulating" (with corresponding $Bi = 0$, we have two "practically" insulating boundaries), there is a long-wave instability with $k_* = 0$ and $Ma_* = 48$ [2]. This instability mode is difficult to observe because in practice the condition of a very poorly conducting low boundary is difficult to realize [3,4].

Increasing the heat transfer, the minimum of the critical Marangoni number increases and shifts deeper into the region of short wavelengths. The value of Ma grows with increasing values of Bi. The limiting case $Bi \to \infty$ implies a transition to an isothermal upper free surface. In this case there is no gradient of surface tension, no Marangoni effect, and thermocapillary instability is ruled out.

II. INSTABILITY OF A LAYER WITH A DEFORMABLE SURFACE

In the previous section we have considered only monotonic disturbances, hence leading to steady patterned convection past a given threshold. Oscillatory instability eventually leading to transverse vibrations, and hence wave motion, becomes possible when the free surface is deformable, as discussed earlier.

A. Two Modes of Oscillatory Instability

Let us consider weak surface deformations, ζ. Then the boundary conditions on the free surface can be written as

$$p + \sigma_0\left(\frac{\partial^2 \zeta}{\partial x^2} + \frac{\partial^2 \zeta}{\partial y^2}\right) = 2\eta \frac{\partial \mathrm{v}_z}{\partial z}$$

$$-\sigma_1\left(\frac{\partial T}{\partial x} + \frac{\partial T_0}{\partial z}\frac{\partial \zeta}{\partial x}\right) = \eta\left(\frac{\partial \mathrm{v}_x}{\partial z} + \frac{\partial \mathrm{v}_z}{\partial x}\right) \tag{16a}$$

$$-\sigma_1\left(\frac{\partial T}{\partial y} + \frac{\partial T_0}{\partial z}\frac{\partial \zeta}{\partial y}\right) = \eta\left(\frac{\partial \mathrm{v}_y}{\partial z} + \frac{\partial \mathrm{v}_z}{\partial y}\right) \tag{16b}$$

$$\frac{\partial \zeta}{\partial t} = \mathrm{v}_z \tag{16c}$$

It was noted earlier that surface deformability opens the way to yet another mode of monotonic instability in the long-wave ($k = 0$) range [2–6]. Consider the role of the deformable open surface. When the liquid surface is inclined with respect to the isotherms of the main temperature field it begins to experience the action of a longitudinal thermocapillary force directed toward the region with lower temperature. If the temperature gradient field is directed inward, a spontaneous protuberance of the surface causes liquid to move toward the top of the hill, thus sustaining the generated surface deformation. On the other hand, if the temperature gradient is directed outward, an accidental bump in the surface generates a backward flow from the top of the hill, suppressing the surface deformation. Thus, a monotonic long-wave instability is indeed expected in a liquid layer heated on the side of its solid boundary support. The influence of the open surface deformation on the critical Marangoni number of the monotonic instability will be discussed in detail later when dealing with the stability of a liquid layer with two free boundaries.

In the case of heating at the open side of the free deformable surface, as discussed in Chapter 2, the above-described interaction of the temperature field with thermocapillary motion is expected to produce an elasticlike force, which could give rise to oscillatory instability in the form of transverse surface vibrations and waves [7–16].

To define the threshold of the oscillatory convection and the structure of the expected flow in better quantitative terms than given earlier in the book, let us write the general solution of Eqs (1) and (2) for neutral oscillatory disturbances with $\lambda = i\omega$ and ($Pr \neq 0$, $Pr \neq 1$)

$$v = c_1 \exp(kz) + c_2 \exp(-kz) + c_3 \exp(lz) + c_4 \exp(-lz)$$

$$\theta = \frac{c_1 \exp(kz) + c_2 \exp(-kz)}{i\omega Pr} - \frac{c_3 \exp(lz) + c_4 \exp(-lz)}{i\omega(Pr - 1)} + c_5 \exp(mz) + c_6 \exp(-mz)$$

$$l = (k^2 - i\omega)^{1/2}, \qquad m = (k^2 - i\omega Pr)^{1/2}$$

The coefficients c_i are specified by the boundary conditions

$$z = 0\text{: } \mathrm{v} = \frac{\mathrm{dv}}{\mathrm{d}z} = 0, \qquad \text{(i) } \theta = 0; \qquad \text{(ii) } \frac{\mathrm{d}\theta}{\mathrm{d}z} = 0$$

$$z = 1\text{: } \frac{\mathrm{d}^3\mathrm{v}}{\mathrm{d}z^3} - (3k^2 - i\omega)\left(\frac{\mathrm{dv}}{\mathrm{d}z}\right) - Ca\, k^4 \xi = 0$$

$$\frac{\mathrm{d}^2\mathrm{v}}{\mathrm{d}z^2} + k^2\mathrm{v} + Ma\, Pr^{-1} k^2 \left(\theta + \frac{\mathrm{d}T_0}{\mathrm{d}z}\xi\right) = 0 \tag{17}$$

$$\frac{\mathrm{d}\theta}{\mathrm{d}z} = -Bi\theta, \qquad \xi = -\mathrm{v}(i\omega)^{-1} \tag{18}$$

where $Ca = \sigma_0 d / \eta\chi$ is the capillary parameter relating surface tension to viscous forces.

The requirement that a homogeneous complex system is solvable for c_i defines the critical Marangoni number. Imposing that the Marangoni number must be real yields the corresponding dispersion equation.

Shown in Fig. 2 are the neutral $Ma(k)$ and dispersion $\omega(k)$ curves for $Pr = 0.1$, $Ca = 10^4$, and $Bi = 0$. The dashed lines show the boundary of stability for oscillatory disturbances. For reference, the dispersion relation for capillary waves on the surface of an inviscid liquid is represented by the dotted–dashed line. Curves under odd numbers correspond to the case of an isothermal solid boundary

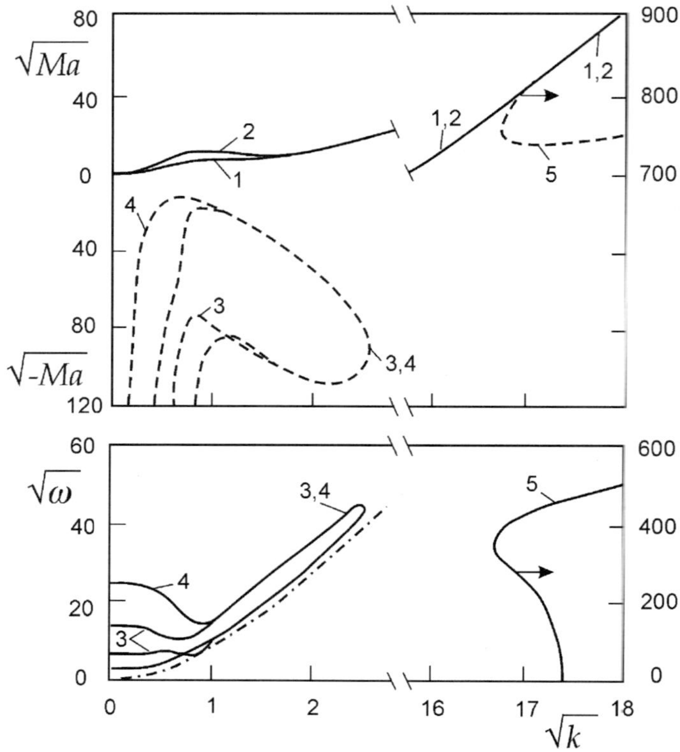

FIG. 2 Surface tension gradient (Marangoni) driven convective instability in a plane layer. Neutral curves, $Ma(k)$, and dispersion relation, $\omega(k)$, for $Pr = 0.1$, $Ca = 10^4$, and $Bi = 0$. Curves labeled with odd numbers have been obtained for the case of an isothermal solid boundary (i), and curves corresponding to even numbers for the case of a given heat flux at the solid boundary (ii). Dashed lines show the stability boundary for oscillatory disturbances. (From Ref. 17.)

(i), and curves under even numbers to the case of given heat flux held fixed at the solid boundary (ii). Note that the dispersion relation, $\omega(k)$, defines two modes of oscillatory surface disturbances: oscillations with the dispersion relation closely fitting the dotted–dashed curve, hence capillary waves sustained by the Marangoni effect, and long-wave oscillations with practically constant frequency at $k \rightarrow 0$ also sustained by the same effect. Also note that surface waves can be excited and eventually sustained for different ranges of wavenumbers under both heating regimes, on the open side of the liquid surface or on the side of the solid support of the liquid layer. Recall the earlier mentioned elasticlike role played by the Marangoni effect and the fact that long waves in liquids suffer virtually no dispersion. The Marangoni effect appears as an agent responsible for such dilation–compression waves at the interface of two incompressible immiscible liquids or at the open free surface of an incompressible liquid.

B. Long-Wavelength Instability

Let us now consider in detail the possibility of thermocapillary instability with a deformable open surface in the range of long waves ($k \to 0$). As mentioned earlier, when heating is carried out on the side of the solid support of the liquid layer it can give rise to monotonic long-wave instability while oscillatory instability is all that is expected when heating on the open side of the liquid surface. Because the gravitational force strongly interferes with the relief formation on the free surface (hydrostatic pressure) this factor must be kept when calculating the threshold for stationary convection. Note that if g is set to zero, and we therefore neglect not only buoyancy but hydrostatic head at the open surface, the capillary length diverges to infinity and long-wavelength disturbances may settle with infinitesimal energy supply, hence at vanishing Marangoni number and no instability is evoked. Smith [2] dealt appropriately with this case and showed that indeed there are two different monotonic instability modes with corresponding finite threshold Marangoni numbers.

1. Monotonic Instability

As noted earlier the evolution of neutral monotonic disturbances obeys the equations

$$\frac{\mathrm{d}^4\mathrm{v}}{\mathrm{d}z^4} - 2k^2\frac{\mathrm{d}^2\mathrm{v}}{\mathrm{d}z^2} + k^4\mathrm{v} = 0 \tag{19}$$

$$\frac{\mathrm{d}^2\theta}{\mathrm{d}z^2} - k^2\theta + Pr\mathrm{v} = 0 \tag{20}$$

$$z = 0 : \mathrm{v} = \frac{\mathrm{d}\mathrm{v}}{\mathrm{d}z} = 0, \qquad \text{(i) } \theta = 0; \qquad \text{(ii) } \frac{\mathrm{d}\theta}{\mathrm{d}z} = 0$$

$$z = 1 : \mathrm{v} = 0, \qquad \frac{\mathrm{d}^3\mathrm{v}}{\mathrm{d}z^3} - 3k^2\frac{\mathrm{d}\mathrm{v}}{\mathrm{d}z} - (Ca\,k^4 + Pr^{-1}Ga\cdot k^2)\xi = 0$$

$$\frac{\mathrm{d}^2\mathrm{v}}{\mathrm{d}z^2} + k^2\mathrm{v} + Ma\,Pr^{-1}k^2(\theta - \xi) = 0, \qquad \frac{\mathrm{d}\theta}{\mathrm{d}z} = -Bi\theta$$

where $Ga = gd^3/\nu\chi$ is the Galileo number.

The neutral curve of the marginal stability for the isothermal solid boundary case with $Bi = 0$, $Ga = 120$, and $Pr\,Ca = 10^4$ has two minima for the same value of the Marangoni number [2]. Figure 3 depicts theoretical predictions. It does

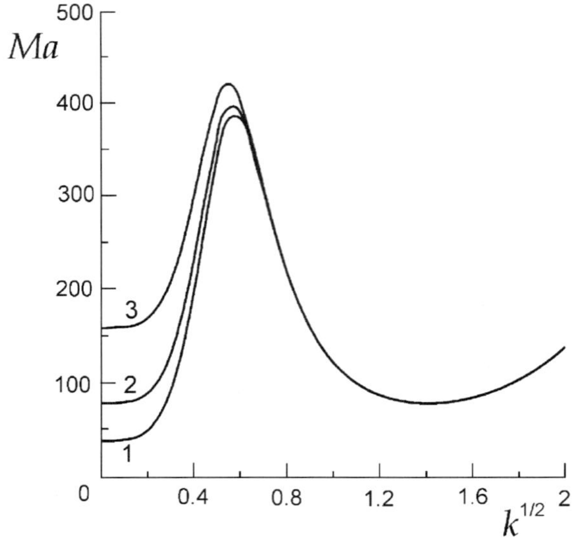

FIG. 3 Surface tension gradient (Marangoni) driven convective instability in a plane layer. Marginal stability curves for $Bi = 0$, $Pr\,Ca = 10^4$ and various values of Ga: (1) 60, (2) 120, and (3) 240. Note that one or the other of the two instability thresholds is possible depending on the value of Ga, although they may also coexist as illustrated with the locus 2. (From Ref. 2.)

appear that for the short-wavelength minimum the Marangoni number does not practically depend on Ga. For the long-wavelength instability the role of gravity, and hence of the Galileo number, can be analyzed in the $k \to 0$ limit. Solving the problem of Eqs. (19) and (20) using power series in k^2, in the 0th order we obtain

$$\frac{d^4 v}{dz^4} = 0, \qquad \frac{d^2 \theta}{dz^2} = -v \tag{21}$$

$$z = 0: v = \frac{dv}{dz} = 0, \qquad \text{(i) } \theta = 0; \qquad \text{(ii) } \frac{d\theta}{dz} = 0$$

$$z = 1: v = 0, \qquad \frac{d^3 v}{dz^3} - Ga\,Ma^{-1}\frac{d^2 v}{dz^2} = 0, \qquad \frac{d\theta}{dz} = -Bi\theta$$

where we have kept the Biot number appearing explicitly.

It follows that the threshold value of the Marangoni number for long-wavelength instability, Ma_l, is

$$Ma_l = \tfrac{2}{3}Ga$$

hence $Ma = 0$ as g vanishes, as earlier noted. The Galileo number depends on the layer thickness as d^3 whereas $Ma \sim d$; thus one may select a value of the layer depth, d, for which $Ma_l = Ma* = 80$. In fact, at $Pr\,Ca > 10^4$ the critical Marangoni number of the short-wave instability is insensitive to surface deformation.

Recently, these two modes of monotonic instability were experimentally observed [3,4]. Shown in Fig. 4 are photographs of the corresponding convective cells. The first photograph (a) shows the motionless heat diffusion regime (the Marangoni number is 10% below threshold) in a layer of thickness $d = 0.033$ cm. A large black spot in the second photograph (b) ($d = 0.011$ cm) corresponds to a long-wave convective cell. Photograph (c) ($d = 0.022$ cm) shows the coexistence of both instability modes, a long cell and hexagonal cells. The supercritical regime in the layer, with $d = 0.033$ cm, is shown in photograph (d) where convection appears only in the form of hexagonal cells. However, the experimenters noted that

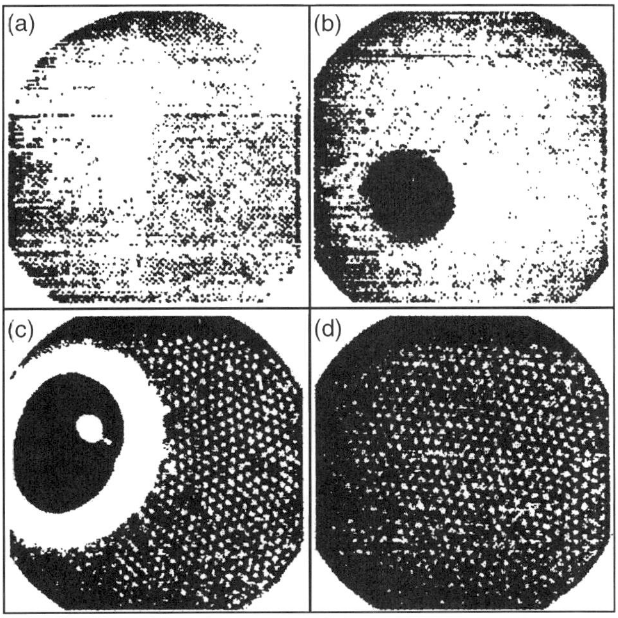

FIG. 4 Surface tension gradient (Marangoni) driven convective instability in a plane layer. Infrared images from experiments with increasing fluid depth. (a) ($d = 0.033$ cm) 10% below onset of instability. (b) ($d = 0.011$ cm). A long-wavelength mode is the primary instability. The dark region is a dry spot. (c) ($d = 0.022$ cm). Long-wavelength and hexagonal modes coexist. A droplet (white circle) is trapped within the dry spot (dark oval). (d) ($d = 0.033$ cm). Hexagons are the primary instability. (From Ref. 3.)

the onset occurs 35% below the value predicted by linear theory for the single layer model. Accordingly, the influence of the air layer depth, or the depth of another liquid layer above, must be taken into account for an appropriate quantitative theory predicting the correct onset and the form of the interface shape or relief. The latter question will be discussed further below.

2. Oscillatory Instability

Let us now consider the possibility of a long-wavelength oscillatory mode. Because long-wavelength thermocapillary oscillations have practically constant finite frequency at $k \to 0$, and therefore do not interact with capillary-gravity waves, the Galileo number can be disregarded, for simplicity, and hence the boundary-value problem for the amplitudes of Fourier normal modes in the 0th order in k^2 is

$$-i\omega\frac{d^2v}{dz^2} = \frac{d^4v}{dz^4} \tag{22}$$

$$-i\omega Pr\theta = \frac{d^2\theta}{dz^2} + Prv \tag{23}$$

$$z = 0: v = \frac{dv}{dz} = 0, \qquad \text{(i) } \theta = 0; \qquad \text{(ii) } \frac{d\theta}{dz} = 0$$

$$z = 1: \frac{d^3v}{d^3z} + i\omega\frac{dv}{dz} = 0, \qquad \frac{d\theta}{dz} = -Bi\theta$$

$$i\omega\frac{d^2v}{d^2z} + i\omega Ma\,Pr^{-1}k^2\theta + Ma\,Pr^{-1}k^2v = 0 \tag{24}$$

Since for neutral oscillations the Marangoni number diverges to infinity when $k \to 0$ (dashed line in Fig. 2) the product $M = k^2\,Ma$ must be kept finite. The solution to the system of Eqs. (22) to (24) can be written as

$$v = c_1(\cosh lz - 1) \tag{25}$$

$$\theta = c_2 \sinh mz + c_3 \cosh mz - c_1 Pr(\cosh lz/(l^2 - m^2) + 1/m^2) \tag{26}$$

with

$$l = (-i\omega)^{1/2}, \qquad m = (-i\omega Pr)^{1/2}$$

and

$$\text{(i)}\quad c_2 = \frac{1}{m\cosh m + Bi\sinh m}\left[c_1 Pr\left(\frac{l\sinh l + Bi\cosh l}{l^2 - m^2} + \frac{Bi}{m^2}\right) - c_3(m\sinh m + Bi\cosh m)\right]$$

$$c_3 = c_1\frac{Prl^2}{m^2(l^2 - m^2)}$$

$$\text{(ii)}\quad c_2 = 0$$

$$c_3 = \frac{c_1 Pr}{m\sinh m + Bi\cosh m}\left(\frac{l\sinh l + Bi\cosh l}{l^2 - m^2} + \frac{Bi}{m^2}\right).$$

The solvability condition of Eqs. (22) to (24) determines the threshold and the frequency of oscillations at threshold. At $Bi = 0$ (heat flux held fixed at the liquid–air interface) a countable number of frequencies corresponding to different modes of instability is obtained. Higher frequencies correspond to larger values of the parameter M (redefined Marangoni number). Table 1 gives two low frequencies of oscillations and respective values of M for $Bi = 0$.

The data listed in the table show that long-wave oscillations occur only when the layer is heated on the open side of the free surface ($Ma < 0$). The oscillations for the isothermal surface have lower frequency, decreasing with the increase of the Prandtl number. The frequencies of critical perturbations significantly depend on the value of the liquid–air Biot number. The dependence of the low-oscillation frequencies on Bi is shown in Fig. 5 for a good heat conducting, isothermal solid lower boundary and in Fig. 6 for a poorly conducting, "insulating" bottom boundary. In fact, this dependence appears rather involved. With the increase of

TABLE 1 Surface Tension Gradient (Marangoni) Driven Convective Instability in a Plane Layer. Low Frequencies of Long Wavelength Oscillations and Respective Values of M for the Liquid–Air $Bi = 0$; (i) Isothermal Solid Bottom Boundary, and (ii) Poorly Conducting Solid Bottom Boundary (Heat Flux Held Fixed)

	Case (i)		Case (ii)	
Pr	ω	M	ω	M
0.1	8.82	−7.46	49.4	−309
0.1	48.2	−323	197	−5136
0.01	16.0	−2.35	493	−2652
0.01	493	−2706	1970	−42,766

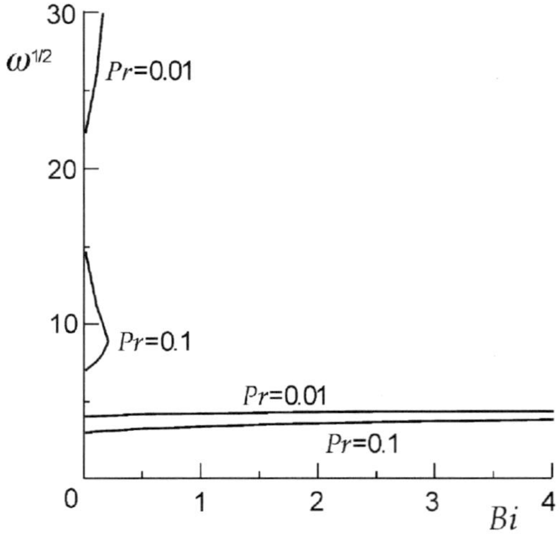

FIG. 5 Surface tension gradient (Marangoni) driven convective instability in a plane layer. Frequency of long-wavelength oscillations as a function of the liquid–air Bi for an isothermal solid bottom boundary.

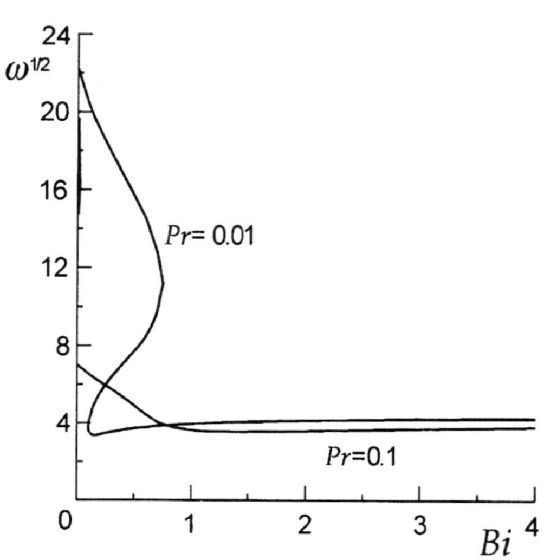

FIG. 6 Surface tension gradient (Marangoni) driven convective instability in a plane layer. Frequency of long-wave oscillations as a function of the liquid–air Bi for a poorly conducting solid bottom boundary.

the liquid–air Bi number the frequencies of some oscillations begin to approach each other and after becoming equal the oscillations disappear. Beyond a certain value of the Biot number, there is only one low level of instability. The frequency of the neutral oscillations is close to the frequency of oscillations at $Bi = 0$ for the isothermal solid boundary case.

The particularity of the insulating lower boundary case is that, for liquids with low values of Prandtl number ($Pr \sim 0.01$), the low frequency mode disappears at $Bi = 0$. But for the Biot number close to zero there appear two oscillation modes with equal frequencies. As it grows the difference between these frequencies increases. However the dispersion curves for high-frequency oscillations are closed and disappear. At $Bi \geq 1$ in all cases only oscillations with frequencies somewhat higher than 10 are observed. As the Biot number lowers (Fig. 6), the low-frequency mode of oscillations vanish, and there is a sudden increase in the frequency. This means that the model of a theoretically insulating heat boundary can hardly differ from a real, poorly conductive one.

As the solution of Eqs. (25) and (26) show the flow exponentially decays going away from the surface with a logarithmic decrement proportional to $\omega^{1/2}$. This means that for high-frequency oscillations the flow is localized in a thin layer near the open surface, and explains why these oscillations disappear with the increase of the Biot number. For $Pr = 0.01$ this fact is illustrated in Fig. 7, where the real and imaginary parts of the amplitudes of the longitudinal velocity components, u_r and u_i, and perturbation of temperature, θ_r and θ_i, are shown as functions of z.

For an isothermal solid bottom boundary and for liquid–air with $Bi = 0$ a critical motion with frequency $\omega = 16$ (lower level of spectrum) penetrates through the layer [Fig. 7(a)]. The velocity and temperature distributions for a solid boundary with liquid–air $Bi = 0$ are displayed in Fig. 7(b) where it is shown that the oscillation with frequency $\omega = 493$ is localized near the open surface, and in most of the layer no motion exists. In Figs 7(c) and (d) appear the distributions of the amplitudes of perturbations corresponding to the case of a lower boundary with $Bi = 0.2$ close to the "nose" of the curve for $Pr = 0.01$ in Fig. 6. For lower frequencies ($\omega = 11.6$) these distributions are close to those of Fig. 7(a) and hence all elements of the liquid layer are involved in the motion. From the upper curve ($\omega = 29.8$) we can see the appearance of a motionless spot near the solid boundary [Fig. 7(d)], though it is less noticeable than for oscillations with frequency $\omega = 493$ [Fig. 7(b)].

III. INSTABILITY OF A LIQUID FILM SUBJECTED TO A TRANSVERSE TEMPERATURE GRADIENT

Let us now consider the instability of a film, or a plane liquid layer with two free boundaries, subjected to a uniform transverse temperature gradient in vanishing

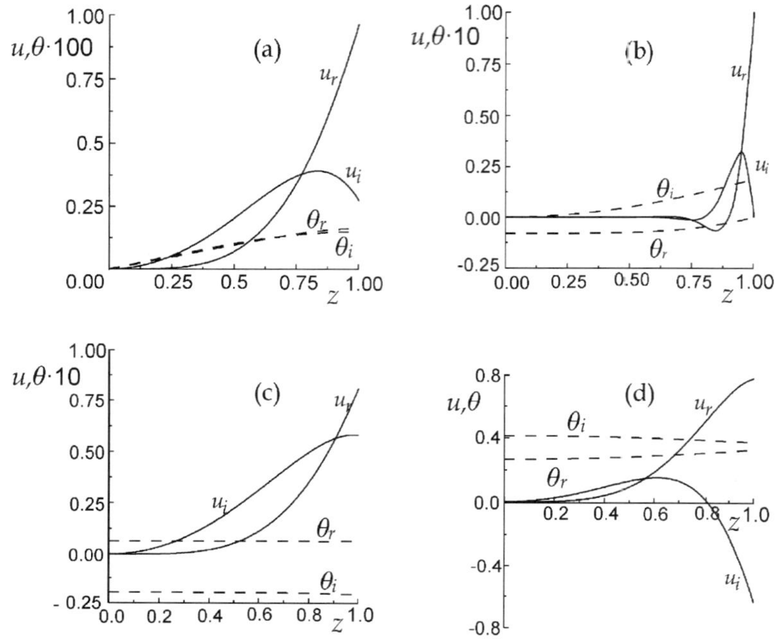

FIG. 7 Surface tension gradient (Marangoni) driven convective instability in a plane layer. Longitudinal velocity component (along the horizontal surface) and perturbation of temperature for oscillations of the main modes: (a) and (b) are for an isothermal solid boundary at liquid–air $Bi = 0$; (c) and (d) are for the two main modes for an "insulation" solid boundary at liquid–air $Bi = 0.2$.

effective gravity conditions. As earlier, here again the surface tension is assumed to depend linearly on the temperature. On the basis of an exact solution of the linear perturbation problem for a layer with deformable boundaries, the instability domains, the dispersion curves, and the shape of the perturbations can be determined. For liquids with low enough Prandtl numbers, waves with predominantly longitudinal flow and capillary-gravity waves can develop on the film as a result of the Marangoni effect.

For a layer with two outer open deformable boundaries, both monotonic and oscillatory instability have been studied [18,19]. In the latter case, the attention has been mostly devoted to the analysis of the neutral curves, while the dispersion relations have remained practically unexplored. In the region of low frequencies and small wavenumbers there are no unstable oscillatory disturbances. Further studies have dealt with heat sources or sinks located in the midplane section (a divider) of the layer as we shall discuss in a subsequent chapter.

A. A Boundary Value Problem for Small Disturbances

Let us consider a film, or a plane layer of viscous liquid with two free boundaries located at $z = 0$ and $z = d$, respectively, with a temperature difference, Θ, imposed from outside (Fig. 8). Disregarding buoyancy, a steady state corresponding to mechanical equilibrium with a linear temperature profile is possible:

$$T_0 = \Theta\left(1 - \frac{z}{d}\right)$$

In order to investigate the stability of this state we shall consider small perturbations of the velocity, $\mathbf{v}$, pressure, p, and temperature, T, whose behavior is described, as in earlier cases, by the continuity, the Navier–Stokes, and the heat (Fourier) diffusion equations, which we recall for completeness here,

$$\operatorname{div} \mathbf{v} = 0 \tag{27}$$

$$\frac{\partial \mathbf{v}}{\partial t} = -\frac{1}{\rho}\nabla p + \nu \Delta \mathbf{v} \tag{28}$$

$$\frac{\partial T}{\partial t} + \mathbf{v} \cdot \nabla T_0 = \chi \Delta T \tag{29}$$

In formulating the boundary conditions, we shall assume that the deviation, ζ, of either liquid surface from the equilibrium position is small compared with the layer thickness, d, the surrounding medium, say air, has no (mechanically) active influence on the liquid flow, and again the surface tension depends linearly on temperature: $\sigma = \sigma_0 - \sigma_1 T$. However, in view of the smallness of the term $\sigma_1 T$, to further simplify the problem the reference value of the surface tension σ_0 is assumed to be the same on both surfaces. Accordingly, the normal and tangential

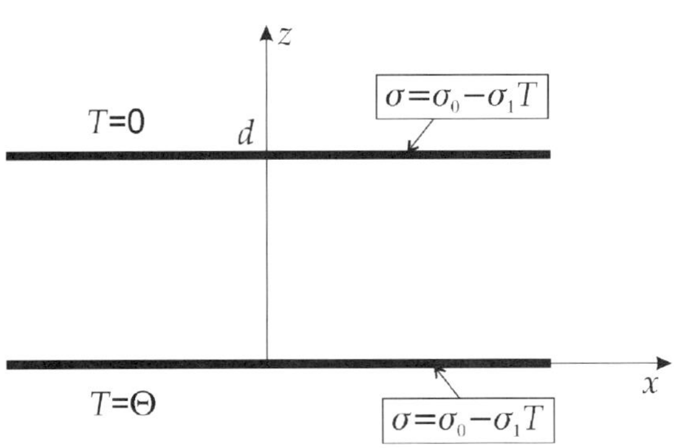

FIG. 8 Surface tension gradient (Marangoni) driven instability of a liquid film or a liquid layer with two open boundaries. Geometry of the problem and the coordinate system.

stress balances and the kinematic conditions in the planes $z = 0$ and $z = d$ are

$$p \pm \sigma_0 \left(\frac{\partial^2 \zeta}{\partial x^2} + \frac{\partial^2 \zeta}{\partial y^2} \right) = 2\eta \frac{\partial \mathrm{v}_z}{\partial z} \tag{30}$$

$$-\sigma_1 \left(\frac{\partial T}{\partial x} + \frac{\partial T_0}{\partial z} \frac{\partial \zeta}{\partial x} \right) = \pm \eta \left(\frac{\partial \mathrm{v}_x}{\partial z} + \frac{\partial \mathrm{v}_z}{\partial x} \right) \tag{31a}$$

$$-\sigma_1 \left(\frac{\partial T}{\partial y} + \frac{\partial T_0}{\partial z} \frac{\partial \zeta}{\partial y} \right) = \pm \eta \left(\frac{\partial \mathrm{v}_y}{\partial z} + \frac{\partial \mathrm{v}_z}{\partial y} \right) \tag{31b}$$

$$\frac{\partial \zeta}{\partial t} = \mathrm{v}_z \tag{32}$$

where x and y are the longitudinal coordinates, and the lower sign corresponds to the boundary $z = 0$. Assuming that the heat flux is held constant at the two outer boundaries, we write

$$z = 0, \qquad d : \frac{\partial T}{\partial z} = 0 \tag{33}$$

As in earlier cases studied, we change to dimensionless quantities, taking d, d^2/ν, ν/d, and Θ, as units of length, time, velocity, and temperature, respectively. Once more we construct the solution of the problem using Fourier normal modes

$$(\mathrm{v}_z, T, \zeta) = (\mathrm{v}, \theta, \xi) \exp[i(k_1 x + k_2 y - \lambda t)],$$

and we formulate the problem for the amplitudes of the z-components of the velocity, $\mathrm{v}(z)$, the temperature, $\theta(z)$, and the surface deformation, ξ.

According to Eqs. (27) to (33) the boundary value problem, with $\lambda = i\omega$, for the disturbance amplitudes has the form:

$$-i\omega \left(\frac{\mathrm{d}^2 \mathrm{v}}{\mathrm{d}z^2} - k^2 \mathrm{v} \right) = \frac{\mathrm{d}^4 \mathrm{v}}{\mathrm{d}z^4} - 2k^2 \frac{\mathrm{d}^2 \mathrm{v}}{\mathrm{d}z^2} + k^4 \mathrm{v} \tag{34}$$

$$-i\omega Pr \theta = \frac{\mathrm{d}^2 \theta}{\mathrm{d}z^2} - k^2 \theta - Pr \left(\frac{\mathrm{d}T_0}{\mathrm{d}z} \right) \mathrm{v} \tag{35}$$

$$k^2 = k_1^2 + k_2^2$$

$$z = 0, 1 : \frac{\mathrm{d}^3 \mathrm{v}}{\mathrm{d}z^3} - (3k^2 - i\omega) \left(\frac{\mathrm{d}\mathrm{v}}{\mathrm{d}z} \right) \pm Cak^4 \xi = 0$$

$$\frac{\mathrm{d}^2 \mathrm{v}}{\mathrm{d}z^2} + k^2 \mathrm{v} \pm Ma\, Pr^{-1} k^2 \left[\theta + \left(\frac{\mathrm{d}T_0}{\mathrm{d}z} \right) \xi \right] = 0$$

$$\frac{\mathrm{d}\theta}{\mathrm{d}z} = 0, \qquad \xi = -\mathrm{v}(i\omega)^{-1} \tag{36}$$

The problem of Eqs. (34) to (36) contains three independent dimensionless parameters. As in earlier studied cases the capillary parameter, *Ca*, characterizes the possible deformability of the boundaries by viscous forces. Recall that plane undeformable boundary corresponds to very high values of the capillary number ($Ca \to \infty$).

B. Neutral Curves and Form of Disturbances

The boundary value problem of Eqs. (34) to (36) admits an exact solution. For illustration, the results of determining the stability boundaries for a liquid with the parameter values $Pr = 0.01$ and $Ca = 10^4$ are summarized in Fig. 9(a). The neutral curves of the oscillatory disturbances and monotonic instability are represented by the broken and solid curves, respectively. Curves 4 correspond to the abovementioned parameter values. In Fig. 9(b) we have plotted the dispersion

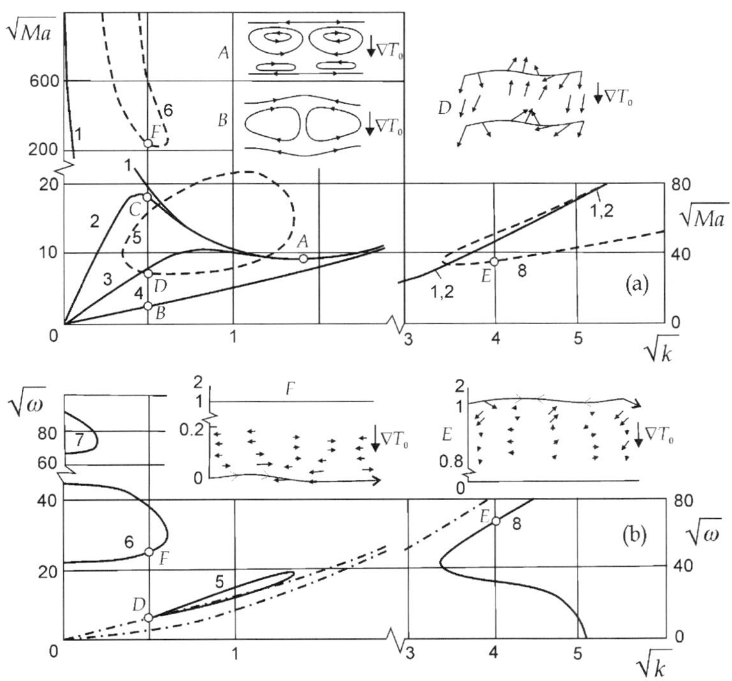

FIG. 9 Surface tension gradient (Marangoni) driven instability of a liquid film with two free boundaries. Neutral curves, $Ma(k)$, and dispersion relations, $\omega(k)$, for $Pr = 0.01$ and $Ca = 10^4$. Insets *A* and *B*: streamlines of the monotonic disturbances for points *A* and *B* on the neutral curves. Insets *D*, *E*, and *F*: instantaneous values of the velocity vector of the oscillatory disturbances at points *D*, *E*, and *F*. (From Ref. 16.)

curves of the oscillatory disturbances. In the insets in Fig. 9 we have sketched the shape of the convective cells and the structure of the flow for characteristic points of the neutral curves. Later on both monotonic and oscillatory instabilities are considered in further detail.

In a layer with two free boundaries, the Marangoni effect in one boundary suppresses the formation of a single-vortex neutral perturbation and, therefore, the critical Marangoni number turns out to be higher ($Ma = 87.4$) than for a layer with only a free surface ($Ma = 80$) [18].

The deformability of the boundaries creates another monotonic instability mechanism associated with the inclination of the layer boundary to the isotherms of the imposed base temperature field. Under conditions of weightlessness for a layer with one free boundary, and for a zero liquid–air Biot number, there is absolute instability with $k* = 0$ [4]. In Fig. 9(a), for a layer with two free boundaries the neutral curves of monotonic instability obtained from the exact solution of the boundary value problem of Eqs (31a,b) and (32) are depicted by curves 1–4 for the parameter value $Pr \cdot Ca \to \infty$, 10^6, 10^4, and 10^2, respectively. In inset *A* in Fig. 9 we have illustrated the disturbance convective structure in the neighborhood of the minimum of the neutral curves (point *A*) where there is instability [1]. There is practically no deformation of the boundary of the layer and the growing longitudinal temperature gradient due to the inflow of hot liquid to the surface leads to the formation of an oppositely directed vortex on the externally heated boundary. In inset *B* we have reproduced the disturbance generated by a different mechanism. On the surface deformed by the motion of the liquid the main temperature field T_0 induces a longitudinal gradient that maintains the motion due to the Marangoni effect. The deformation of the liquid surface is maintained by the flow in the neighborhood of the cold surface, and hence the latter can be considered an active surface. However, the shape of the cell is such that the motion (swelling–compression) is maintained on both surfaces.

Let us consider the disturbance structure in more details. The transverse and longitudinal velocity components, v_z and v_x, are

$$v_z = v(z) \cos kx$$

$$v_x = -\frac{1}{k}\left(\frac{dv}{dz}\right) \sin kx \equiv u(z) \sin kx$$

In Fig. 10, curves 1, 2, and 3 reproduce the distribution of the amplitudes of u and v over the layer thickness for the points *A*, *B*, and *C* in Fig. 9(a). The solid and broken lines represent the amplitudes of u and v, respectively. The solution of the problem has been normalized with respect to the amplitude of the maximum longitudinal velocity component. For all monotonic disturbances the temperature disturbance depends only slightly on the transverse coordinate and can be characterized by its value at the boundary.

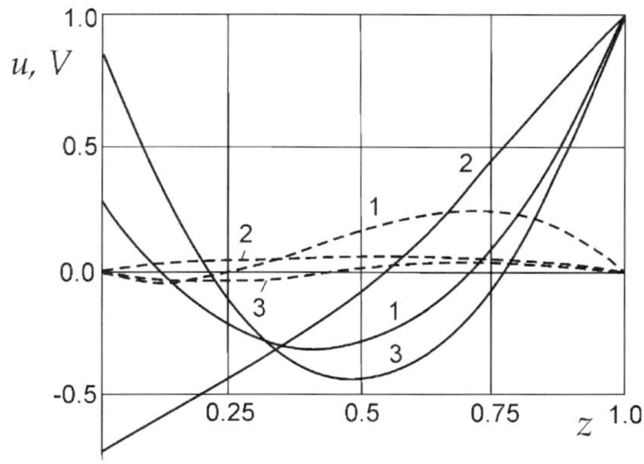

FIG. 10 Surface tension gradient (Marangoni) driven instability of a liquid film with two free boundaries. Amplitude distribution of the transverse and longitudinal velocity components, $v(z)$ and $u(z)$, of the monotonic disturbances at points A, B, and C on the neutral curves [curves 1, 2, and 3, respectively: Solid lines, $u(z)$; broken lines, $v(z)$]. (From Ref. 16.)

In Table 2 we present the amplitudes of the surface deformation, ξ, and the temperature disturbance, θ, at the boundaries. It appears that significant deformation of the boundary occurs only at point B where instability originates in the inclination of the layer.

Now let us study the behavior of the oscillatory perturbations as determined from the solution of the boundary value problem of Eqs. (34) to (36). For $Pr \neq 0$ and $Pr \neq 1$ the general solution of Eqs. (34) and (35) has the form:

$$\mathrm{v} = c_1 \exp(kz) + c_2 \exp(-kz) + c_3 \exp(lz) + c_4 \exp(-lz) \tag{37}$$

$$\theta = \frac{c_1 \exp(kz) + c_2 \exp(-kz)}{i\omega Pr} - \frac{c_3 \exp(lz) + c_4 \exp(-lz)}{i\omega (Pr - 1)} + c_5 \exp(mz) + c_6 \exp(-mz) \tag{38}$$

$$l = (k^2 - i\omega)^{1/2}, \qquad m = (k^2 - i\omega Pr)^{1/2}$$

TABLE 2 Surface Tension Gradient (Marangoni) Driven Instability of a Liquid Film: Amplitudes of the Boundary Deformation, ξ, and the Temperature Disturbance, θ, on the Boundaries.

Points	k	Ma	$\xi\,(0)$	$\xi\,(1)$	$\theta\,(0)$	$\theta\,(1)$
A	2.0	87.7	0.91×10^{-6}	-0.66×10^{-6}	0.018	0.032
B	0.25	6.37	1.12	-1.05	0.570	0.570
C	0.25	321	0.69×10^{-3}	-0.69×10^{-3}	0.065	0.069

Source: From Ref. 16.

The coefficients c_i are determined from the boundary conditions. The solvability condition of the system of equations for c_i determines a critical Marangoni number for fixed k. The requirement that the Marangoni number be real determines the oscillation frequency ω.

In Fig. 9(a) the neutral curves for oscillatory disturbances are represented by the broken lines 5, 6, and 8 for $Pr = 0.01$ and $Ca = 10^4$ (the instability domain is inside the curves). The corresponding dispersion curves are reproduced in Fig. 9(b). For curve 7 the neutral curve lies in the range of quite large Marangoni numbers ($Ma* = 1.3 \times 10^8$) and is not reproduced in Fig. 9(a).

In Fig. 9(b) the dotted–dashed lines give the dispersion relations for capillary waves on the surface of an isothermal inviscid liquid. The upper curve, $\omega^2 = Ca\, k^3 \operatorname{cotanh}(k/2)$, corresponds to flexural antisymmetric/stretching waves on the film whereas the lower one, $\omega^2 = Cak^3 \tanh(k/2)$, corresponds to squeezing waves with symmetric surface deflections (swelling–compression of the film). The oscillations described by curve 5 have a frequency similar to the flexural frequency. The existence domain of these waves is bounded for small wave numbers. There are no undamped oscillations as $k \to 0$ and $\omega \to 0$ [19].

The structure of the oscillatory flow disturbances for points on the various neutral curves is reproduced in insets D, E, and F of Fig. 9. The arrows show the instantaneous velocity at the characteristic points of the wave. An investigation of the form of the critical motion at point D on the neutral curve 5 shows that the two outer surfaces of the layer oscillate practically in phase at a frequency close to the frequency of the flexural waves. The transverse velocity component varies only slightly with depth, whereas the longitudinal velocity component depends significantly on the transverse coordinate.

In Fig. 11 lines 1 account for the amplitude distribution of these velocities

$$\mathrm{v}_x = u_r(z)\sin(kx - \omega t) + u_i(z)\cos(kx - \omega t),$$
$$\mathrm{v}_z = \mathrm{v}_r(z)\cos(kx - \omega t) + \mathrm{v}_i(z)\sin(kx - \omega t),$$

(the longitudinal and transverse components are represented by the solid and broken lines, respectively).

On the hotter surface of the layer the longitudinal velocity component exceeds the velocity on the colder surface approximately by a factor of 2 and is of the same order as the transverse velocity. This indicates that on the hotter surface, energy is pumped into the capillary wave as a result of the Marangoni effect.

In view of the structure of the flow and the characteristic dependence of ω on k, the oscillations described by the lower branch of curve 8 in Fig. 9(a) should be considered as capillary waves. In the abovementioned units the oscillations have a small wavelength and are localized in the neighborhood of the cold surface of the liquid. The structure of the motion and the liquid velocity distribution are shown in inset E in Fig. 9 and by lines 2 in Fig. 11.

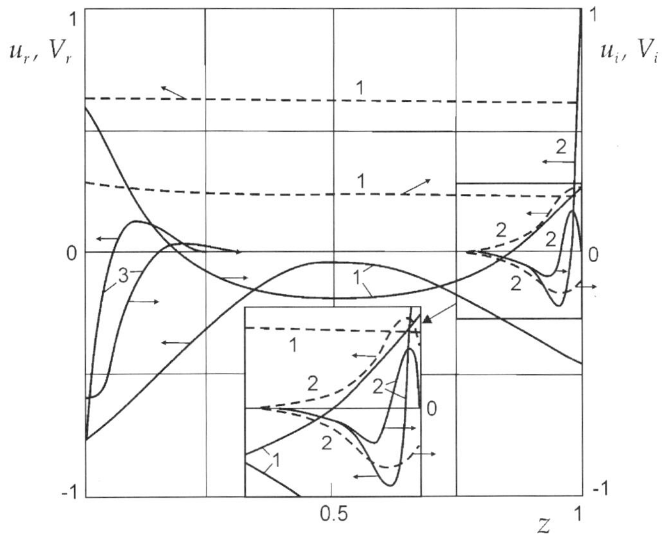

FIG. 11 Surface tension gradient (Marangoni) driven instability of a liquid film with two free boundaries. Amplitude distribution of the transverse (v_r, v_i) and longitudinal (u_r, u_i) velocity components for oscillatory disturbances at points *D*, *E*, and *F* on the neutral curves (curves 1, 2, and 3): solid lines, (u_r, u_i); broken lines, (v_r, v_i). (From Ref. 16.)

In Fig. 9(a) the neutral curve 6 depicts neutral oscillations of a qualitatively different structure. The oscillation frequency remains finite as $k \rightarrow 0$, which, as in an earlier studied case, distinguishes these oscillations from transverse capillary-gravity waves. The motion of the liquid is characterized by intense longitudinal flow in the neighborhood of the hotter surface of the layer (see inset *F* in Fig. 9 and curves 3 in Fig. 11). In Fig. 11 we have not reproduced the transverse velocity of this disturbance because of its smallness. As in the case of long-wavelength monotonic instability, the excitation of longitudinal waves is associated with the relative inclination of the layer boundary to the isotherms of the equilibrium temperature field and the generation of motion due to the Marangoni effect. Once more we note that the oscillations are excited on a liquid surface heated externally, whereas the monotonic long-wavelength instability is generated when the surface is heated from inside the liquid layer.

The neutral curves and the form of the critical motion are fairly sensitive to variation in the values of the parameters of the problem. For example, for $Ca = 10^4$ a closed, baglike region of oscillatory instability (curve 5) appears at $Pr = 0.0022$. At lower values of the Prandtl number the Marangoni effect does not support flexural waves. With the increase in Pr to 0.05 the instability domains, bounded by curves 5 and 6, approach each other and then merge. After that it appears difficult to classify the oscillatory disturbances into transverse and longitudinal. For overstability, the minimum critical Marangoni number increases from 16 at $Pr = 0.0022$ to 540 at $Pr = 0.05$, and the range of values of the wave numbers of the undamped disturbances is enlarged.

IV. THERMOCAPILLARY INSTABILITY OF A HANGING LAYER WITH INTERNAL HEAT SOURCES

In the earlier considered cases it has been assumed that the liquid has a linear temperature distribution due to heat diffusion in a film or a layer heated on either of its outer boundaries. Internal heat sources offer a different example of temperature distribution in a liquid layer. If the layer boundaries, $z = \pm d$, are maintained at the same temperature, taken as a reference value, and the heat sources at all points have the same power, the temperature distribution can be written as

$$T_0 = 0.5\Theta\left(1 - \frac{z^2}{d^2}\right)$$

where the temperature in the midplane of the layer center is defined by the source power; the factor 0.5 is introduced to ensure that the temperature gradient at the open surface is the same as in the case of a linear temperature distribution. The problem of small disturbances is identical to the problem discussed in the previous section except for a different dependence of T_0 on the transverse coordinate. Results exist about the thermocapillary instability of a layer with a homogeneous heat source and an open surface [20] but here we shall consider a film or layer with two free boundaries [17,21].

As given earlier, the equations for the Fourier normal modes

$$-i\omega\left(\frac{\mathrm{d}^2 \mathrm{v}}{\mathrm{d}z^2} - k^2 \mathrm{v}\right) = \frac{\mathrm{d}^4 \mathrm{v}}{\mathrm{d}z^4} - 2k^2 \frac{\mathrm{d}^2 \mathrm{v}}{\mathrm{d}z^2} + k^4 \mathrm{v} \tag{39}$$

$$-i\omega Pr\theta = \frac{\mathrm{d}^2 \theta}{\mathrm{d}z^2} - k^2 \theta + Prz\mathrm{v} \tag{40}$$

have exact solutions

$$\mathrm{v} = c_1 \exp(kz) + c_2 \exp(-kz) + c_3 \exp(lz) + c_4 \exp(-lz) \tag{41}$$

$$\begin{aligned}\theta = {} & \frac{c_1 \exp(kz)}{i\omega Pr}\left(z - \frac{2k}{i\omega Pr}\right) + \frac{c_2 \exp(-kz)}{i\omega Pr}\left(z + \frac{2k}{i\omega Pr}\right) \\ & + \frac{c_3 \exp(lz)}{i\omega(Pr-1)}\left(z - \frac{2l}{i\omega(Pr-1)}\right) + \frac{c_4 \exp(-lz)}{i\omega(Pr-1)}\left(z + \frac{2l}{i\omega(Pr-1)}\right) \\ & + c_5 \exp(mz) + c_6 \exp(-mz)\end{aligned} \tag{42}$$

$$l = (k^2 - i\omega)^{1/2}, \qquad m = (k^2 - i\omega Pr)^{1/2}$$

The solvability condition of the problem provides the critical Marangoni number and the frequency of neutral oscillations. Figure 12 illustrates neutral curves for a film heated or cooled by internal, homogeneously distributed, sources. Monotonic instability of the layer is predicted for both earlier examined mechanisms. Neutral curve 1 describes instability caused by the mid-layer symmetric inclination of the free surface (an odd mode with respect to the transverse velocity) to isotherms of the imposed base temperature field. Curve 2 corresponds to the case studied by Pearson [1]. The motion in this mode is even in velocity and odd in temperature and is maintained only for strong heating conditions ($Ma > 0$).

Curves 3 and 5 (squeezing wave) and 4 and 6 (flexural/stretching waves) represent the neutral oscillations. As expected, for very short wavelengths (curves 5 and 6) the surface wave modes in the layer with linear distribution and in the layer with internal homogeneously distributed heat sources are nearly the same. The long-wavelength disturbances freely propagate through the whole layer with a behavior strongly affected by the curvature of the temperature profile. For $Pr = 10^{-2}$ no oscillations having nonzero frequency at k tending to 0 have been

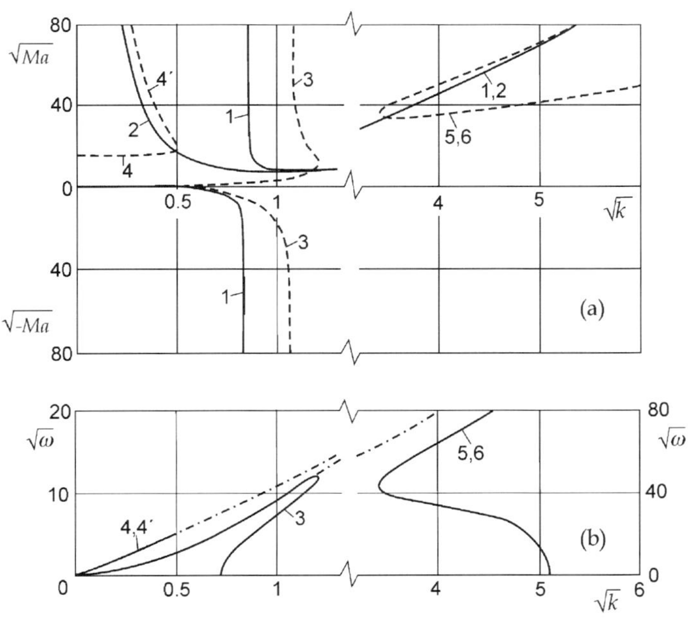

FIG. 12 Surface tension gradient (Marangoni) driven instability of a liquid film with internal heat sources and two free boundaries. Neutral curves and frequencies of neutral disturbances for $Ca = 10^4$ and $Pr = 10^{-2}$. (From Ref. 21.)

obtained. It is most likely that the decrease of the temperature gradient across the layer thickness causes the decay of the thermocapillary elasticlike mechanism of instability leading to longitudinal waves. All oscillatory modes could be reasonably referred to the category of capillary-gravity waves sustained by the Marangoni effect.

Later on we shall show that symmetric cooling of a liquid layer with two free outer surfaces by means of a thin partition can also lead to the generation of long-wavelength waves.

V. COMBINED THERMOCAPILLARY AND BUOYANCY INSTABILITY OF A PLANE LIQUID LAYER

Let us now consider the influence of buoyancy upon the results found for the stability of a horizontal liquid layer subjected to the Marangoni effect. In this case, we need to introduce into Eq. (1) an additional term accounting for buoyancy

$$-\lambda\left(\frac{\mathrm{d}^2\mathrm{v}}{\mathrm{d}z^2}-k^2\mathrm{v}\right)=\frac{\mathrm{d}^4\mathrm{v}}{\mathrm{d}z^4}-2k^2\frac{\mathrm{d}^2\mathrm{v}}{\mathrm{d}z^2}+k^4\mathrm{v}-Rak^2\theta \tag{43}$$

$$-\lambda Pr\theta=\frac{\mathrm{d}^2\theta}{\mathrm{d}z^2}-k^2\theta+\mathrm{v} \tag{44}$$

where $Ra=\alpha g\Theta d^3/\nu\chi$ is the Rayleigh number, α is the thermal expansion coefficient, and the other parameters have their earlier given meaning. To make quantities dimensionless we have used as scales d, d^2/ν, χ/d, and θ for length, time, velocity, and temperature, respectively, as usually done for buoyancy-driven convection.

Assuming that the free surface ($z=1$) is flat, the boundary condition on the upper boundary can be written as

$$z=1:\mathrm{v}=0,\qquad \frac{\mathrm{d}^2\mathrm{v}}{\mathrm{d}z^2}=-k^2Ma\theta,\qquad \frac{\mathrm{d}\theta}{\mathrm{d}z}=-Bi\theta \tag{45}$$

while on the lower solid boundary we set

$$z=0:\mathrm{v}=\frac{\mathrm{d}\mathrm{v}}{\mathrm{d}z}=0,\qquad \theta=0 \tag{46}$$

We solve Eqs. (43) and (44) for all boundary conditions [Eqs. (45) and (46)] with the exception of the Marangoni stress condition [second expression in Eq. (45)]. Then, using the latter yields the solvability condition in the form of a dispersion relation, which can be written as $Ma=f(\lambda, k, Pr, Bi, Ra)$. As in earlier cases treated in this book, the growth rate of disturbances, λ, is a complex quantity. If $\lambda=0$ then there is a real value of Ma (and Ra) for neutral monotonic stability. If

$\mathrm{Im}\lambda = \omega \neq 0$ the roots of the equation $\mathrm{Im}f(\omega, k, Pr, Bi, Ra) = 0$ provide the frequency of neutral oscillations (overstability) for the corresponding values of *Ma* (and *Ra*) [22].

A. Monotonic Instability: Theory

Let us first recall results for monotonic instability [23]. Figure 13 illustrates the variation of the neutral curves for the Marangoni–Rayleigh problem. Solid lines indicate the stability boundary for pure thermocapillary convection in the case when the Biot number at the liquid–air interface equals zero or unity. The dashed lines define the boundaries of the instability region for a Rayleigh number half the critical value, which leads to instability in the absence of thermocapillarity.

The Marangoni and Rayleigh numbers for the combined problem obey the equation $Ma/Ma_* + Ra/Ra_* = 1$ [23]. This relation shows the cooperative contribution of both mechanisms leading to instability and their tight coupling as they appear with about equal wavenumber at the onset of convection either side ($Ra = 0$ and $Ma = 0$, respectively). The validity of the above relation is supported by Fig. 14, which gives the results of the numerical investigation of the linear stability problem. Two solid lines in the plane (*Ma*, *Ra*) define the boundaries of the instability region (the region above the curve) for the limiting cases $Bi = 0$ (insulating outer open surface) and $Bi \to \infty$ (conducting liquid–air boundary).

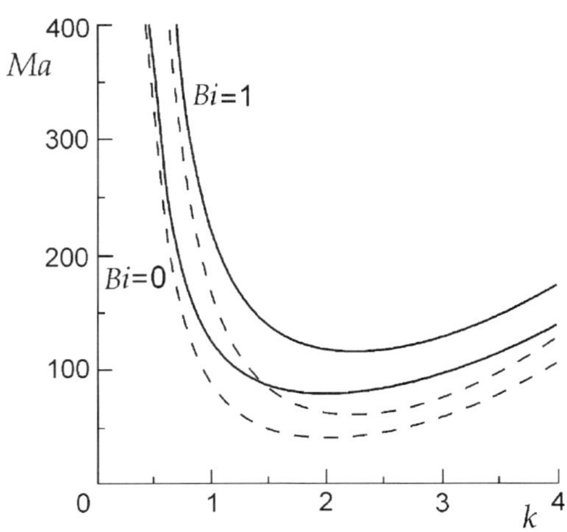

FIG. 13 Neutral curves for surface tension gradient (Marangoni) driven convection (solid lines) and combined thermocapillary and buoyancy Rayleigh driven convection (dashed lines) when $Bi = 0$ and $Bi = 1$ at the liquid–air interface. The Rayleigh number in combined convection is half the critical value leading to instability in the absence of Marangoni effect.

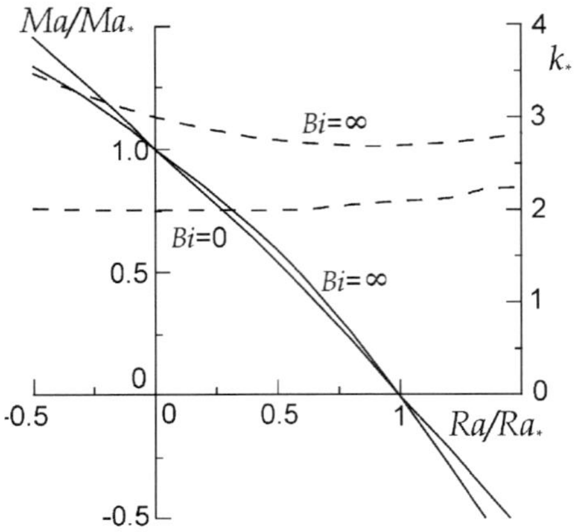

FIG. 14 Combined Marangoni–Rayleigh convection. Stability diagram showing the neutral stability boundary (solid lines) and corresponding critical wavenumber (broken lines) for two values of Bi at the liquid–air interface ($Bi = 0$ and $Bi \to \infty$).

The Marangoni and Rayleigh numbers are normalized by their corresponding values when instability occurs separately (at $Ra = 0$ and $Ma = 0$, respectively). Broken lines show the variation of the corresponding wavenumbers at threshold. The critical values of Marangoni, Ma_*, and Rayleigh numbers, Ra_*, and the corresponding wave numbers for various values of Bi at the liquid–air interface are given in Table 3. These values show that with the increase of Bi the Marangoni number grows unbounded and asymptotically becomes proportional to the Biot number, $Ma_* = 32.073Bi$, whereas the Rayleigh number tends to a finite limit, $Ra_* \to 1100$.

TABLE 3 Combined Marangoni–Rayleigh Convection. Critical Values of Marangoni Number, Ma_*, and Rayleigh Number, Ra_*, and the Corresponding Wave Numbers for Various Values of the Liquid–Air Bi.

Bi	Ma_*	k_*	Ra_*	k_*
0	79.607	1.993	669.00	2.086
0.1	83.427	2.028	682.36	2.117
0.2	87.195	2.060	694.78	2.144
0.5	98.256	2.142	727.42	2.212
1	116.127	2.246	770.57	2.293
2	150.679	2.386	831.27	2.393
5	250.598	2.598	925.51	2.519
10	413.440	2.743	989.49	2.589
100	3303.83	2.976	1085.90	2.672

B. Monotonic Instability: Experiment

Let us recall results from the paradigmatic, careful experimental study of combined thermocapillary-driven and buoyancy-driven convection carried out by Koschmieder [24–26]. A schematic section of his set-up is shown in Fig. 15. A shallow layer of silicone oil on a plane, circular copper plate was uniformly heated from below. The air on its surface was kept at constant temperature by a uniformly cooled glass plate close to the oil, which at the same time tended to inhibit air motions. Motions began with concentric circular rolls, which, after the center ring had formed, broke down into a hexagonal pattern. A rather accurate determination of the wavelength of the motions was carried out. The wavelength was found to be variable with the depth of the fluid layer in qualitative agreement with the theory [23].

1. Description of the Apparatus

Details of the experimental set-up are reproduced verbatim from Ref. 24 (pp. 9–10). A 12-mm thick copper plate of 200-mm diameter served as bottom and was heated from below by a disk made of brass with ten concentric resistance wires in series. The copper plate and the brass plate were separated by a shallow water tank in which a stirrer was rotated to homogenize the temperature distribution in the brass plate. The lateral wall of the oil layer was a 1-mm thick lucite ring from which, 7.5-mm above the copper plate, a shoulder extended horizontally outwards, on which the lid rested. The lucite ring was surrounded by 50-mm foam rubber for thermal insulation. The bottom of the lid was a 1.9-mm

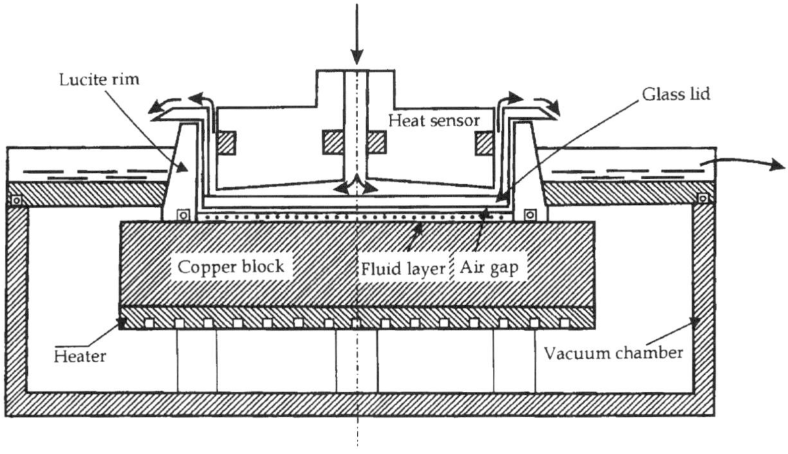

FIG. 15 Combined Marangoni–Rayleigh convection. Schematic section of the set-up. (From Ref. 25.)

thick glass plate of 216-mm diameter onto which a 60-mm high Bakelite® wall was attached. The glass plate was cooled by a water flow of around 200 cm^3/s from a controlled bath. The water jet impinged on the center of the plate and was then forced to spread evenly to the sides under a lucite disk. This arrangement provides a good approximation to a uniformly cooled lid save the axisymmetry of the flow, which played no significant role. The lid and the oil on the copper plate were separated by an air layer, which was sealed in at the circle where the glass plate rests on the shoulder of the rim. The minimum distance from the oil to the glass was determined by the need to avoid contact of the oil, whose level was raised at the rim due to the adhesion of the oil to the lucite ring. This closest distance was 0.75 mm. With such a narrow gap, the air upon the oil virtually stayed at rest and obtained a reasonable uniform temperature from the glass plate. The silicone oil was Dow Corning 200 fluid of kinematic viscosity $\nu = 10^{-4}\,m^2/s$, density $\rho = 0.968\,g/cm^3$, thermal expansion coefficient $\alpha = 0.96 \times 10^{-3}/K$, specific heat $C_p = 0.35\,cal/g°C$ $(= 1.46\,J/kgK)$, thermal conductivity $\kappa = 0.37 \times 10^{-3}\,cal/cm\,s°C$ $(= 1.54 \times 10^{-1}\,J/msK)$ and surface tension $\sigma_0 = 20.9$ dyne/cm $(= 20.9\,mN/m)$ at 25°C. The dependence of surface tension on temperature was measured and found to be $-\sigma_1 = 0.058\,dyne/cm°C$ $(= 0.058\,mN/mK)$ $\pm 5\%$. The overall temperature difference between the cooling water and the bottom plate was measured with copper–constantan thermocouples ($1\,\mu V = 2.5 \times 10^{-2}\,°C$). The motions of the oil were made visible with aluminum flakes and direct illumination of the surface of the liquid layer. Dark lines in the pictures obtained indicate vertical motions, while areas of predominantly horizontal motions appear bright as light was reflected by the flakes in almost horizontal motion. Pictures were taken with the lid removed for about 20 s. The copper bottom plate was not blackened.

2. Further Details About the Onset of Patterned Convection

The establishment of a hexagonal pattern under an air surface, as observed by Koschmieder [24,26], is worth recalling as an example of previous careful study and is shown in Fig. 16. The leveled copper plate was in this case covered by a 5.32 ± 0.02 mm deep oil layer, which was gradually heated from zero vertical temperature difference. Rings formed consecutively inward from the rim as expected in view of the cylindrical axisymmetry imposed by the container. Incidentally, introducing various symmetry breaking elements that favor rolls, as done by Velarde and Normand [27], illustrates how the geometry tends to impose a convective pattern due to the influence of boundaries but subsequently yields to hexagons.

The following sequence of events is taken almost verbatim from Ref. 24. Four hours after heating started there were eight rings at a temperature difference corresponding to 370 μV. If the vertical temperature difference was maintained

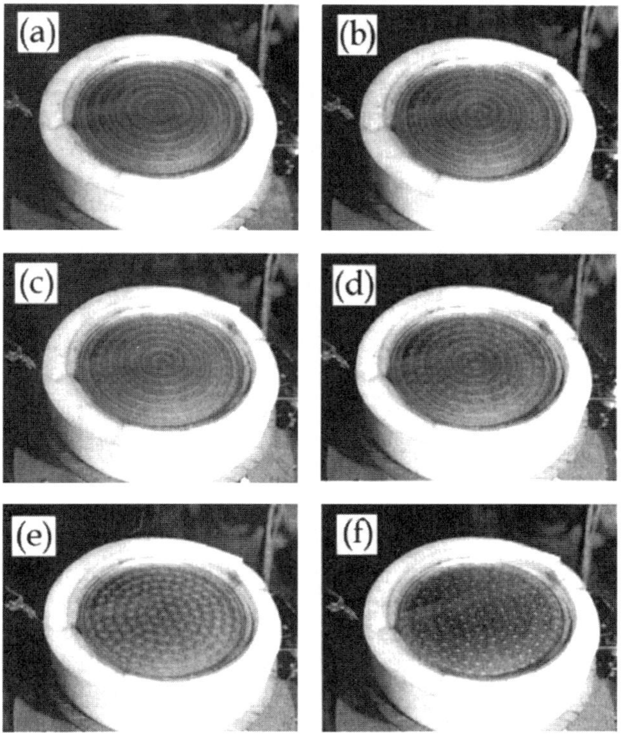

FIG. 16 Combined Marangoni–Rayleigh convection. (a) Concentric ring pattern, $\Delta T = 380\,\mu V$, t = 0; Breakdown of axial symmetry, (b) $\Delta T = 390\,\mu V$, t = 8 min, (c) $\Delta T = 400\,\mu V$, t = 15 min; Hexagonal pattern, (d) $\Delta T = 400\,\mu V$, t = 35 min, (e) $\Delta T = 400\,\mu V$, t = 85 min; Supercritical hexagonal pattern, (f) $\Delta T = 960\,\mu V$, t = 315 min. (From Ref. 24.)

constant the rings were observed to be unchanged for hours. Twelve rings could be observed at 380 μV 15 minutes later. After five more minutes, 15 concentric rings had just established [see Fig. 16(a)]. The lid had not been lifted so far and the motions were still almost axially symmetric. The motions had developed precisely the same way as with the lid in contact with the fluid, but the horizontal dimensions of the rings differed considerably. While in the latter case the concentric rings stayed up to five times the critical temperature difference, under an air surface the rings broke down into hexagons immediately.

The beginning of the breakdown of the axial symmetry can be seen in Fig. 16(b), taken 8 minutes after Fig. 16(a) at a temperature difference corresponding to 390 μV. The conversion into hexagons proceeded in a well-defined sequence. The originally smooth rolls became bulgy in a very well-coordinated way for a pair of rolls with ascending motions in the middle. When the bulges of the undulating rolls met at the centerline of the pair of rolls they narrowed rapidly to form a "pinch." Where the pinches were located, the rolls soon became disconnected by fine, transparent, straight lines, perpendicular to the original axis of the rolls. The cells, which thus had established themselves, slowly adopted a more circular form and finally filled the available space with proper hexagons. Figure 16(c) shows how the first cells have already become established in the outer rings, while the center rings have begun to decay. In Fig. 16(d), the hexagonal pattern is established, although some irregularities remain. The two rolls adjacent to the rim stayed virtually unaffected. Almost an hour later at the same temperature difference [Fig. 16(e)] the cells had slightly more regular boundaries. There was one hexagon right at the center of the plate and then in rings consecutively from the center there were 6–12–18–24–29–33 not always perfect cells. The outermost cell ring never managed to align perfectly to the cells further inwards. Finally, Fig. 16(f) shows, at considerably increased temperature difference, the motions still in a hexagonal pattern. The number of cells in the cell rings are 1–6–12–18–23–31–35. Both rolls adjacent to the rim then broke into cells shaped like pentagons, to accommodate the shape of the rim. The falling out of aluminum flakes accumulated tiny heaps at the centers of the cells, thereby possibly stabilizing the pattern.

The appearance of the concentric rings prior to the breakdown into hexagons permitted an accurate determination of the wavelength of the convective motion, matching the theory of Nield [23]. At first a series of experiments were made to determine the wavelength of the motions in the case where 13 concentric rolls occurred with exactly the same set-up as described above. From 16 consecutive runs, which produced 13 rings, the depth d for 13 rings was determined to be $d = 5.145 \pm 0.04$ mm ($\pm$ standard deviation). These experiments covered all possible diameters of the center ring, which consequently produce different arrangements of the cells in the cell rings. The average d coincided with the depth for the most regular distribution of the cells in the different rings, in other words with harmonic rings on the whole plate. From the knowledge of h it follows that the wavelength for one roll is $\lambda = \pi/\kappa_c = r/nd = 1.495 \pm 0.012$ for 13 rings (κ_c is the critical wavenumber). The magnitude of the error ($\pm 0.8\%$) is not surprising if one keeps in mind that the largest possible error here is one-half wavelength, corresponding to $\pm 4\%$ for 13 rings. The λ measured here is compatible with Nield's values, which range up to 1.57 for the case of "insulation" at the upper free surface open to air. It is difficult to assess how effective the insulation on top of the oil is, since as earlier said the heat conductivity is expressed in terms of the Biot number $Bi = q_0 d/\kappa_c$. The quantity q_0 is the rate of change with temperature of the heat flux from the fluid surface and κ_c its heat conductivity. The value of q_0 is not known. Because for 13 rings, Bi is of the order 0.4 according to the measured λ ($Bi = 0$ corresponds to heat flux held fixed), and since κ_c^{-1} is of the order 10^3, q_0

has to be very small here. The most conveniently variable quantity in Bi is the depth. The wavelengths were therefore determined for several values of d, with corresponding variations in the thickness of the air layer, since the lid position could not easily be changed. The average depth for 15 rings [Fig. 16(a)–(e)] was found to be 4.28 ± 0.04 mm, corresponding to $\lambda = 1.56$ and hence the wavelength increased 4% with the decrease of d. The measured λ came very close to the maximal possible in Nield's theory, namely $\lambda_{max} = 1.576$. Koschmieder's attempts to produce 17 concentric rings at $d = 3.8$ mm failed. Instead of the innermost rings there appeared clusters of hexagonal cells. The wavelength inferred from the final hexagonal pattern scattered and therefore these experiments were discontinued. On the other hand, 11 harmonic rings appeared at 6.76 mm, corresponding to $\lambda = 1.35$, a decrease of 8% due to the increased depth. The air layer was then 0.75 mm thick, which was the narrowest ever tried by Koschmieder [26]. He also cross-checked results with another experimental set-up using silicone oil of 10 Stokes viscosity ($1\,\mathrm{St} = 10^{-4}\,\mathrm{m}^2/\mathrm{s}$). The onset of convection and the establishment of the hexagonal pattern were exactly of the same type as described above. The wavelength varied a little, 11 harmonic rings appeared at a depth of 6.44 mm, corresponding to $\lambda = 1.41$. This change might be caused by a variation of q_0 or could have been due to a 3% increasing in the thermal conductivity of the liquid relative to that of 1 Stokes viscosity. Further recent details about theory, numerics, and experiments leading to Benard cells under various circumstances of the thermal constraints and boundary conditions, role of defects and parameters, etc., can be found in the book by Colinet et al. [28] and in the review paper by Schatz and Neitzel [29].

C. Oscillatory Instability

Let us now return to results predicted for a liquid layer, as in Koschmieder's experiment, but when the heating is done from the air side, a subject already discussed in Chapter 2. Although surface deformation is significant for a complete description of Benard cells, it is not so in accounting for some salient features like wavelength, etc., of instability in the case of the combined Marangoni–Rayleigh instability. We shall disregard here surface deformation, thus eliminating from our consideration the possibility of capillary-gravity waves. However, we shall show that in the combined problem, with the Marangoni effect and buoyancy coupled together, oscillatory motions and waves are also expected.

As shown earlier, for standard liquids with surface tension decreasing with the increase of temperature, when heating from the ambient air above in the presence of the Marangoni effect, longitudinal surface waves are possible. Buoyancy-driven steady convection does not arise when heating the layer from above. However, oscillatory disturbances (internal waves) may appear in such a stably stratified layer with a frequency scaling with the square root of the (negative) Rayleigh number (Brunt–Väisälä frequency). Note that although longitudinal surface waves and internal waves may be damped if excited separately, the outcome is drastically changed if both are excited simultaneously, which will be the case for high enough (in absolute value) Marangoni and Rayleigh numbers [22]. For example, for $Pr = 6$ and liquid–air $Bi = 0$, undamped oscillations appear at $Ra = -1.21 \times 10^6$.

Neutral curves, at fixed Rayleigh number, are shown in Fig. 17(a), while the frequency for those states is plotted in Fig. 17(b). The point 1 corresponds to $Ra = -1.21 \times 10^6$, below which (in absolute value) no neutral states exist. The solid lines 2 and 3 correspond to $Ra = -1.5 \times 10^6$ and $Ra = -3 \times 10^6$, respectively. As in an earlier studied case the neutral stability curves appear in the form of a "bag" or a "bubble" whose interior corresponds to the region of instability. The bubble and its shape largely depend on the value given to the Rayleigh number. Indeed, as the absolute value of Ra is increased the bubble is seen to grow at point 1 and is then shifted to higher absolute values of Ma.

The general scenario is that at $Ra \to -\infty$, a countable number of bubbles should appear, corresponding to a countable number of vertical modes (internal waves). The bubble shown in Fig. 17 corresponds to the fundamental mode. Broken lines in Fig. 17 show the asymptotic results (in the limit of high, in absolute value, Marangoni and Rayleigh numbers) for curve 3. In this range it

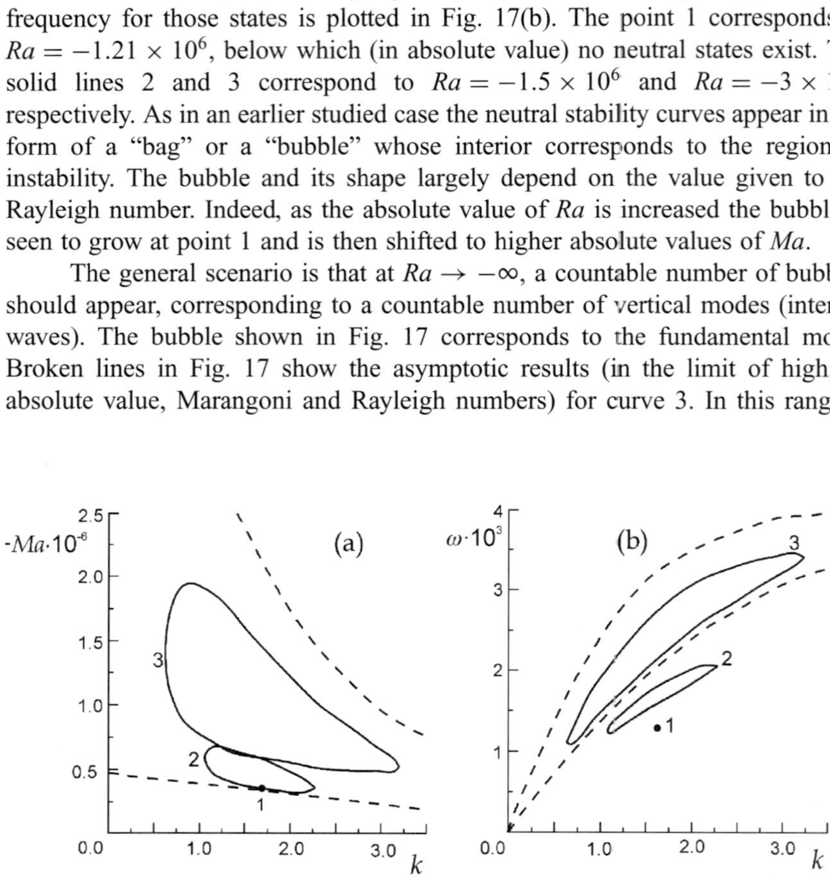

FIG. 17 Combined Marangoni–Rayleigh oscillatory convection in a liquid layer with open undeformable surface. (a) Marginal stability curves and (b) critical frequencies for the lowest unstable mode at fixed Rayleigh number, $Pr = 6$ and $Bi = 0$: solid line 2 for $Ra = -1.5 \times 10^6$, and solid line 3 for $Ra = -1.5 \times 10^6$. The point 1 corresponds to $Ra = -1.21 \times 10^6$, below which (in absolute value) no marginal states exist. Broken lines show asymptotic results for line 3 when Ma, $Ra \to -\infty$. The upper and lower curves correspond to surface and internal waves, respectively. (From Ref. 22.)

appears that the upper and lower branches of the bubble correspond to longitudinal surface waves and internal waves, respectively.

As the Marangoni and Rayleigh numbers are both proportional to the temperature gradient, it is therefore more convenient to use the *dynamic* Bond number, $Bd = Ra/Ma = \alpha\rho g d^2/\sigma_1$, which does not change with changing temperature difference. Figure 18 shows the marginal curves (a) and the corresponding frequencies (b) at $Pr = 6$ and $Bi = 0$ for $Bd = 1.5$, 2, 4, 6, 10 and 15. The broken line is the locus of the minima as Bd is varied. With increasing Bd, the minimum shifts from long ($k \to 0$) to short waves ($k \to \infty$). The lowest absolute value of the critical Marangoni number is achieved for $Bd \approx 4$, which

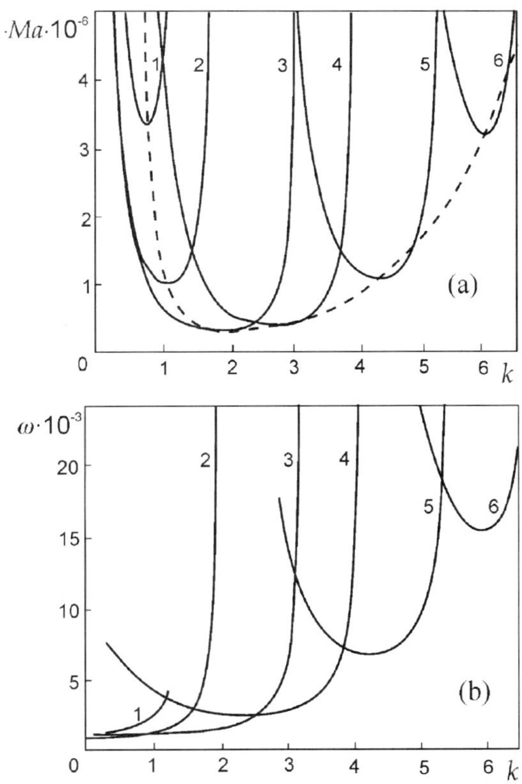

FIG. 18 Combined Marangoni–Rayleigh oscillatory convection in a liquid layer with open undeformable surface. (a) Marginal stability curves and (b) critical frequency for the lowest unstable mode at fixed *dynamic* Bond number, $Pr = 6$ and $Bi = 0$: curves 1–6 correspond to $Bd = 1.5$, 2, 4, 6, 10, and 15, respectively. When Bd varies, the minimum of the marginal curve moves along the broken line. (From Ref. 22.)

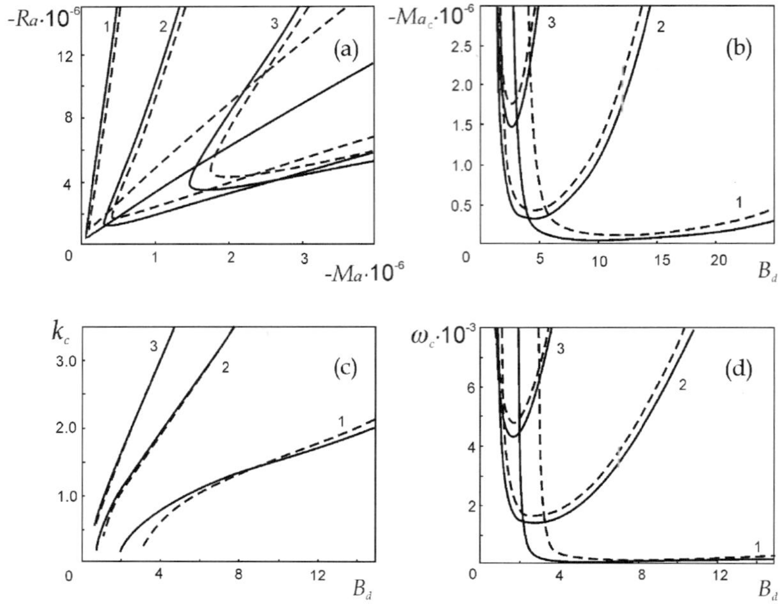

FIG. 19 Combined Marangoni–Rayleigh oscillatory convection in a liquid layer with open undeformable surface. (a) Stability boundary in the plane $\{-Ma, -Ra\}$, (b) critical Marangoni number, (c) wavenumber, and (d) frequency vs. *dynamic* Bond number at various values of the Prandtl and Biot numbers. The three pairs of curves corresponds to $Pr = 0.1$, $Pr = 6$, $Pr = 20$, marked 1, 2 and 3, respectively. Within each pair, the solid and broken lines are for $Bi = 0$ and $Bi = 1$, respectively. (From Ref. 22.)

represents the optimal balance between the Marangoni effect and buoyancy. $Bd = 0.715$ is the lowest possible value, below which the neutral stability curves disappear at infinity with $k \rightarrow 0$.

Figure 19 illustrates the influence of the Prandtl and Biot numbers on the stability diagram (a), and on the variations of the critical Marangoni number (b), the wavenumber (c), and the frequency (d) with the *dynamic* Bond number. In particular, one can see that a decrease in *Pr* (lowering thermal inertia or enhancing mechanical inertia) results in lower instability thresholds. The growth of *Bi* (an increase of heat transfer from the free surface) generally leads to an enhancement of the stability of the system.

As already indicated in Chapter 2, Velarde and Linde and their collaborators have explored, theoretically and experimentally, the various oscillatory motions and waves excited and, eventually, sustained by the Marangoni effect, coupled or not to buoyancy ([10–14,22]; see also further references in Chapter 2).

REFERENCES

1. Pearson, J.R.A. On convection cells induced by surface tension. J. Fluid Mech. **1958**, *4*, 489–500.
2. Smith, K.A. On convective instability induced by surface-tension gradients. J. Fluid Mech. **1966**, *24*, 401–414.
3. VanHook, S.J., Schatz, M.F., McCormick, W.D., Swift, J.B., Swinney, H.L. Long-wavelength instability in surface-tension-driven Benard convection. Phys. Rev. Lett. **1995**, *75*, 4397–4400.
4. VanHook, S.J., Schatz, M.F., Swift, J.B., McCormick, W.D., Swinney, H.L. Long-wavelength surface-tension-driven Benard convection: experiment and theory. J. Fluid Mech. **1997**, *345*, 45–78.
5. Takashima, M. Surface tension driven instability in a horizontal liquid layer with a deformable free surface. I. Stationary convection. J. Phys. Soc. Japan **1981**, *50*, 2745–2750.
6. Goussis, D.A., Kelly, R.E. On the thermocapillary instabilities in a liquid layer heated from below. Int. J. Heat Mass Transfer **1990**, *33*, 2237–2245.
7. Levchenko, E.B., Chernyakov, A.L. Instability of surface waves in a nonuniformly heated liquid. Soviet Phys. JETP **1981**, *54*, 102–105.
8. Takashima, M. Surface tension driven instability in a horizontal liquid layer with a deformable free surface. II. Overstability. J. Phys. Soc. Japan **1981**, *50*, 2751–2756.
9. Garcia-Ybarra, P.L., Velarde, M.G. Oscillatory Marangoni–Benard interfacial instability and capillary-gravity waves in single- and two-component liquid layers with or without Soret thermal diffusion. Phys. Fluids **1987**, *30*, 1649–1660.
10. Chu, X.-L., Velarde, M.G. Sustained transverse and longitudinal waves at the open surface of a liquid. Physicochem. Hydrodyn. **1988**, *10*, 727–737.
11. Chu, X.-L., Velarde, M.G. Transverse and longitudinal waves induced and sustained by surfactant gradients at liquid-liquid interfaces. J. Colloid and Interface Sci. **1989**, *131*, 471–484.
12. Chu, X.-L., Velarde, M.G. Dissipative hydrodynamic oscillators. I. Marangoni effect and sustained longitudinal waves at the interface of two liquids. Il Nuovo Cimento 11 **1989**, D, 709-716.
13. Chu, X.-L., Velarde, M.G. Dissipative hydrodynamic oscillators. V. Interfacial oscillation in finite containers. Il Nuovo Cimento **1989**, *11* D, 1615–1629.
14. Chu, X.-L., Velarde, M.G. Dissipative hydrodynamic oscillators. VI. Transverse interfacial oscillations not obeying the Laplace–Kelvin law. Il Nuovo Cimento **1989**, *11* D, 1631–1643.
15. Ryabitskii, E.A. Thermocapillary instability of a plane layer with a vertical temperature gradient. Izv. Akad. Nauk SSSR, Mech. Zhidk. Gaza **1992**, *3*, 19–23 (in Russian).
16. Birikh, R.V., Rudakov, R.N. Thermocapillary instability of a deformable liquid film. Izv. Akad. Nauk SSSR, Mech. Zhidk. Gaza **1996**, *5*, 30–36.
17. Birikh, R.V., Briskman, V.A., Rudakov, R.N., Velarde, M.G. Marangoni–Benard convective instability of a floating layer with an internal, permeable, heated or cooled

divider and two deformable open surface. Int. J. Heat Mass Transfer **1995**, *38*, 2723–2731.

18. Funada, T. Marangoni instability of thin liquid sheets. J. Phys. Soc. Japan **1986**, *55*, 2191–2202.
19. Funada, T., Kubo, T., Matsuura, S. Marangoni instability of unsteady mode for a liquid sheet. Numazu College Technol. Research Annual **1994**, *28*, 53–59.
20. Ryabitskii, Ye.A. Thermocapillary instability of liquid layer with internal heat generation. Microgravity Sci. Technol. **1994**, *VII/1*, 20–23.
21. Birikh, R.V., Briskman, V.A., Rudakov, R.N., Velarde, M.G. On the Overstability of Nonisothermal Deformable Layer; 33rd Aerospace Sciences Meeting and Exhibit. Reno. NV., USA, Paper AIAA 95-0818, 1–9, 1995.
22. Rednikov, A.Ye., Colinet, P., Velarde, M.G., Legros, J.C. Rayleigh-Marangoni oscillatory instability in a horizontal liquid layer heated from above: coupling and mode-mixing of internal and surface dilational waves. J. Fluid Mech. **2000**, *405*, 57–77.
23. Nield, D.A. Surface tension and buoyancy effects in cellular convection. J. Fluid Mech. **1964**, *19*, 341–352.
24. Koschmieder, E.L. On convection under an air surface. J. Fluid Mech. **1967**, *30*, 9–15.
25. Koschmieder, E.L., Biggerstaff, M.I. Onset of surface-tension-driven Benard convection. J. Fluid Mech. **1986**, *167*, 49–64.
26. Koschmieder, E.L. *Benard Cells and Taylor Vortices*; Cambridge University Press: Cambridge, 1993 (and references therein).
27. Velarde, M.G., Normand, Ch. Convection. Sci. Am. **1980**, *243* (1), 92–108.
28. Colinet, P., Legros, J.C., Velarde, M.G. *Nonlinear Dynamics of Surface-Tension-Driven Instabilities*. Wiley-VCH: Berlin, 2001.
29. Schatz, M.F., Neitzel, G.P. Experiments on thermocapillary instabi-lities. Annu. Rev. Fluid Mech. **2001**, *33*, 93–129.

6

Convective Instability of a Liquid Layer with a Permeable Partition

In this chapter we shall analyze the role played by an internal permeable, pervious divider or partition grid placed inside a liquid layer parallel to the outer boundaries. First we shall consider the influence of the divider properties upon buoyancy (Rayleigh) driven convection and, subsequently, upon thermocapillary convection. The latter will be considered first when the outer surfaces are undeformable and then when surface deformation is taken into account. The onset of either monotonic or oscillatory instability (overstability) will also be studied. We shall identify convective thresholds and the dependence of critical disturbance modes on the boundary conditions of the layer as well as on the properties of the grid.

I. BUOYANCY (RAYLEIGH) DRIVEN INSTABILITY OF A HORIZONTAL LIQUID LAYER

We consider now the instability of an infinitely extended horizontal layer of viscous liquid heated from below in the presence of gravity and partitioned by a thin (relative to liquid depths) horizontal permeable divider (grid) into two liquid sublayers.

A. A Layer with Outer Stress-Free Boundaries

Let us start with a film or a liquid layer with two slippery, stress-free outer horizontal boundaries localed at $z = \pm d$. We shall scale length with d and so in dimensionless form $z = \pm 1$, at fixed temperatures $\pm\Theta$ (the lower boundary has a higher temperature). The temperature difference at the boundaries responsible for the possible onset of convective flow, with wave number $\mathbf{k} = (k_x,\ k_y)$, determines the critical value of the Rayleigh number, as defined in the preceding chapter, which is the eigenvalue of the boundary value problem for monotonic instability,

$$\frac{d^4 v}{dz^4} - 2k^2 \frac{d^2 v}{dz^2} + k^4 v = Ra\, k^2 \theta \tag{1}$$

$$\frac{d^2 \theta}{dz^2} - k^2 \theta = -v \tag{2}$$

$$z = \pm 1 : v = \frac{d^2 v}{dz^2} = \theta = 0$$

where v(z) and $\theta(z)$ are the amplitudes of normal mode disturbances of vertical velocity component and temperature, respectively:

$$\{v_z, T\} = \{v(z), \theta(z)\} \cdot \exp\left[i(k_x x + k_y y)\right].$$

A dimensionalization was carried out using as scales (besides d for length) χ/d and Θ for velocity and temperature, respectively. All parameters here used have, generally, the meaning given in the preceding chapter.

Equations (1) and (2) yield the critical Rayleigh number $Ra = (\pi^2/4 + k^2)^2/k^2$. The fundamental mode of the convective flow is a roll whose vertical velocity component is $v_x \to V \sim \cos(\pi z/2)$. The first harmonic mode is a two-vortex motion with the vertical velocity $v_z \to W \sim \sin \pi z$. Note that the components in this double cell rotate in opposite directions with the horizontal velocity component in the middle of the cell being at its maximum.

Let a thin permeable divider be placed in the center section, midplane of the liquid layer ($z = 0$). Neglecting its thickness we may assume [1] that flow through the divider maintains the continuity of temperature T, heat flux, and transverse velocity v_z, i.e.,

$$z = 0 : T_- = T_+, \quad \frac{\partial T_-}{\partial z} = \frac{\partial T_+}{\partial z}, \quad v_{z-} = v_{z+} \tag{3}$$

Here and in the following the subscripts + and − denote the values of the function above and below the divider.

The divider is defined by its resistance to longitudinal, α_r, and to transverse, α_n, flows. In this section for the longitudinal velocity component at the divider we shall apply the no-slip condition (assuming that the tangential resistance is quite large; $\alpha_r \to \infty$). The normal resistance of the divider will be assumed finite; $\alpha_n = \alpha < \infty$. Then the boundary condition for the velocity on the divider is

$$z = 0 : v_z = -\alpha^{-1}(p_+ - p_-), \quad v_{x\pm} = v_{y\pm} = 0 \tag{4}$$

Accordingly, α is the only phenomenological coefficient characterizing the divider. Note that here we use the symbol α to characterize the partition resistance.

A permeable divider with the above-described properties permits the equilibrium state of the layer heated from below. However, let us see how it may affect the instability threshold and the expected steady convective pattern. We introduce in the boundary-value problem [Eqs. (1) and (2)] the boundary conditions at the divider. Using the continuity, the Navier–Stokes, and the heat (Fourier) diffusion equations we may write the conditions of Eqs. (3) and (4) for the amplitudes of Fourier normal mode disturbances of the temperature and the vertical velocity component:

$$z = 0 : \theta_- = \theta_+, \qquad \frac{d\theta_-}{dz} = \frac{d\theta_+}{dz}$$

$$v_- = v_+, \qquad \frac{d^3 v_-}{dz^3} - \frac{d^3 v_+}{dz^3} = \alpha k^2 v, \qquad \frac{dv_-}{dz} = \frac{dv_+}{dz} = 0 \tag{5}$$

Thus, Eq. (5), together with Eqs. (1) and (2), are sufficient to determine the critical value of the Rayleigh number at which the equilibrium in the horizontal liquid layer becomes unstable, now with the inserted permeable divider and heated from below.

The boundary-value problem of Eqs. (1), (2), and (5) is symmetric about $z = 0$ and hence the solutions fall into even and odd modes. For convenience we eliminate temperature. Then an even solution for v can be written in the form

$$v = \sum_{j=1}^{3} c_j(\sin\mu_j \cos\mu_j z \pm \cos\mu_j \sin\mu_j z) \tag{6}$$

where the upper sign refers to the region $z \geq 0$ and the lower one to the region $z \leq 0$; the coefficients μ_j are the roots of characteristic Eqs. (1) and (2)

$$\mu_1 = (r - k^2)^{1/2}, \qquad \mu_{2,3} = \sqrt{0.5(-1 \pm i\sqrt{3})r - k^2} \equiv \mu_r \pm i\mu_i,$$
$$r \equiv (k^2 Ra)^{1/3}$$

The solution of Eq. (6) satisfies Eqs (1) and (2) and from the condition of Eq. (5) on the divider we obtain the equation

$$\frac{\alpha k^2}{6r}\left(\frac{tg\mu_1}{\mu_1} + \frac{(\sqrt{3}\mu_r - \mu_i)\sinh 2\mu_i - (\sqrt{3}\mu_i + \mu_r)\sin 2\mu_r}{(\mu_r^2 + \mu_i^2)(\cos 2\mu_r + \cosh 2\mu_i)}\right) = -1 \tag{7}$$

An odd solution is constructed, similarly to Eq. (6), and its corresponding transcendental relation is

$$\frac{(\mu_i + \sqrt{3}\mu_r)\sinh 2\mu_i + (\mu_r - \sqrt{3}\mu_i)\sin 2\mu_r}{\cosh 2\mu_i - \cos 2\mu_r} = \frac{\mu_1}{\tan \mu_1} \tag{8}$$

Note that Eq. (7) shows that the critical value of the Rayleigh number for even modes depends on the divider resistance while this is not the case for the odd modes [Eq. (8)]. As expected, at $\alpha = 0$, Eq. (7) reduces to the equation determining the even modes in the spectrum of the Rayleigh number for the problem without divider, and hence a permeable divider with zero normal resistance does not affect the stability to even disturbances. For low enough values of the resistance ($\alpha \ll 1$) the minimum (critical) Rayleigh number, $Ra*$, and the corresponding wavenumber are approximately:

$$Ra* = \frac{27}{64}\pi^4\left(1 + \frac{8\alpha}{9\pi^2}\right), \qquad k* = \frac{\pi}{2\sqrt{2}}\left(1 - \frac{2\alpha}{9\pi^2}\right) \tag{9}$$

For arbitrary values of divider resistance the numerical values of $Ra*$ and $k*$ are obtained by solving Eq. (7) and are given in Fig. 1. It appears that by increasing α, the critical Rayleigh number grows monotonically and for $\alpha \to \infty$ reaches the value of 816.7. With the increase of α the wavenumber of the most dangerous disturbance first diminishes and then, at and above $\alpha = 80$, it grows monotonically to the limiting value of 2.215. This behavior can be explained by analyzing the convective modes expected at onset.

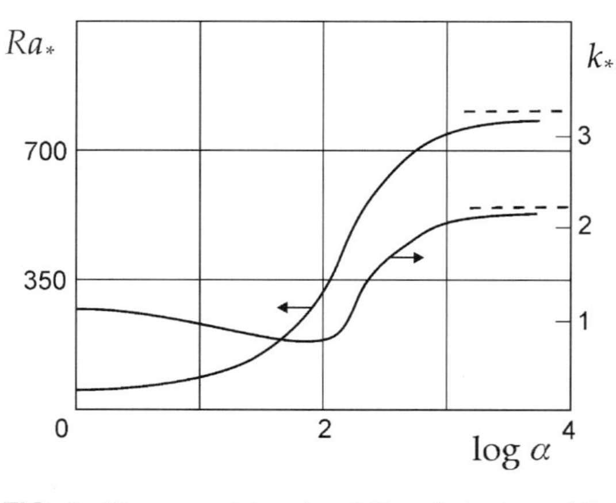

FIG. 1 Buoyancy-driven instability of a horizontal liquid layer with two free surfaces and an internal divider. Semilog plot of critical Rayleigh numbers and corresponding wavenumbers as a function of divider resistance. (From Ref. 1.)

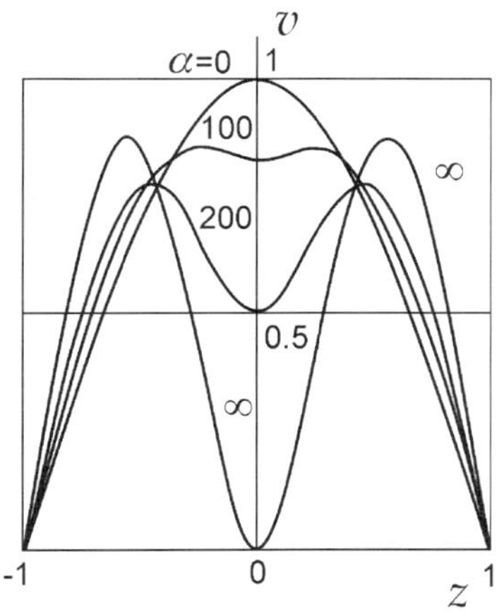

FIG. 2 Buoyancy-driven instability of a horizontal liquid layer with two free surfaces and an internal divider. Amplitude of transverse velocity, v, vs. transverse coordinate, z, for different values of the divider resistance from 0 to α. (From Ref. 1.)

In Fig. 2 the amplitude of the vertical velocity component is plotted against the transverse coordinate at different values of the resistance, α. When this resistance vanishes, the convective flow develops in the form of cells, which spread over both sublayers. As α increases an intense convective flow progresses deeper into the layer towards the layer surfaces, which causes a decrease in the wavenumber of the most dangerous disturbance. From $\alpha = 80$ onwards, closed vortices develop on the background of this flow in both parts of the liquid layer and as $\alpha \to \infty$ the flow develops as a two-layer structure, thus justifying the growth of the critical wavenumber.

The transcendental equation for odd disturbances [Eq. (8)] shows that the critical Rayleigh number does not depend on the normal divider resistance. However, under the assumption of infinite tangential resistance it is necessary to ensure that in the middle of the layer the longitudinal velocity component vanishes. In the absence of hydrodynamic interaction between two vortices the minimum critical Rayleigh number in the odd solution increases until the value 1100.6 is reached. The wavenumber of the most dangerous disturbances is equal to 2.68. Comparison with Fig. 1 shows that for all values of α, a lower even mode is the most dangerous.

B. Layer with Outer Solid Boundaries

Let us now consider that the liquid layer is confined between two solid flat surfaces ($z = \pm 1$), again at corresponding temperatures $\pm\Theta$. Let the permeable divider

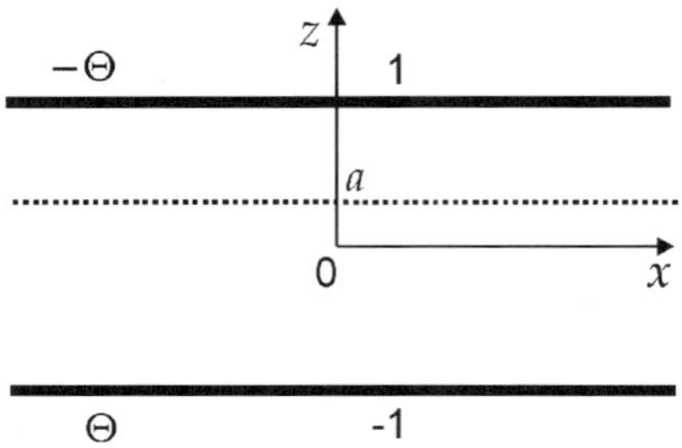

FIG. 3 Liquid layer with an internal divider off-midplane: Layer geometry and coordinate system.

now be placed inside the layer at a distance a from the midplane (Fig. 3). One may expect that the critical value of the Rayleigh number leading to instability significantly depends on the location of the divider and its properties [2].

As before, we now assume that the physical properties of the divider are described by the following relations:

$$z = a : T_- = T_+, \qquad \frac{\partial T_-}{\partial z} = \frac{\partial T_+}{\partial z} \tag{10}$$

$$\mathbf{v}_- = \mathbf{v}_+, \qquad p_- - p_+ = \alpha_n \mathrm{v}_z,$$

$$\tau_{zx+} - \tau_{zx-} = \alpha_\tau \mathrm{v}_x, \qquad \tau_{zy+} - \tau_{zy-} = \alpha_\tau \mathrm{v}_y \tag{11}$$

The conditions of Eq. (10) suggest that the divider has no thermal inertia and according to Eq. (11) the velocity is continuous. The velocity projection is proportional to the sum of corresponding stresses acting on both sides of the divider.

The neutral disturbances responsible for instability in such a horizontal layer satisfy Eqs. (1) and (2). Those equations should be augmented with conditions on the solid boundaries

$$z = \pm 1 : \mathrm{v} = \frac{\mathrm{dv}}{\mathrm{dz}} = \theta = 0 \tag{12}$$

and boundary conditions on the divider, which, using the equations of motion, become

$$z = a : \theta_- = \theta_+, \qquad \frac{\mathrm{d}\theta_-}{\mathrm{dz}} = \frac{\mathrm{d}\theta_+}{\mathrm{dz}}, \qquad \mathrm{v}_- = \mathrm{v}_+, \qquad \frac{\mathrm{dv}_-}{\mathrm{dz}} = \frac{\mathrm{dv}_+}{\mathrm{dz}} \tag{13}$$

$$\frac{\mathrm{d}^3\mathrm{v}_-}{\mathrm{dz}^3} - \frac{\mathrm{d}^3\mathrm{v}_+}{\mathrm{dz}^3} = \alpha_n k^2 \mathrm{v}, \qquad \frac{\mathrm{d}^2\mathrm{v}_+}{\mathrm{dz}^2} - \frac{\mathrm{d}^2\mathrm{v}_-}{\mathrm{dz}^2} = \alpha_\tau \frac{\mathrm{dv}}{\mathrm{dz}} \tag{14}$$

This boundary-value problem was solved using the Runge–Kutta method to find independent solutions satisfying Eq. (12). A linear combination of independent solutions meeting the condition of Eq. (14) leads to a 6th-order homogeneous algebraic system for mode amplitudes. The solvability condition for this system yields the Rayleigh number for neutral disturbances, $Ra = Ra(k, a, a_n, a_\tau)$.

Neutral curves and motion modes have been obtained for a divider with high tangential resistance ($\alpha_\tau \to \infty$). In Figs. 4 and 5 the critical value of the Rayleigh number, $Ra*$, and the critical wave number, $k*$, are plotted as a function of the normal resistance for different divider locations. When the position of the divider coincides with the midplane of the layer ($a = 0$) the critical Rayleigh number rapidly grows with increasing divider resistance.

Experiments [3] have been conducted to test the theoretical predictions. The interlaced grid used in these experiments was made of a kapron filament 0.02 mm thick with 75% porosity. For a convective cell 0.85 mm thick in water, the divider dimensionless resistance was estimated at about 200 units. A multilayer grid was used to increase the divider resistance. In Fig. 4 the experimentally observed values of $Ra*$ are depicted by dots. Theory shows that practically the maximum stabilization of equilibrium is reached already at $\alpha_n \sim 10^3–10^4$.

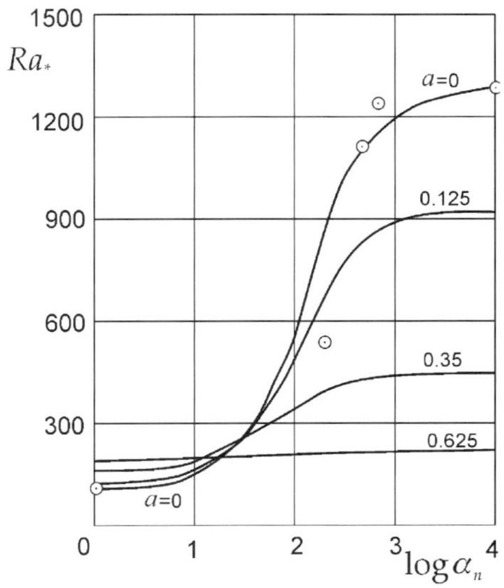

FIG. 4 Buoyancy-driven instability of a horizontal liquid layer with two solid boundaries and an internal divider. Semilog plot of the lowest critical value of the Rayleigh number vs. resistance for different divider locations inside the layer [2]. The dots denote experimental values [3] for a divider placed at the layer midplane. (From Ref. 2.)

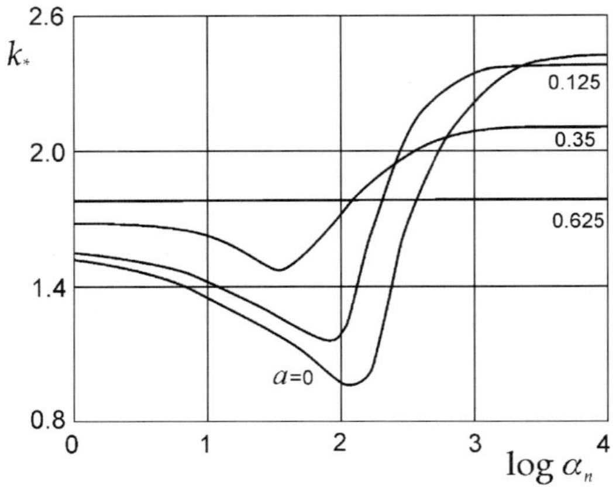

FIG. 5 Buoyancy-driven instability of a horizontal liquid layer with two solid boundaries and an internal divider. Semilog plot of the wavenumber of the most dangerous disturbances vs. divider resistance for a divider placed at different locations off-midplane. (From Ref. 2.)

As in the case of a film, or a layer with one or two free boundaries, the wavenumber of the most dangerous disturbance first diminishes with the increase of the divider resistance and then, beginning from $\alpha_n = 120$, it monotonically grows. Such a behavior of k_* can be attributed to a change of the transverse velocity distribution in the critical disturbance mode.

The amplitude of the disturbance vertical velocity for a symmetrically located divider is given in part (a) of Fig. 6 for different values of divider resistance. It can be seen that in the lowest resistance region ($\alpha_n \sim 100$) of the curve $k_*(\alpha_n)$ the effective vertical dimension of the convective cell is the largest. At small resistance values a displacement of the divider away from the midplane layer to $a = 0.625$ slightly increases the stability domain while at large resistance values the stability is drastically reduced.

For $a > 0.625$ the normal divider resistance has practically no effect on stability and, for $a \to 1,\ Ra \to 106.7$. In this region the dependence of the critical Rayleigh number and the wavenumber on a is solely due to variation of the effective layer thickness.

The modes of critical disturbances for $a = 0.125$, 0.350, and 0.625 at different α_n are shown in Fig. 6(b)–(d). For low normal resistance, and small displacements of the divider, the cell center is still at the divider. Further vertical displacement of the divider causes the cell center to move away in a wider channel region. Increasing the divider resistance, the flow changes to a two-layer flow with weak intensity in the narrow part of the channel. At high resistance values these changes in the flow mode significantly increase the stability domain.

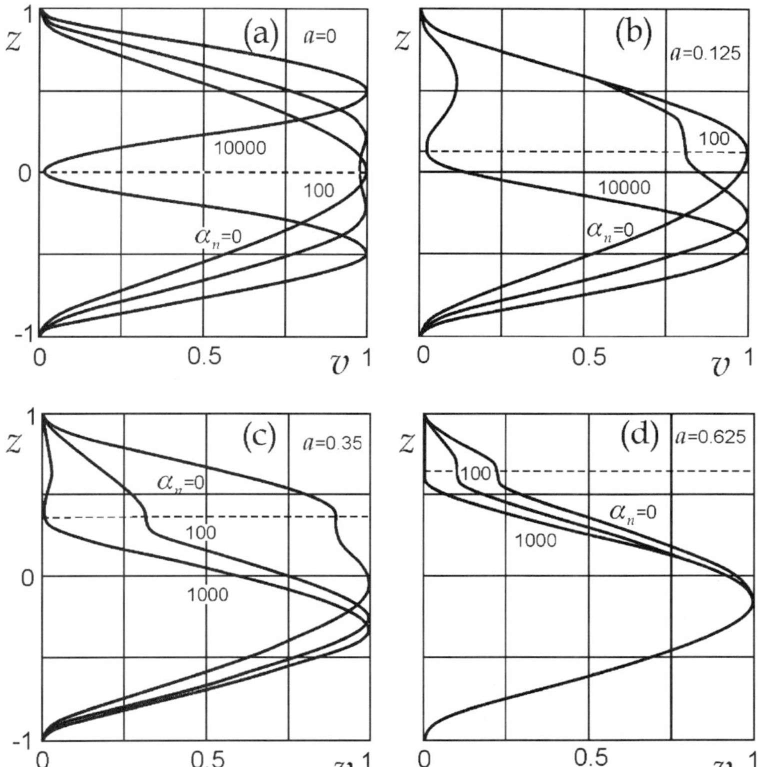

FIG. 6 Buoyancy-driven instability of a horizontal liquid layer with an internal divider. Amplitude of transverse velocity as a function of transverse coordinate for different values of divider resistance and at different locations of the divider: (a) $a = 0$, (b) $a = 0.125$, (c) $a = 0.35$, and (d) $a = 0.625$. (From Ref. 2.)

II. SURFACE TENSION GRADIENT (MARANGONI) DRIVEN INSTABILITY OF A FILM WITH TWO OPEN, UNDEFORMABLE FREE SURFACES AND HEATED OR COOLED DIVIDER

Consider a film, or a suspended liquid layer ($-1 < z < 1$), with two external free level boundaries open to air (considered mechanically passive) located at $z = \pm 1$. Again at $z = 0$, i.e., midway between the two open boundaries, we consider inserted a permeable divider, thickness $2h(h \ll d)$, taken as a partition that on average is uniformly heated. Disregarding surface deformability and buoyancy in the liquid layer, here we shall consider the result of the Marangoni effect acting at the two opposite open boundaries [4].

In the quiescent state the temperature profile is assumed to be initially linear

$$T_0 = \Theta\left(1 - \frac{|z|}{d}\right) \tag{15}$$

with Θ defined by the power at the divider, as a heat source or sink, and the heat flux at open surfaces. Let (T, p, v) denote infinitesimal temperature, pressure, and velocity disturbances. At the open boundaries we assume

$$z = \pm 1 : \mathrm{v}_z = 0 \tag{16}$$

together with the tangential stress balance. If, as usual, the surface tension at the open boundaries depends linearly on temperature, then the boundary conditions are

$$z = \pm 1 : \eta \frac{\partial \mathrm{v}_x}{\partial z} = \pm \sigma_1 \frac{\partial T}{\partial x} \tag{17}$$

where, as earlier, here $\sigma_1 \geq 0$.

The disturbed heat flux at both outer surfaces is assumed such that

$$z = \pm 1 : \frac{\partial T}{\partial z} = 0 \tag{18}$$

Note that as in earlier cases studied these are the conditions most favorable for the onset of thermocapillary convection.

At the divider, we assume continuity of the velocity, with the tangential velocity component taken proportional to the sum of the tangential stresses on both sides of the partition. The normal component is taken proportional to the pressure jump across it. Thus we have at $z = 0$:

$$\mathrm{v}_x^+ = \mathrm{v}_x^-, \qquad \mathrm{v}_z^+ = \mathrm{v}_z^- \tag{19a}$$

$$\eta\left(\frac{\partial \mathrm{v}_x^+}{\partial z} - \frac{\partial \mathrm{v}_x^-}{\partial z}\right) = \alpha_\tau \mathrm{v}_x, \qquad p^- - p^+ = \alpha_n \mathrm{v}_z \tag{19b}$$

Thermally, the divider has the characteristics

$$T_+ = T_-, \qquad \kappa_\mathrm{f}\left(\frac{\partial T_+}{\partial z} - \frac{\partial T_-}{\partial z}\right) = \kappa_\mathrm{p} 2d\Delta_1 T \tag{20}$$

with + and − subscripts denoting, as earlier, the half-layer above and below the divider, respectively. $\Delta_1 = (\partial^2/\partial x^2 + \partial^2/\partial y^2)$, while κ_f and κ_p are the heat conductivity of the fluid and the divider, respectively. Here, the thermal boundary condition on the partition takes into account its longitudinal heat conductivity.

We now introduce dimensionless variables using suitable scales: length, d; time, d^2/ν; velocity, χ/d; partition resistance, η/d; and temperature, Θ. Using once more the Fourier normal mode decomposition, $\{\mathrm{v}_z = \mathrm{v}(z)\exp(ikx),\ T = \theta(z)\exp(ikx)\}$, and restricting consideration to monotonic instability without buoyancy we have:

$$\frac{\mathrm{d}^4\mathrm{v}}{\mathrm{d}z^4} - 2k^2\frac{\mathrm{d}^2\mathrm{v}}{\mathrm{d}z^2} + k^4\mathrm{v} = 0 \tag{21}$$

$$\frac{\mathrm{d}^2\theta}{\mathrm{d}z^2} - k^2\theta \pm \mathrm{v} = 0 \tag{22}$$

$$z = 0: \mathrm{v}_- = \mathrm{v}_+, \qquad \frac{\mathrm{d}\mathrm{v}_-}{\mathrm{d}z} = \frac{\mathrm{d}\mathrm{v}_+}{\mathrm{d}z}, \qquad \frac{\mathrm{d}^3\mathrm{v}_-}{\mathrm{d}z^3} - \frac{\mathrm{d}^3\mathrm{v}_+}{\mathrm{d}z^3} = \alpha_n k^2 \mathrm{v}$$

$$\frac{\mathrm{d}^2\mathrm{v}_+}{\mathrm{d}z^2} - \frac{\mathrm{d}^2\mathrm{v}_-}{\mathrm{d}z^2} = \alpha_\tau \frac{\mathrm{d}\mathrm{v}}{\mathrm{d}z} \tag{23}$$

$$\theta_+ = \theta_-, \qquad \frac{\mathrm{d}\theta_+}{\mathrm{d}z} - \frac{\mathrm{d}\theta_-}{\mathrm{d}z} = 2k^2\kappa\theta_+, \qquad \kappa = \kappa_\mathrm{p} d/\kappa_\mathrm{f} \tag{24}$$

$$z = \pm 1: \mathrm{v} = 0, \qquad \frac{\mathrm{d}^2\mathrm{v}}{\mathrm{d}z^2} = \pm k^2 Ma\theta, \qquad \frac{\mathrm{d}\theta}{\mathrm{d}z} = 0 \tag{25}$$

As expected, besides the wavenumber, k, this boundary-value problem involves the Marangoni number, defined in the preceding chapter, as well as three dimensionless divider parameters: normal and tangential resistances, α_n and α_τ, and its longitudinal heat conductivity, κ.

Again, solutions of the boundary-value problem of Eqs. (22)–(25) can be represented as a linear combination of even $\{\mathrm{v}(-z) = \mathrm{v}(z),\ \theta(-z) = -\theta(z)\}$ and odd $\{\mathrm{v}(-z) = -\mathrm{v}(z),\ \theta(-z) = \theta(z)\}$ modes. With such representation we may construct the solution for the region $z > 0$, after appropriately rewriting the boundary conditions of Eq. (23). At the midplane $z = 0$ we must find from the condition of Eq. (23) the jumps of the odd functions. We have

(i) For the even solution at $z = 0^+$:

$$z = +0: \frac{\mathrm{d}\mathrm{v}}{\mathrm{d}z} = 0, \qquad \frac{\mathrm{d}^3\mathrm{v}}{\mathrm{d}z^3} = -0.5\alpha_n k^2\mathrm{v}, \qquad \theta = 0 \tag{26}$$

(ii) For the odd solution at $z = 0^+$:

$$z = +0: \mathrm{v} = 0, \qquad \frac{\mathrm{d}^2\mathrm{v}}{\mathrm{d}z^2} = 0.5\alpha_\tau\frac{\mathrm{d}\mathrm{v}}{\mathrm{d}z}, \qquad \frac{\mathrm{d}\theta}{\mathrm{d}z} = k^2\kappa\theta \tag{27}$$

It appears that the even solution depends on a single parameter, namely α_n, while the odd mode depends on two parameters, α_τ and κ.

The general form of both solutions can be written in the region $z > 0$ as

$$v = c_1 \sinh kz + c_2 \cosh kz + c_3 z \cosh kz + c_4 z \sinh k_z \tag{28a}$$

$$\theta = c_5 \sinh kz + c_6 \cosh kz + \left(\frac{c_4}{4k^2} - \frac{c_2}{2k}\right) z \sinh kz + \left(\frac{c_3}{4k^2} - \frac{c_1}{2k}\right) z \cosh kz - (c_3 \sinh kz + c_4 \cosh kz)\frac{z^2}{4k} \tag{28b}$$

where $c_i (i = 1, \ldots, 6)$ are constants to be fixed by the boundary conditions.

The problem is now reduced to solving a determinantal equation, which, upon separation of the solution into its even and odd parts, yields:

(i) Even modes

$$Ma = \frac{32k^2 + 8\alpha_n(k^2t^2 + kt - k^2)}{4(k^2 - 2kt + t^2 - k^2t^2 + 2kt^3) + \alpha_n(t^3/k - k^2 + k^2t^2)} \tag{29}$$

where $t = \tanh k$. Asymptotically we have:

(i.a) $k \to 0$:

$$Ma = \frac{480 + 80\alpha_n k^2}{60k^2 + \alpha_n k^4} \tag{30a}$$

(i.b) $k \to \infty$:

$$Ma = 8k^2 \tag{30b}$$

Thus the threshold values correspond to the minima of the curves shown in Fig. 7. At $\alpha_n = 0$ we have $Ma* = 21.8$ and $k* = 1.1$, while at $\alpha_n = 10^4$ (considered as practical infinity) we have $Ma* = 79.5$ and $k* = 2.0$. The results found for an impervious divider correspond to the earlier known finding for a *good* heater [4,5].

Figure 8 provides the vertical velocity in the upper half layer and related flow profiles as an increases from zero to (practical) infinity. Very much like in the case of buoyancy-driven convection earlier discussed we see the crucial role played by α_n in establishing a *single cell* convective pattern all over the two halves of the liquid layer. Indeed, for the even solution, when α_n tends to zero the divider plays little mechanical role. In the opposite case, when α_n tends to infinity, the partition truly divides the liquid into two separate layers. For completeness, Fig. 9 depicts the expected vertical velocity component at vanishing α_n for various values of the wavenumber at neutral stability conditions.

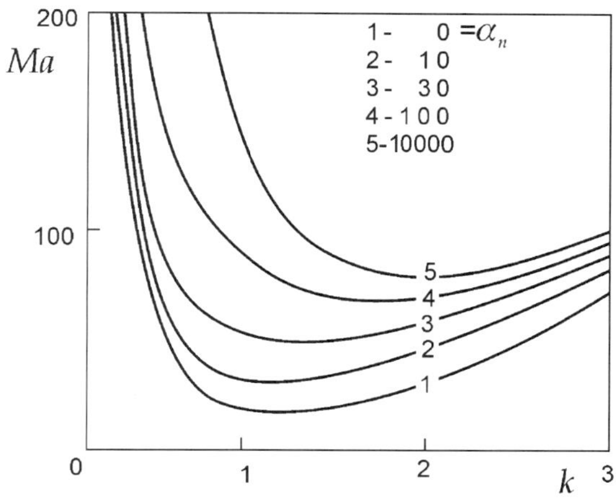

FIG. 7 Surface tension gradient (Marangoni) driven instability of a film or a floating liquid layer with an internal divider. Neutral stability curves (even solution) at fixed values of the transverse resistance of the divider, α_n ($\alpha_n = 0,\ 10,\ 30,\ 10^2$, and 10^4). (From Ref. 4.)

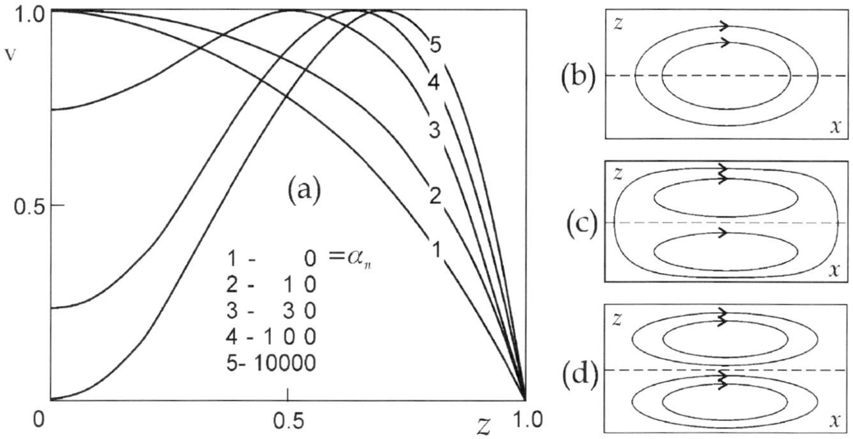

FIG. 8 Surface tension gradient (Marangoni) driven instability of a film or a floating liquid layer with an internal divider. Even solution: (a) vertical velocity component of the expected flow at threshold for several values of the transverse resistance of the divider, α_n, and qualitative sketch of the corresponding flow profiles (b) for $\alpha_n = 0$, (c) for $\alpha_n = 30$, and (d) for $\alpha_n = 10^4$. Note that at large enough values of the resistance the divider splits the layer into two separately convecting halves. (From Ref. 4.)

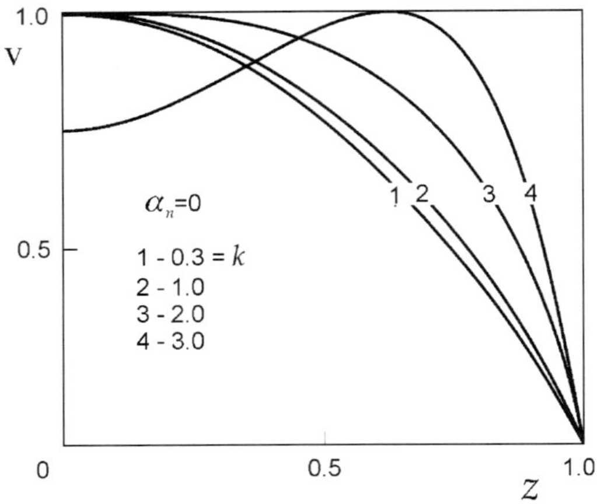

FIG. 9 Surface tension gradient (Marangoni) driven instability of a film or a floating liquid layer with an internal divider. Even solution: vertical velocity component of neutral disturbances for vanishing α_n and different values of k correspond to critical values at threshold. (From Ref. 4.)

(ii) Odd modes

$$Ma = \frac{16k^2t^3 + 4\alpha_\tau(k^2t^3 + kt^2 - k^2t) + 16\kappa k^3t^2 + 4\alpha_\tau\kappa(k^3t^2 + k^2t - k^3)}{\begin{array}{r}2(t^3 - k^2t + k^2t^3) + 0.5\alpha_\tau(k + t^2/k - 2t + 2t^3 - k^2t + k^2t^3 - kt^2) \\ +2\kappa(2k^3t^2 - k^2t^3 + k^2t - 2k^3 + kt^2) + 0.5\alpha_\tau\kappa(t^3 + k^3t^2 - k^3\end{array}} \tag{31}$$

Asymptotically, we have:

(ii.a) $k = 0$:

$$Ma = \frac{(288 + 48\alpha_\tau)(1 + \kappa)}{12 + \alpha_\tau}, \tag{32a}$$

(ii.b) $k \to \infty$:

$$Ma = \frac{16k^2 + 4a_\tau k + 16\kappa k^3 + 4a_\tau \kappa k^2}{2 + 0.5a_\tau/k + 2\kappa k + 0.5a_\tau\kappa} \tag{32b}$$

The threshold values are the minima of the curves shown in Fig. 10, which, for illustration, correspond to $\kappa = 0$, $\kappa = 1$, and $\kappa = 5$, respectively. For $\kappa = 0$, at $\alpha_\tau = 0$ we have $Ma* = 24$ and $k* = 0$, while at $\alpha_\tau = 10^4$ we have $Ma* = 48$ and $k* = 0$. The results for a rigid poor conducting divider recover findings already known. When κ does not vanish, e.g., when $\kappa = 5$, the threshold values do not correspond to a vanishing wavenumber and are: at $\alpha_\tau = 0$, $Ma* = 54.8$ with $k* = 1.6$; and at $\alpha_\tau = 10^4$; $Ma* = 77.8$ with $k* = 1.9$. Figure 11 depicts the

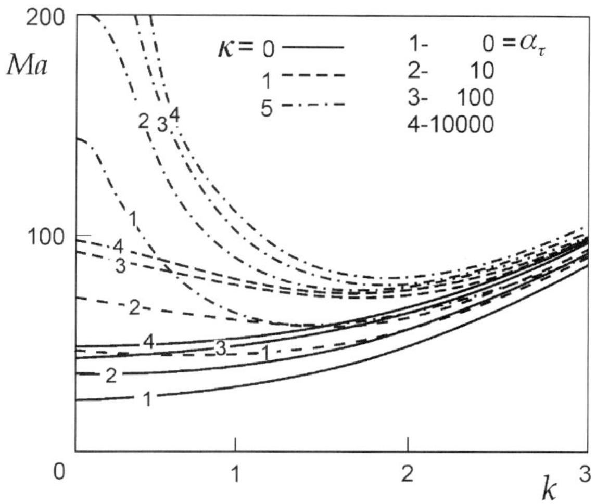

FIG. 10 Surface tension gradient (Marangoni) driven instability of a film or a floating liquid layer with an internal divider. Neutral stability curves (odd solution) at fixed values of the parallel resistance of the divider, α_τ, and different values of divider longitudinal heat conductivity, κ. (From Ref. 4.)

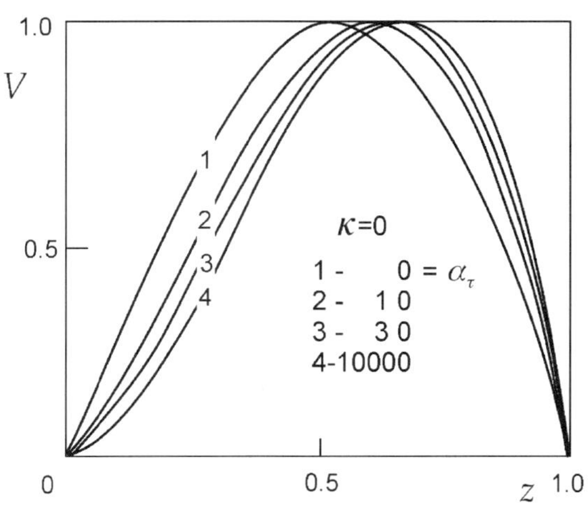

FIG. 11 Surface tension gradient (Marangoni) driven instability of a film or a floating liquid layer with an internal divider. Odd solution: vertical component of the expected flow at threshold for several values of the parallel resistance of the divider, α_τ, and vanishing heat conductivity. (From Ref. 4.)

predicted vertical velocity component for the odd solution. It clearly shows that, at variance with the even mode, for the odd mode both halves develop separate convection cells as may intuitively be guessed.

On the other hand, for the later case it appears that the flow pattern shows only a slight dependence on divider heat conductivity, κ. Both even and odd convective modes show velocity profiles strongly dependent on the corresponding divider resistance. For small values of α_n, i.e., in the case of a good permeability of the divider, we see that the two initially unrelated thermocapillary instabilities yield flows that merge the two expected steady cell patterns into a single convective structure. In the opposite extreme case, a rigid partition trivially yields Benard cells in the two halves.

Comparison between the expected behavior of even and odd modes of convective instability is illustrated in Fig. 12. For the particular case of vanishing divider longitudinal heat conductivity, $\kappa = 0$, it appears that in the long-wave region odd disturbances have a lower threshold while in the short-wave region, for small resistance values, the even modes are the most dangerous. Note that for even modes, under appropriate conditions, the two halves of the layer behave like a

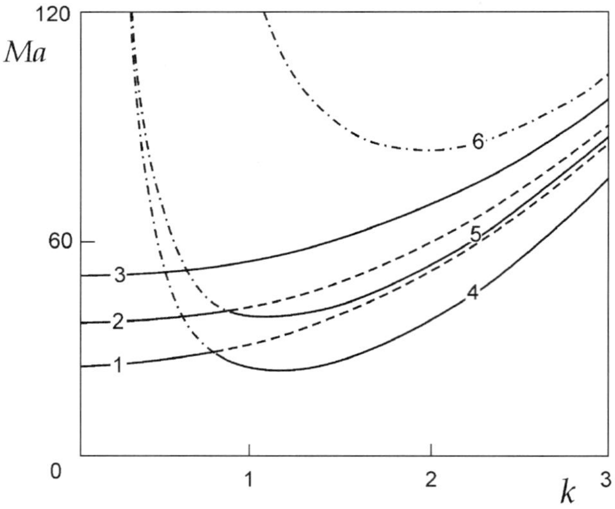

FIG. 12 Surface tension gradient (Marangoni) driven instability of a film or a floating liquid layer with an internal divider. Neutral curves (even and odd disturbances) at vanishing κ. Solid line portions denote the lower threshold, most dangerous instability mode at equal transverse and parallel resistances, $\alpha_n = \alpha_\tau = 0$ (lines 1, 4), 10 (lines 2, 5), and 10^{-4} (lines 3, 6). (From Ref. 4.)

single convective cell while otherwise both halves may convect in a rather unrelated way. Overall, for large enough values of the resistances, α_n and α_τ, the odd modes are the most dangerous. From the behavior of all neutral stability curves, at vanishing wavelengths it can be shown that all critical Marangoni numbers, $Ma*$, asymptote to $8k^2$.

Figure 13 depicts the dependence of the critical Marangoni number, $Ma*$, on the corresponding hydrodynamic resistance of the divider, α: for even modes (curve 1) $\alpha = \alpha_n$, and for odd modes curves (2–6) $\alpha = \alpha_\tau$. There is a rapid variation of $Ma*$ in the region $1 < \alpha < 100$. This shows that the upper value of this interval fits well with a divider with not so small holes. Even though such a divider suppresses through-flow, it still allows for molecular diffusion from one half to the other in the liquid layer.

In conclusion, we can say that in view of the drastic influence of the divider on the expected thermocapillary flows, the use of a heated permeable divider, for example, as a grid catalyst, shows potential usefulness to control flows in variable effective gravity and microgravity conditions.

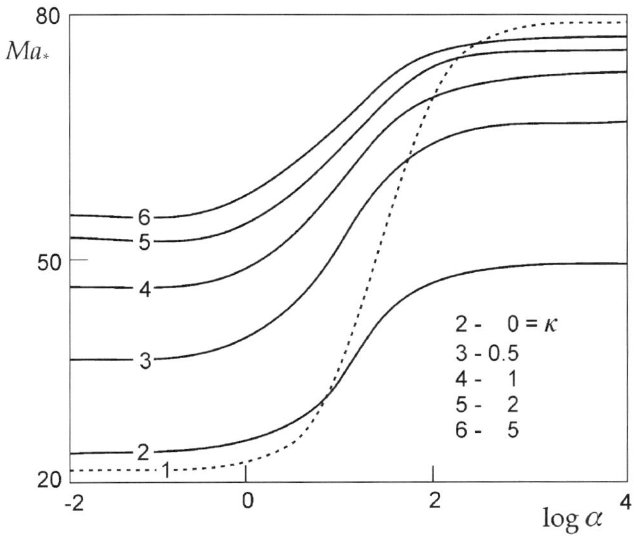

FIG. 13 Surface tension gradient (Marangoni) driven instability of a film or a floating liquid layer with an internal divider. Semilog plot (using the appropriate resistance α: for even modes line 1, $\alpha = \alpha_n$; and for odd modes lines 2–6, $\alpha = \alpha_\tau$) of the critical Marangoni numbers at different values of the divider heat conductivity. κ. (From Ref. 4.)

III. THERMOCAPILLARY INSTABILITY OF A FILM OR LIQUID LAYER WITH TWO OUTER FREE DEFORMABLE SURFACES AND HEATED OR COOLED DIVIDER

Let us now consider that both surfaces bounding the floating layer are deformable, and as earlier examined, have a heated or cooled permeable divider. An infinitely high hydrodynamic resistance of the divider reduces to the case of a two-layer system divided by a solid impermeable or impervious boundary [6]. The opposite case of a heated divider with vanishingly small resistance may be realized by heating with a laser sheet.

At $z = 0$, i.e., midway between the two open surface boundaries ($z = \pm 1$), we insert the permeable divider, which on average is uniformly heated or cooled. Disregarding gravitational effects we shall consider the result of the Marangoni effect acting at the two opposite surface boundaries of the liquid layer eventually coupled through flows in the bulk. In the quiescent state the temperature profile is assumed to be initially linear

$$T_0 = \Theta(1 - |z|) \tag{33}$$

We have set the zero reference value at both outer boundaries of the liquid layer. Note that although we start with isothermal boundary conditions we assume that due to the outer ambient air having a largely lower heat conductivity than the liquid the actual boundary condition for disturbances is adiabatic for heat disturbances, hence heat flux is held fixed. The temperature of the divider Θ may be either positive or negative (above or below the ambient temperature).

Let T, p, and $\mathbf{v}$ denote infinitesimal temperature, pressure, and velocity disturbances, respectively. Assuming that the transverse displacements of the free liquid surfaces, ζ, are small quantities, for a linear approximation the boundary condition may be applied at $z = \pm 1$. At the open boundaries, with a linear temperature dependence of the surface tension, we also assume that, mechanically, air plays no active role, hence

$$z = \pm 1 : \frac{\partial \zeta}{\partial t} = \mathrm{v}_z, \qquad p \pm \sigma_0 \frac{\partial^2 \zeta}{\partial x^2} = 2\eta \frac{\partial \mathrm{v}_z}{\partial z}$$
$$-\sigma_1 \left(\frac{\partial T}{\partial x} + \frac{\partial T_0}{\partial z}\frac{\partial \zeta}{\partial x} \right) = \pm \eta \left(\frac{\partial \mathrm{v}_x}{\partial z} + \frac{\partial \mathrm{v}_z}{\partial x} \right) \tag{34}$$

The disturbed heat flux at both free outer surfaces is as in Eq. (18).

At the permeable divider, i.e., at level $z = 0$ we assume continuity of the velocity. The tangential velocity component is proportional to the sum of the tangential stresses on either side of the partition, while the normal component is proportional to the pressure jump across it as in Eqs. (19a) and (19b).

Thermal conditions at the divider are such that disturbances of temperature and heat flux are continuous:

$$T_+ = T_-, \qquad \frac{\partial T_+}{\partial z} = \frac{\partial T_-}{\partial z} \tag{35}$$

that is, we consider the case with vanishing divider longitudinal heat conductivity.

A. Monotonic Instability

As for the case of undeformable outer boundaries, again we take Fourier normal modes

$$(\mathrm{v}_z,\ T,\ \zeta) = (\mathrm{v},\ \theta,\ \xi)\exp[i(k_1 x + k_2 y)]$$

obeying the continuity, Navier–Stokes, and heat (Fourier) diffusion equations, which here reduce to

$$\frac{\mathrm{d}^4\mathrm{v}}{\mathrm{d}z^4} - 2k^2\frac{\mathrm{d}^2\mathrm{v}}{\mathrm{d}z^2} + k^4\mathrm{v} = 0 \tag{36}$$

$$\frac{\mathrm{d}^2\theta}{\mathrm{d}z^2} - k^2\theta + \mathrm{v} = 0 \tag{37}$$

$$k^2 = k_1^2 + k_2^2$$

with boundary conditions

$$z = \pm 1: \frac{\mathrm{d}^3\mathrm{v}}{\mathrm{d}z^3} - 3k^2\frac{\mathrm{d}\mathrm{v}}{\mathrm{d}z} \pm Pr\,Ca k^4\xi = 0$$

$$\frac{\mathrm{d}^2\mathrm{v}}{\mathrm{d}z^2} + k^2\mathrm{v} \pm Mak^2\left(\theta + \xi\frac{\mathrm{d}T_0}{\mathrm{d}z}\right) = 0$$

$$\frac{\mathrm{d}\theta}{\mathrm{d}z} = 0, \qquad \mathrm{v} = 0 \tag{38}$$

$$z = 0: \mathrm{v}_- = \mathrm{v}_+, \qquad \frac{\mathrm{d}\mathrm{v}}{\mathrm{d}z} = \frac{\mathrm{d}\mathrm{v}_+}{\mathrm{d}z}, \qquad \frac{\mathrm{d}^3\mathrm{v}_-}{\mathrm{d}z^3} - \frac{\mathrm{d}^3\mathrm{v}_+}{\mathrm{d}z^3} = \alpha_n k^2\mathrm{v}$$

$$\frac{\mathrm{d}^2\mathrm{v}_+}{\mathrm{d}z^2} - \frac{\mathrm{d}^2\mathrm{v}_-}{\mathrm{d}z^2} = \alpha_\tau\frac{\mathrm{d}\mathrm{v}}{\mathrm{d}z}$$

$$\theta_+ = \theta_-, \qquad \frac{\mathrm{d}\theta_+}{\mathrm{d}z} = \frac{\mathrm{d}\theta_-}{\mathrm{d}z} \tag{39}$$

Five dimensionless groups appear in this case: the earlier defined capillary number, *Ca*, Prandtl number, *Pr*, and Marangoni number, *Ma*, and the transverse/normal and tangential/parallel divider resistances, α_n and α_τ.

Again, as in the case treated in the preceding section, the solutions of Eqs (37)–(39) naturally split into even, $\{v(z) = v(-z),\ \theta(-z) = -\theta(z)\}$, and odd, $\{v(-z) = -v(z),\ \theta(-z) = \theta(z)\}$, modes. We obtain

(i) For the even solution at $z = 0^+$:

$$\frac{dv}{dz} = 0, \qquad \frac{d^3v}{dz^3} = -0.5\alpha_n k^2 v, \qquad \theta = 0 \tag{40}$$

(ii) For the odd solution at $z = 0^+$:

$$v = 0, \qquad \frac{d^2v}{dz^2} = 0.5\alpha_\tau \frac{dv}{dz}, \qquad \frac{d\theta}{dz} = 0 \tag{41}$$

Hence the even solution depends only on the transverse resistance, α_n, while the odd mode depends on the tangential one, α_τ.

The general form of both solutions can be written, in the region $z > 0$, as

$$\begin{aligned} v &= c_1 \sinh kz + c_2 \cosh kz + c_3 z \cosh kz + c_4 z \sinh k_z \\ \theta &= c_5 \sinh kz + c_6 \cosh kz + \left(\frac{c_4}{4k^2} - \frac{c_2}{2k}\right) z \sinh kz \\ &\quad + \left(\frac{c_3}{4k^2} - \frac{c_1}{2k}\right) z \cosh kz - (c_3 \sinh kz + c_4 \cosh kz)\frac{z^2}{4k} \end{aligned} \tag{42}$$

where c_i are constants to be determined by the boundary conditions.

For the even solution the solvability condition of the problem yields the neutral stability curve, i.e., it provides the Marangoni number as a function of wavenumber and the dimensionless groups of the problem. We have

$$Ma = \frac{32k^2 + 8\alpha_n(k^2t^2 + kt - k^2)}{4(k^2 - 2kt + t^2 - k^2t^2 + 2kt^3) + \alpha_n(t^3/k - k^2 + k^2t^2) + (Pr\,Ca)^{-1} 8k^2(t^2 - 1)(4 - \alpha_n)} \tag{43}$$

with $t = \tanh k$. Clearly, for an undeformable liquid surface ($Pr\,Ca \to \infty$) Eq. (43) reduces to the earlier obtained value of Eq. (28). Asymptotically, we have:

(i.a) $k \to 0$:

$$Ma = \frac{480 + 80\alpha_n k^2}{60k^2 + \alpha_n k^4 + 120(Pr\,Ca)^{-1}(\alpha_n - 4 + 4k^2)} \tag{44a}$$

and for $\alpha_n > 4$ the onset of instability occurs at $Ma > 0$ (heated divider). At $k = 0$ the Marangoni number is equal to $4Pr\,Ca/(\alpha_n - 4)$. Thus for $\alpha_n < 4$ there is a region of small k where instability exists at $Ma < 0$ (cooled divider or heated ambient air). For large values of the combination $Pr\,Ca$ the region is $0 \leq k^2 < 2(4 - \alpha_n)/Pr\,Ca$ with $Ma* = -4Pr\,Ca/(4 - \alpha_n)$.

(i.b) $k \to \infty$:

$$Ma = 8k^2 \tag{44b}$$

i.e., the instability appears only for a heated divider.

Typical neutral stability curves for the even mode in the case $\alpha_n > 4$ are shown in Fig. 14. Curve 1 corresponds to an impervious divider ($\alpha_n \to \infty$) and an undeformable surface ($Ca \to \infty$). The curve has only one minimum at $k* = 2.0$ with $Ma* = 80$. For a deformable surface (finite values of Ca) and an impermeable divider the lowest threshold for instability appears at $k = 0$ (curves 2 and 3). For finite values of the resistance the latter instability disappears and the critical Marangoni number increases as α_n is decreased. In some range of α_n and $Pr\,Ca$ the neutral curves possess two minima (e.g., curve 5). Further decreasing the resistance makes the disturbances with finite wavenumber become the most dangerous. For $\alpha_n = 4$ the vertical axis is the asymptote of the neutral curve at $k = 0$.

The inset in Fig. 14 qualitatively illustrates the shape of critical disturbances and the distorted free surfaces. It is readily seen that bulging of the fluid surface is caused by the fluid inflow from the regions with higher surface temperature. The index "+" denotes the regions of the surface in which the Marangoni effect sustains the disturbances. In the valleys, the heated fluid is carried by the velocity

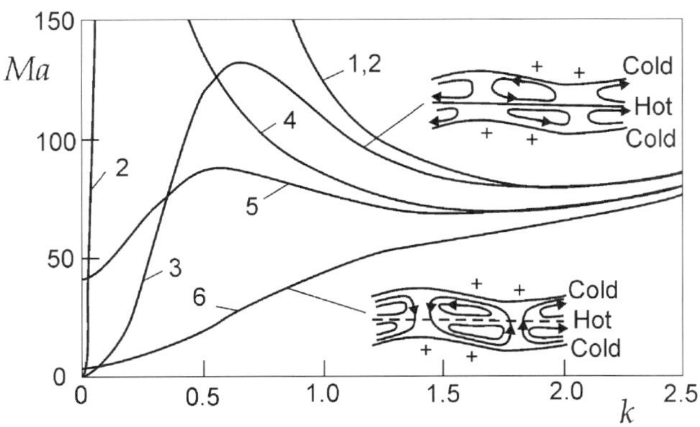

FIG. 14 Surface tension gradient (Marangoni) driven instability of a film or a floating liquid layer with an internal divider. Neutral curves for even monotonic disturbances at large divider resistances and different values of the combination $Pr\,Ca$: 1, $\alpha_n \to \infty$, $Pr\,Ca \to \infty$; 2, ∞, 10^5; 3, ∞, 10^5; 4, 10^2, ∞; 5, 10^2, 10^3; and 6, 10^2, 10^2. (From Ref. 6.)

disturbance to the surface, hence Pearson's instability result for a layer with a flat open surface [5]. Mismatch in the slope of the layer surface with respect to the isotherms of the imposed base temperature field provides another mechanism sustaining disturbances. Here this mechanism operates efficiently in the region of long waves and for the case with an impervious divider, contributes to the onset of the lowest threshold instability at $k = 0$.

For $\alpha_n < 4$ the asymptote is the straight line $k_a \approx [2(4 - \alpha_n)/Pr\,Ca]^{1/2}$ and a new instability range appears when the layer is heated from outside or there is a cooled divider. Typical neutral stability curves for this case are shown in Fig. 15. Curve 1 represents the case of a layer with undeformable boundaries and a permeable divider. As in the case $\alpha_n \to \infty$ this curve asymptotes to the vertical axis at $k = 0$. Curves 2 and 3 ($Pr\,Ca = 100$, $\alpha_n = 0$ and 1, respectively) have discontinuities at $k = k_a$ and illustrate the presence of instability for both ways of heating.

Instability in the short-wavelength region occurs with a heated divider. A typical flow pattern is depicted in the upper inset of Fig. 15. Contrary to the cases displayed in Fig. 14, distortion of the fluid surface in the layer with the permeable divider should be attributed to transverse flows. The Marangoni effect at the inclined section of the free surface slows down the initiated motion (this point is

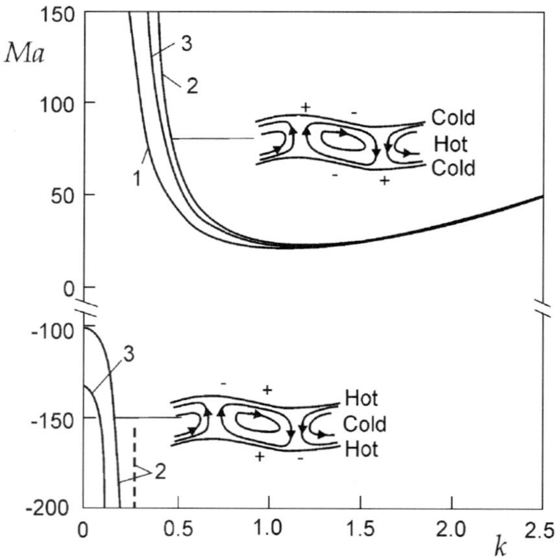

FIG. 15 Surface tension gradient (Marangoni) driven instability of a film or a floating liquid layer with an internal divider. Neutral curves for even monotonic disturbances at low divider resistances and different values of $Pr\,Ca$: 1, $\alpha_n = 0$, $Pr\,Ca \to \infty$; 2, 0, 10^2; and 3, 1, 10^2. (From Ref. 6.)

marked by index "−") and suppresses it in the region of long waves. However, with the layer heated from outside, this mechanism leads to a long-wavelength instability, as shown at the bottom of Fig. 15. Here, as the depicted pattern of the flow shows, the surface tension gradient prevents the growth of disturbances. Heating from outside or with a cooled divider, the disturbances with $k = 0$ are the most dangerous. Since a new mechanism of instability makes itself evident only at low divider resistance and within the long-wavelength range, it seems to be related to the interaction of disturbances occurring at the two opposite outer free surfaces.

For the odd solutions the solvability condition yields

$$\begin{aligned} Ma = {} & 8[4k^2t^3 + \alpha_\tau(k^2t^3 + kt^2 - k^2t)]/[4(t^3 - k^2t + k^2t^3) \\ & + \alpha_\tau(k + t^2/k - 2t + 2t^3 - k^2t + k^2t^3 - kt^2) \\ & + 8k^2(Pr\,Ca)^{-1}(1 - t^2)(4 + \alpha_\tau)], \end{aligned} \tag{45}$$

which asymptotically provides

$$k \to 0: \qquad Ma = \frac{288(1 - k^2) + 16\alpha_\tau(3 - 4k^2)}{12(1 - 2k^2) + \alpha_\tau(1 - 4k^2) + 72k^2(Pr\,Ca)^{-1}(4 - \alpha_\tau)} \tag{46a}$$

$$k \to \infty: \qquad Ma = 8k^2 + 2\alpha_\tau k \tag{46b}$$

Instability appears only when we have a heated divider. Typical neutral curves for the odd mode are shown in Fig. 16. For undeformable surfaces ($Pr\,Ca \to \infty$) the instability occurs with $k = 0$ at finite critical Marangoni numbers (curves 1, 4). With a deformable surface the odd mode always yields the lowest instability threshold ($Ma = 0$ at $k = 0$). This is typical of microgravity conditions. With non-negligible (effective) gravitational acceleration our boundary condition [in particular Eq. (34)] needs to be augmented with the hydrostatic contribution, which brings the threshold away from $k = 0$, and as buoyancy may play a non-negligible role due care must also be given to the possible Rayleigh–Taylor instability at the lower surface.

The inset of Fig. 16 depicts the convective flow in the case of a solid boundary (the upper inset) and a divider with $\alpha_\tau = 0$. The solid divider suppresses fluid motion in the middle of the layer and moves the convective rolls towards the outer free surface. This provides stability of the layer to disturbances of finite k as the divider resistance is increased. As the figure shows the convective flow is sustained by the two earlier mentioned instability mechanisms. The lowest threshold for instability is related to motions generated at the part of the surface not aligned with the isotherms of the basic temperature field.

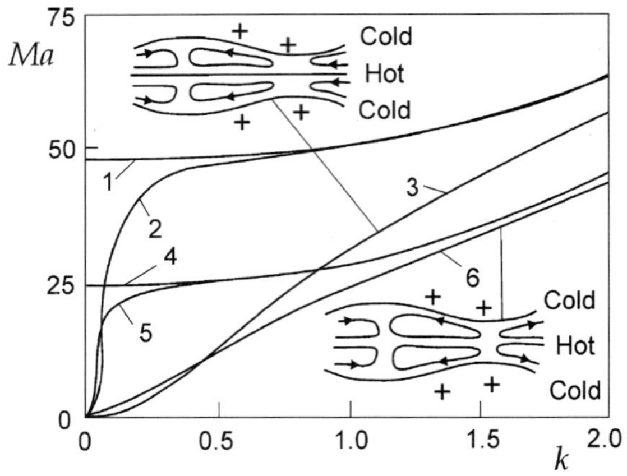

FIG. 16 Surface tension gradient (Marangoni) driven instability of a film or a floating liquid layer with an internal divider. Neutral curves for odd monotonic disturbances at different values of α_τ and $Pr\,Ca$:1, $\alpha_n \to \infty$, $Pr\,Ca \to \infty$; 2, ∞, 10^4; 3, ∞, 10^2; 4, 0, ∞; 5, 0, 10^4; 6, 0, 10^2. (From Ref. 6.)

B. Overstability and Oscillatory Modes

Let us now turn our attention to oscillatory modes of convection. Using once more the Fourier normal mode decomposition for neutral disturbances (with $\lambda = i\omega$)

$$(\mathrm{v}_z,\ T,\ \zeta) = (\mathrm{v},\ \theta,\ \xi)\exp\left[i(k_1 x + k_2 y - \omega t)\right]$$

we have

$$\frac{\mathrm{d}^4\mathrm{v}}{\mathrm{d}z^4} - 2k^2\frac{\mathrm{d}^2\mathrm{v}}{\mathrm{d}z^2} + k^4\mathrm{v} + i\omega\left(\frac{\mathrm{d}^2\mathrm{v}}{\mathrm{d}z^2} - k^2\mathrm{v}\right) = 0 \tag{47}$$

$$\frac{\mathrm{d}^2\theta}{\mathrm{d}z^2} - k^2\theta - \left(\frac{\mathrm{d}T_0}{\mathrm{d}z}\right)\mathrm{v} + i\omega Pr\theta = 0 \tag{48}$$

$$k^2 = k_1^2 + k_2^2$$

$$z = \pm 1: \frac{\mathrm{d}^3\mathrm{v}}{\mathrm{d}z^3} - (3k^2 - i\omega)\left(\frac{\mathrm{d}\mathrm{v}}{\mathrm{d}z}\right) \pm (Pr\,Ca)k^4\xi = 0 \tag{49a}$$

$$\frac{\mathrm{d}^2\mathrm{v}}{\mathrm{d}z^2} + k^2\mathrm{v} \pm Mak^2\left[\theta + \left(\frac{\mathrm{d}T_0}{\mathrm{d}z}\right)\xi\right] = 0 \tag{49b}$$

$$\frac{\mathrm{d}\theta}{\mathrm{d}z} = 0, \qquad \xi = -\mathrm{v}(i\omega Pr)^{-1} \tag{49c}$$

together with the boundary conditions at the divider:

$$z = 0: \mathrm{v}_- = \mathrm{v}_+, \qquad \frac{\mathrm{dv}_-}{\mathrm{d}z} = \frac{\mathrm{dv}_+}{\mathrm{d}z} \tag{50a}$$

$$\frac{\mathrm{d}^3\mathrm{v}_-}{\mathrm{d}z^3} - \frac{\mathrm{d}^3\mathrm{v}_+}{\mathrm{d}z_+^3} = \alpha_n k^2 \mathrm{v}, \qquad \frac{\mathrm{d}^2\mathrm{v}_+}{\mathrm{d}z^2} - \frac{\mathrm{d}^2\mathrm{v}_-}{\mathrm{d}z^2} = \alpha_\tau\left(\frac{\mathrm{dv}}{\mathrm{d}z}\right) \tag{50b}$$

$$\theta_+ = \theta_-, \quad \frac{\mathrm{d}\theta_+}{\mathrm{d}z} = \frac{\mathrm{d}\theta_-}{\mathrm{d}z} \tag{50c}$$

The solutions of Eqs. (47) and (48) with boundary conditions of Eq. (48) again separate into even and odd modes, with boundary conditions at $z = 0^+$, Eqs. (40) and (41). For both modes, the general solution, when $Pr \neq 0$ and $Pr \neq 1$, can be written for the region $z > 0$ as

$$\mathrm{v} = c_1 \exp(kz) + c_2 \exp(-kz) + c_3 \exp(\nu z) + c_4(-\nu z) \tag{51a}$$

$$\theta = \frac{c_1 \exp(kz) + c_2 \exp(-kz)}{i\omega Pr} - \frac{c_3 \exp(\nu z) + c_4 \exp(-\nu z)}{i\omega(Pr - 1)} + c_5 \exp(\beta z) + c_6 \exp(-\beta z) \tag{51b}$$

where $\nu = (k^2 - i\omega)^{1/2}$ and $\beta = (k^2 - i\omega Pr)^{1/2}$. The solvability condition for the coefficients c_i determines the overstable neutral curves, providing the corresponding frequency, ω, and Marangoni number, Ma. We have

$$Ma = \mathrm{Re}\left(\frac{\Delta_1}{\Delta}\right) \tag{52a}$$

and

$$\mathrm{Im}\left[\frac{\Delta_1(\omega)}{\Delta(\omega)}\right] = 0 \tag{52b}$$

where Δ_1 and Δ, for the even mode, are

$$\Delta_1, \ \Delta \equiv \det\begin{pmatrix} A_+ & A_- & B_+ & B_- & 0 & 0 \\ C_+ & C_- & D_+ & D_- & 0 & 0 \\ E_+ & E_- & F_+ & F_- & G_+ & G_- \\ I_+ & I_- & J_+ & J_- & 0 & 0 \\ K_+ & K_- & L_+ & L_- & M_+ & M_- \\ N_+ & N_- & O_+ & O_- & P_+ & P_- \end{pmatrix}$$

For Δ_1 and even mode

$$A_\pm = \pm k, \qquad B_\pm = \pm\nu, \qquad C_\pm = \pm k^3 + 0.5k^2\alpha_n,$$
$$D_\pm = \pm\nu^3 + 0.5k^2\alpha_n, \qquad E_\pm = \frac{1}{i\omega Pr}, \qquad F_\pm = \frac{1}{i\omega(Pr-1)},$$
$$G_\pm = -1, \qquad I_\pm = (\pm\omega^2 k \pm 2k^3 i\omega - k^4 Ca)\exp(\pm k),$$
$$J_\pm = [\pm\omega^2\nu \pm i\omega\nu(3k^2-\nu^2) - k^4 Ca]\exp(\pm\nu),$$
$$K_\pm = 2k^2 i\omega\exp(\pm k),$$
$$L_\pm = i\omega(\nu^2+k^2)\exp(\pm\nu), \qquad M_\pm = 0, \qquad N_\pm = \pm k\exp(\pm k),$$
$$O_\pm = \pm\nu\exp(\pm\nu), \qquad P_\pm = \pm\beta\exp(\pm\beta).$$

For Δ the corresponding elements of the determinant must be replaced by

$$K_\pm = 0, \qquad L_\pm = k^2\left[\frac{(1/Pr)-1}{(Pr-1)}\right]\exp(\pm\nu),$$
$$M_\pm = i\omega k^2\exp(\pm\beta).$$

For the odd mode some elements of Δ_1 and Δ must be replaced by

$$A_\pm = B_\pm = 1, \qquad C_\pm = k^2 \pm 0.5k^3\alpha_T, \qquad D_\pm = \nu^2 \pm 0.5k^2\nu\alpha_T,$$
$$E_\pm = \frac{\pm k}{i\omega Pr}, \qquad F_\pm = \frac{\pm\nu}{i\omega(Pr-1)}, \qquad G_\pm = \pm\beta.$$

1. Impermeable Divider

An impermeable or impervious divider separates the layer into two halves with vanishing hydrodynamic interaction. In this case the problem for the even mode is equivalent to that of a plane layer with a solid isothermal support, while the odd mode problem reduces to the case of a solid boundary with zero heat flux for disturbances; the heat flux is held fixed. The difference in behavior of even and odd disturbances ought to be clearer in the long-wavelength region, whereas in the region of short waves it is not seen. The onset of thermocapillary instability in a plane layer with one free boundary can be easily reobtained using low Prandtl number liquids. For illustration let us consider the case of a solid divider for a layer with fixed capillary number, $Ca = 10^4$, which is typical for layers of molten semiconductors.

Variations of the critical Marangoni number and the frequency of (neutral) oscillations with wavenumber k are plotted in Fig. 17 for $Pr = 10^{-2}$. The solid line at the top of the figure provides the neutral curve for monotonic disturbances. For the chosen parameter values the curves describing even and odd disturbances coincide on the scale used in our plot. The neutral curves for oscillatory

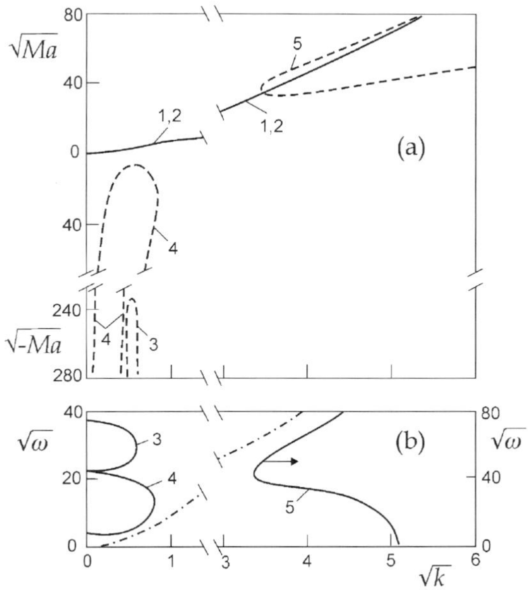

FIG. 17 Surface tension gradient (Marangoni) driven instability of a film or a floating liquid layer with an internal divider. (a) Neutral curves ($Ma,\ k$) and (b) frequency of neutral disturbances for a solid divider ($\omega,\ k$) at $Pr = 10^{-2}$ and $Ca = 10^4$. The even and odd numbers refer, respectively, to even and odd disturbances. The broken lines depict neutral overstable curves. (From Ref. 6.)

disturbances are depicted by broken lines. The even and odd numbers of curves correspond, respectively, to even and odd modes of disturbances. In the region of long waves (small k), oscillatory instability occurs under external heating of the free surface or with a cooled divider. The dispersion relation of these waves differs significantly from the Laplace–Kelvin law ($\omega^2 = Cak^3 \tanh k$) [7] for capillary waves in a liquid layer under isothermal conditions (broken line in the figure). The frequency of oscillation of the observed waves remains finite at low k and as these waves are sustained by the outward-directed temperature gradient of the fluid they are longitudinal waves.

As noted earlier, the discrepancy of neutral curves and frequencies for solutions with different symmetry is caused by the disparity of temperature conditions imposed at the divider. The most dangerous is the even mode (curve 4 in Fig. 17), involving nonzero heat flux through the divider. In the region of short waves ($k > 12$) the oscillations are induced by the internal heating of the fluid (curve 5). The dispersion relation for the waves at the lower branch is close to $\omega^2 = Cak^3$. We have the excitation of capillary waves by the Marangoni effect.

Because of their short length, these waves remain close to the outer free surface and hence are insensitive to the boundary conditions at the divider. Refering to Fig. 17, the frequency at the upper branch of the neutral curve 5 decreases with increasing wavenumber and goes to zero at $k = 26$ where the neutral curve of overstability merges with the neutral curve for monotonic disturbances (curves 1, 2). With the increase of the Prandtl number, and hence an increased contribution of viscous dissipation or increased lifetime of thermal disturbances, the region of short capillary waves is shifted towards still shorter waves. Apart from this, one may observe generation of capillary waves in the intermediate range of wavenumbers ($k \sim 1$) under external heating of the free liquid surface. This region is adjacent to the region of the long-wavelength thermocapillary instability.

Figure 18 gives the neutral curves and frequency of neutral oscillations at $Pr = 0.1$ where the representation of curves is identical to that of Fig. 17. The neutral curves and frequencies of even and odd disturbances, as depicted in Fig. 17 and Eq. (37) for $k > 1$, are essentially the same. However, in the liquid layer with $Pr = 0.1$, relative to the case $Pr = 0.01$, we may distinguish between three regions

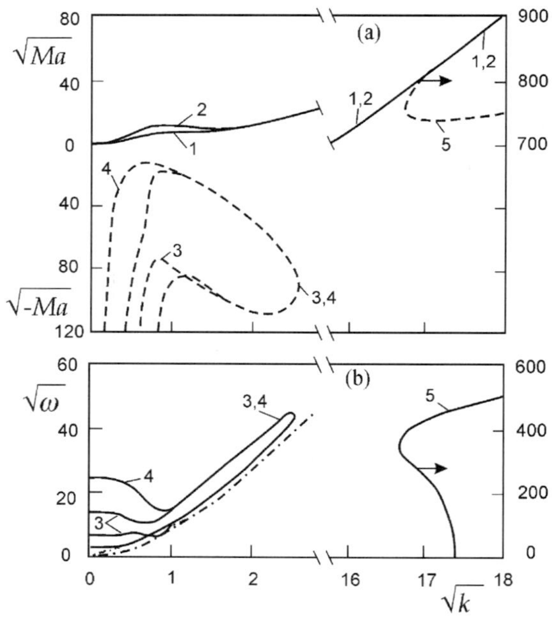

FIG. 18 Surface tension gradient (Marangoni) driven instability of a film or a floating liquid layer with an internal divider. (a) Neutral curves (Ma, k) and (b) frequency (ω, k) of neutral disturbances for a solid divider at $Pr = 0.1$ and $Ca = 10^4$. Symbols are identical to those used in Fig. 17. (From Ref. 6.)

of excitable waves: longitudinal waves ($0 \leqslant k \leqslant 0.5$) and capillary waves ($0.5 < k < 6.3$), initiated by heating from outside, and short capillary waves ($k > 276$), sustained by heating at the divider of the liquid layer. The difference between the layer with a solid divider and that with a solid continuous boundary becomes clear only in the region of long waves, in which one may expect the appearance of waves according to the various thermal conditions on the solid boundary.

2. Divider of Negligible Mechanical Effect

Let us now turn our attention to the case where the divider exerts none or negligible mechanical action on the fluid motion. Accordingly, the divider is an easily pervious partition and serves only as a heat source or sink. The former may

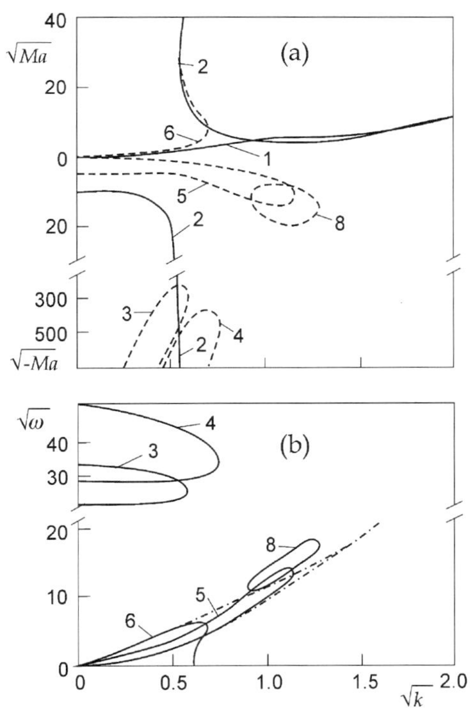

FIG. 19 Surface tension gradient (Marangoni) driven instability of a film or a floating liquid layer with an internal divider. (a) Neutral curves (Ma, k) and (b) frequency (ω, k) of neutral disturbances for divider of negligible mechanical role ($\alpha_n = 0$ and $\alpha_\tau = 0$) at $Pr = 10^{-2}$ and $Ca = 10^4$. Symbols are identical to those used in Fig. 17. (From Ref. 6.)

be the case when heating the liquid with a thin laser sheet or it may be considered as a simplified model of a device with internal heat sources.

In the region of short waves ($k > 10$) the situation is very much like the case of excited surface waves with a solid divider, because such surface disturbances really do not feel the divider.

In the region of intermediate and long waves the oscillations generated at opposite layer outer surfaces are expected to undergo strong hydrodynamic

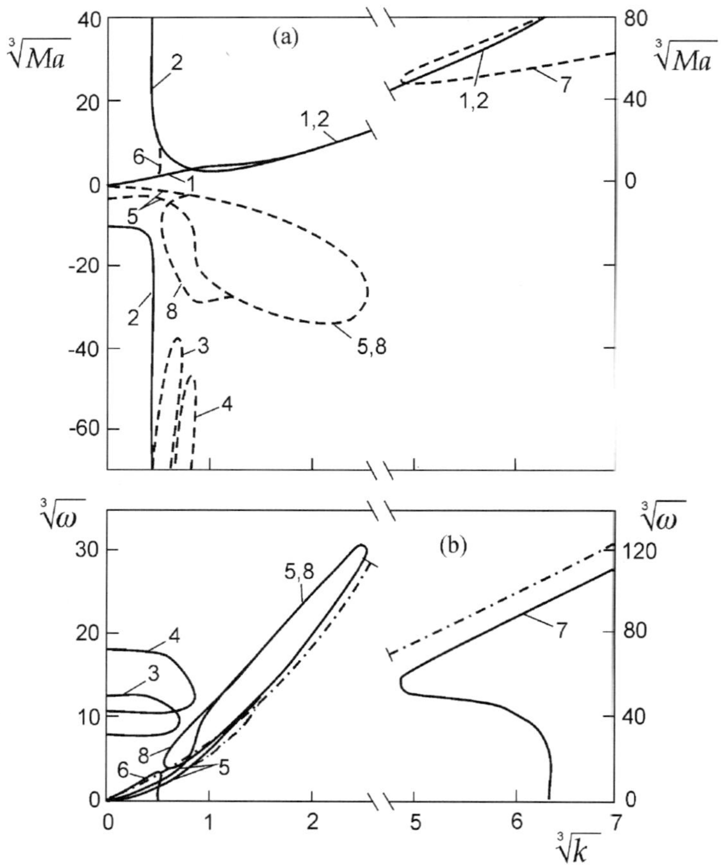

FIG. 20 Surface tension gradient (Marangoni) driven instability of a film or a floating liquid layer with an internal divider. (a) Neutral curves (Ma, k) and (b) frequency (ω, k) of neutral disturbances for a divider of negligible mechanical influence ($\alpha_n = 0$ and $\alpha_\tau = 0$) at $Pr = 10^{-2}$ and $Ca = 10^5$. Symbols are identical to those used in Fig. 17. (From Ref. 6.)

interaction. For a thin isothermal layer, as earlier noted, waves may be classified into two groups: antisymmetrical/flexural waves (even mode) with dispersion relation $\omega^2 = Cak^3 \operatorname{cotanh} k$ and symmetrical/squeezing waves (odd mode) with dispersion relation $\omega^2 = Cak^3 \tanh k$.

The neutral curves for disturbances in the liquid layer with easily pervious partition ($\alpha_n = 0$ and $\alpha_\tau = 0$) are plotted in Fig. 19 for $Pr = 10^{-2}$ and $Ca = 10^4$. The curves describing even and odd disturbances are labeled, respectively, by even and odd numbers. The solid lines at the top of the figure represent the neutral curves for monotonic disturbances. As already noted, at low values of the divider resistance, the neutral curve for even disturbances, at low values of the divider resistance, has two branches corresponding to a long-wavelength region at negative Marangoni numbers and to a short-wavelength region at positive values, respectively. These curves are labeled 2.

The even and odd waves (curves 3, 4) with nonzero frequency at $k = 0$, appear in the region of long waves at high enough (in absolute value) negative Marangoni numbers (external heating). On the other hand, this region also involves the lowest instability threshold for capillary waves (curves 5, 6). The dispersion curves for these waves are consistent with the curves for capillary waves (dashed–dotted lines). Note that the symmetric wave (curve 5) occurs in a rather narrow range of Marangoni numbers, when the liquid layer is heated from outside or there is a cooled divider. In contrast, the antisymmetric waves are observed at positive Marangoni numbers (above the neutral curve 6) and nearly the same wavenumbers, k, corresponding to the onset of monotonic instability for the even mode under internal heating. The upper branch of curve 6 ends at the curve of monotonic instability, labeled 2.

With a liquid layer heated from outside there is also a baglike region of antisymmetric capillary waves with wavenumber $k \sim 1$, for limited values of the Marangoni and wavenumbers (the inner region of curve 8). At $Pr = 0.01$ this region exists only at fairly high capillary numbers ($Ca > 10^4$–10^3), i.e., at relatively small surface deformation. As the capillary number increases the region of excitable capillary waves grows as seen in Fig. 20, which shows neutral curves for $Ca = 10^5$.

It should be emphasized that at $Pr = 10^{-2}$ the capillary and longitudinal waves appear in separate regions. The situation is different with high Prandtl number liquids. Neutral curves for liquids with $Pr = 0.1$ and low divider resistance are displayed in Fig. 21. Here a radically new feature can be observed, namely the appearance of a wide region of sustained capillary waves in the intermediate wavenumber range when heating the layer from outside or when there is a cooled divider ($Ma < 0$). This region overlaps with the region of long longitudinal waves. Note, however, that the lowest instability threshold for long capillary waves governed by the Laplace–Kelvin relation is still valid.

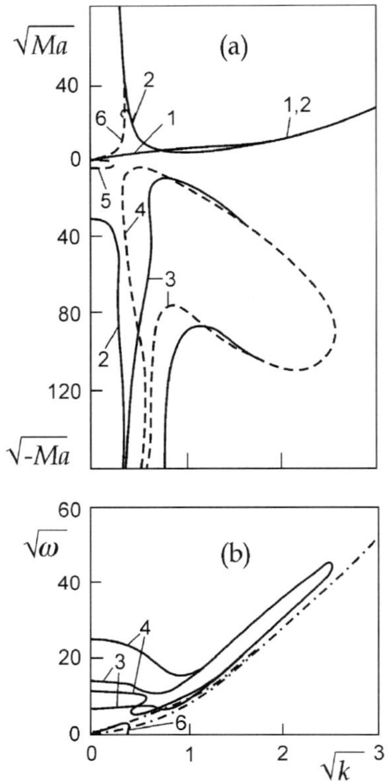

FIG. 21 Surface tension gradient (Marangoni) driven instability of a film or a floating liquid layer with an internal divider. (a) Neutral curves (Ma, k) and (b) frequency (ω, k) of neutral disturbances for a transparent divider ($\alpha_n = 0$ and $\alpha_\tau = 0$) at $Pr = 0.1$ and $Ca = 10^5$. Symbols are identical to those used in Fig. 17. (From Ref. 6.)

REFERENCES

1. Birikh, R.V., Rudakov, R.N. Influence of a permeable partition on convective instability of a horizontal liquid layer. Izv. Akad. Nauk SSSR, Mech. Zhidk. Gaza **1977**, *1*, 157–159 (in Russian).
2. Birikh, R.V., Rudakov, R.N. Convective stability of a horizontal liquid layer with a thin permeable arbitrarily disposed partition. In *Convective Flows*; Zhukhovitsky, E.M., Ed.; PGPI: Perm, 1981; 12–17 (in Russian).
3. Larkin, M.Yu., Sorokin, M.P. Effect of permeable partition on convective stability of a flat horizontal fluid layer. In *Convective Flows*; Zhukhovitsky, E.M., Ed.; PGPI: Perm, 1981; 18–21 (in Russian).

4. Birikh, R.V., Briskman, V.A., Rudakov, R.N., Velarde, M.G. Marangoni–Benard convective instability driven by a heated divider. Int. J. Heat Mass Transfer **1994**, *37*, 493–498.
5. Pearson, J.R.A. On convective cells induced by surface tension. J. Fluid Mech. **1958**, *4*, 489–500.
6. Birikh, R.V., Briskman, V.A., Rudakov, R.N., Velarde, M.G. Marangoni–Benard convective instability of a floating layer with an internal, permeable, heated or cooled divider and two deformable open surface. Int. J. Heat Mass Transfer **1995**, *38*, 2723–2731.
7. Taylor, G.I. The dynamics of thin sheets of fluid. II. Waves on fluid sheets. Proc. Royal Soc. London **1959**, *A253*, 296–312.

7

Thermocapillary Instability of Two-Layer Systems with Liquid–Liquid or Liquid–Gas Interfaces

In earlier chapters we have seen how in a liquid layer, with an undeformable outer surface open to air and temperature-dependent surface tension, a sufficiently high transverse temperature gradient can give rise to stationary convection past an instability threshold as a result of the Marangoni effect, with or without buoyancy. In a two-layer system new flow modes and instabilities are expected not to be encountered in the case of a single free surface instability [1–3], but with too many parameters involved a transparent understanding of the phenomena may be difficult. Thus, to uncover a few features of this complicated problem we shall study, in this chapter, the relatively simple model-problem of thermocapillary instability at the undeformable interface of two liquid layers with similar properties. In fact, as earlier noted, the complete description of the single layer case demands due consideration of the air layer above.

I. INSTABILITY THRESHOLDS IN MODEL SYSTEMS

A. Base State and Equations for Disturbances

Let two layers of immiscible liquids, with respective thicknesses h_1 and h_2, be confined between two solid boundaries at temperatures Θ_1 and $-\Theta_2$ and separated by a flat interface maintained at a temperature taken as a reference value (liquid "1" or "first" layer is placed below). As in earlier cases, the interfacial tension is assumed to depend linearly on the temperature, decreasing with an increase in temperature.

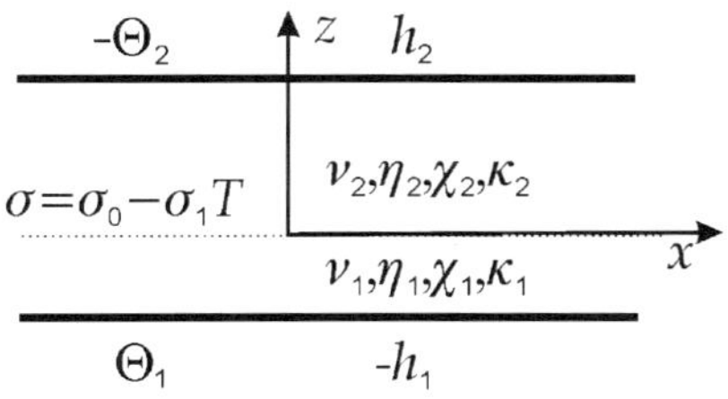

FIG. 1 Thermocapillary instability of a liquid–liquid interface in a two-layer system. Geometry of the problem and the coordinate system.

The layer geometry and the coordinate system are shown in Fig. 1. The base motionless state with linear temperature distribution is

$$\begin{aligned}&\mathbf{v}_0 = 0\\&z < 0: T_0 = -\Theta_1 \cdot z/h_1 = -A_1 z\\&z > 0: T_0 = -\Theta_2 \cdot z/h_2 = -A_2 z\end{aligned} \tag{1}$$

The temperature gradients in the layers $-A_1$ and A_2, respectively, are related as $A_1\kappa_1 = A_2\kappa_2$, where κ denotes liquid heat conductivity.

It is reasonable to scale all quantities using the global characteristics of the system d, ν, η, χ, κ, and Θ defined as $d = h_1 + h_2$, $\nu = \nu_1 + \nu_2$, $\eta = \eta_1 + \eta_2$, $\chi = \chi_1 + \chi_2$, $\kappa = \kappa_1 + \kappa_2$, $\Theta = \Theta_1 + \Theta_2$. We make such a choice to ensure that the description of physical properties for each layer be symmetric and the relative characteristics of the layers, $\tilde{g}_s = g_s/g$ $(s = 1,2)$, vary from 0 to 1. This choice yields the following dimensionless quantities

$$\tilde{t} = \frac{t}{(d^2/\nu)}, \qquad \tilde{\mathrm{v}} = \frac{\mathrm{v}}{(\nu/d)}, \qquad \tilde{T} = \frac{T}{\Theta}, \qquad \tilde{A}_1 = \frac{\tilde{\kappa}_2}{(\tilde{h}_1\tilde{\kappa}_2 + \tilde{h}_2\tilde{\kappa}_1)},$$

$$\tilde{A}_2 = \frac{\tilde{\kappa}_1}{(\tilde{h}_1\tilde{\kappa}_2 + \tilde{h}_2\tilde{\kappa}_1)}, \qquad Pr = \frac{\nu}{\chi}, \quad \text{and} \quad Ma = -\sigma_1\Theta\frac{d}{n\chi}$$

where t is time, v is the liquid velocity, T is the temperature disturbance, Pr is the Prandtl number, Ma is the Marangoni number, which is positive for a system heated on the side of the first layer and negative when cooled there. All other parameters and symbols have their expected meaning in accordance with earlier chapters. To simplify the notation, and as we shall deal with only the dimensionless variables, in what follows the tilde will be omitted.

Small disturbances on the base state are, as usual, expressed as Fourier normal modes,

$$(\mathrm{v}_z, T) = [\mathrm{v}(z), \theta(z)] \exp[i(k_x x + k_y y) - \lambda t)]$$

and obey the following equations (with $\lambda = i\omega$ for neutrally stable, $Re\,\lambda = 0$, modes)

$$-i\omega\left(\frac{\mathrm{d}^2\mathrm{v}}{\mathrm{d}z^2} - k^2\mathrm{v}\right) = \nu_s\left(\frac{\mathrm{d}^4\mathrm{v}}{\mathrm{d}z^4} - 2k^2\frac{\mathrm{d}^2\mathrm{v}}{\mathrm{d}z^2} + k^4\mathrm{v}\right) \tag{2}$$

$$-i\omega\theta = Pr^{-1}\chi_s\left(\frac{\mathrm{d}^2\theta}{\mathrm{d}z^2} - k^2\theta\right) + A_s\mathrm{v} \tag{3}$$

where $k^2 = k_x^2 + k_y^2$, $s = 1, 2$, and with boundary conditions

$$z = -h_1, h_2 : \mathrm{v} = \frac{\mathrm{d}\mathrm{v}}{\mathrm{d}z} = \theta = 0$$

$$z = 0 : \mathrm{v}_1 = \mathrm{v}_2 = 0, \qquad \frac{\mathrm{d}\mathrm{v}_1}{\mathrm{d}z} = \frac{\mathrm{d}\mathrm{v}_2}{\mathrm{d}z}, \qquad \theta_1 = \theta_2, \qquad \kappa_1\frac{\mathrm{d}\theta_1}{\mathrm{d}z} = \kappa_2\frac{\mathrm{d}\theta}{\mathrm{d}z}$$

$$\eta_1\frac{\mathrm{d}^2\mathrm{v}_1}{\mathrm{d}z^2} - \eta_2\frac{\mathrm{d}^2\mathrm{v}_2}{\mathrm{d}z^2} = -k^2 Pr^{-1} Ma\,\theta_1 \tag{4}$$

To determine the critical Marangoni number, the corresponding critical wavenumber, and the shape of the neutral stability curve, an exact solution of the boundary value problem of Eqs. (2)–(4) has been found, and also a step-by-step integration method was used to solve the problem.

B. Solution of the Boundary-Value Problem

The step-by-step method of integration for the boundary problem of Eqs. (2)–(4) is further described in Chapter 11. Although this numerical procedure uses more computer time it leads to a well-posed algebraic system of equations and it may be used to cross-check analytical results.

Here we give the exact solution of the boundary problem [4]. The general solution of Eqs. (2) and (3) satisfying the boundary conditions of Eq. (4) may be written as

$$-h_1 \leq z < 0:$$

$$\mathrm{v}_1 = c_1\left(\frac{\sinh k(z+h_1)}{k} - \frac{\sinh l_1(z+h_1)}{l_1}\right) + c_2[\cosh k(z+h_1) - \cosh l_1(z+h_1)]$$

$$\theta_1 = -c_1\frac{a_1}{i\omega}\left(\frac{\sinh k(z+h_1)}{k} + \frac{Pr}{\chi_1/\nu_1 - Pr}\frac{\sinh l_1(z+h_1)}{l_1}\right) - c_2\frac{a_1}{i\omega}\left(\cosh k(z+h_1) + \frac{Pr}{\chi_1/\nu_1 - Pr}\cosh l_1(z+h_1) - \frac{\chi_1/\nu_1}{\chi_1/\nu_1 - Pr}\cosh m_1(z+h_1)\right) + c_3 \sinh m_1(z+h_1) \tag{5}$$

$$0 < z \leq h_2:$$

$$\mathrm{v}_2 = c_4\left(\frac{\sinh k(z-h_2)}{k} - \frac{\sinh l_2(z-h_2)}{l_2}\right) + c_5[\cosh k(z-h_2) - \cosh l_2(z-h_2)],$$

$$\theta_2 = -c_4\frac{a_2}{i\omega}\left(\frac{\sinh k(z-h_2)}{k} + \frac{Pr}{\chi_2/\nu_2 - Pr}\frac{\sinh l_2(z-h_2)}{l_2}\right) - c_5\frac{a_2}{i\omega}\left(\cosh k(z-h_2) + \frac{Pr}{\chi_2/\nu_2 - Pr}\cosh l_2(z-h_2) - \frac{\chi_2/\nu_2}{\chi_2/\nu_2 - Pr}\cosh m_2(z-h_2)\right) + c_6 \sinh m_2(z-h_2) \tag{6}$$

The coefficients c_i are determined by the boundary conditions of Eq. (4). The solvability condition of the homogeneous system of equations for the complex unknowns, c_i, provides a determinental equation involving the Marangoni number. The requirement that the Marangoni number be real defines the frequency of neutral oscillations. Once the critical Marangoni number and the frequency are obtained, the system of equations for the c_i coefficients was solved and the disturbance amplitudes v(z) and θ(z) were calculated.

The transverse, V, and longitudinal liquid velocity, U, and temperature distribution, T, for monotonic instability ($\omega = 0$) in the layer are then calculated using the following relations:

$$V = \text{v} \cos kx$$

$$U = -\frac{1}{k}\text{v}' \sin kx = u \sin kx$$

$$T = \theta \cos kx.$$

The structure of oscillatory disturbances is most conveniently examined for the case of standing waves obtained by putting together the two (left and right) running waves

$$\begin{aligned} V &= Re(\text{v}_z) = Re\{\text{v} \exp[i(kx - \omega t)] + \text{v} \exp[i(-kx - \omega t)]\} \\ &= A \sin(\omega t + \varphi_\text{v}) \cos kx \end{aligned}$$

$$\begin{aligned} U &= Re(\text{v}_x) = Re\left\{-\frac{1}{ik}\text{v}' \exp[i(kx - \omega t)] + \frac{1}{ik}\text{v}' \exp[i(-kx - \omega t)]\right\} \\ &= B \sin(\omega t + \varphi_\text{u}) \sin kx \end{aligned}$$

$$T = Re\{\theta \cdot \exp[i(kx - \omega t)] + \theta \cdot \exp[i(-kx - \omega t)]\} = C \sin(\omega t + \varphi_\theta) \cos kx$$

where the amplitudes and phases are

$$A = \sqrt{\text{v}_\text{r}^2 + \text{v}_\text{i}^2}, \qquad B = \frac{1}{k}\sqrt{\text{v}_\text{r}'^2 + \text{v}_\text{i}'^2}, \qquad C = \sqrt{\theta_\text{r}^2 + \theta_\text{i}^2},$$

$$\sin \varphi_\text{v} = \text{v}_\text{r}/A, \qquad \cos \varphi_\text{v} = \text{v}_\text{i}/A, \qquad \sin \varphi_u = \frac{-\text{v}_\text{r}'}{kA},$$

$$\cos \varphi_u = \frac{-\text{v}_\text{i}'}{kA}, \qquad \sin \varphi_\theta = \frac{\theta_\text{r}}{C}, \qquad \cos \varphi_\theta = \frac{\theta_\text{i}}{C}$$

C. Neutral Curves and Structure of Disturbances

For simplicity we consider a two-layer model in which the liquids in the layers have similar properties and hence the only dimensionless parameter specifying the difference between two layers is the difference in layer thickness. Figure 2 shows the critical Marangoni number, Ma_*, minimized with k, and the wavenumber, k_*, for monotonic disturbances (curves 1) and Ma_*, k_* and frequency, ω, for oscillatory disturbances (broken line and lines 2, $Pr = 0.4$) plotted against the first layer thickness h_1.

At $h_1 = 0.5$ the layers are identical and there is no instability. At $h_1 \neq 0.5$ both modes of instability may appear depending on the direction of heating. For example, the neutral curves show that there is monotonic instability at small h_1, when heating on the side of the thicker layer 2 ($Ma < 0$), while there is oscillatory

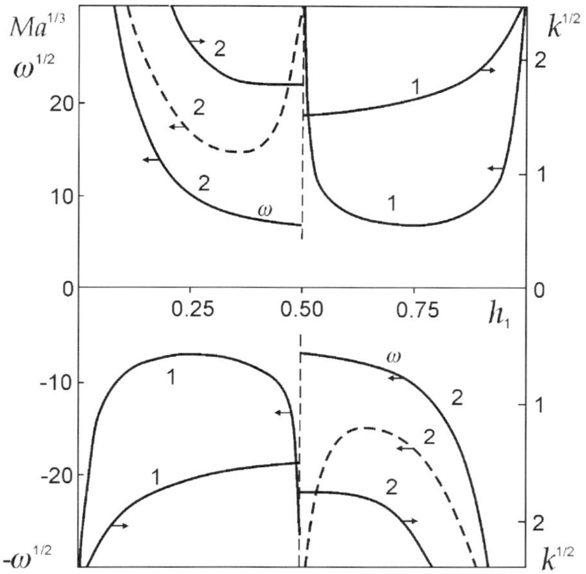

FIG. 2 Thermocapillary instability of a liquid–liquid interface in a two-layer system. Critical Marangoni number (for oscillatory disturbances, broken line), wavenumber (ordinate on the right), and frequency vs. thickness, h_1, of the lower layer for monotonic (curves 1) and oscillatory (curves 2) disturbances. (From Ref. 5.)

instability when heating on the thinner layer side 1 ($Ma > 0$). This corresponds to a mechanism of interaction between the thermal and hydrodynamic disturbances in the liquid, in which the temperature at the interface is determined by the motion in the thicker layer. The mechanism of monotonic instability in a two-layer system is similar to that in a single-layer system with free surface. Therefore we shall concentrate only on the case of oscillatory instability.

The structure of oscillatory disturbances is conveniently investigated focusing on the standing wave problem. Figure 3 presents the structure of the convective flow in different phases of the stationary wave. The arrows denote the liquid velocity vectors (the length of the arrows is proportional to the value of the velocity).

The phase $\varphi = 0^\circ$ corresponds to a maximum value of the longitudinal temperature gradient at the interface and the maximum surface tension gradient-driving force. In this phase flow motions develop in both layers very near to the interface and have practically the same intensity, which is not yet maximum. A purely inertial motion (the interface is isothermal) is shown in the figure for $\varphi = 90^\circ$.

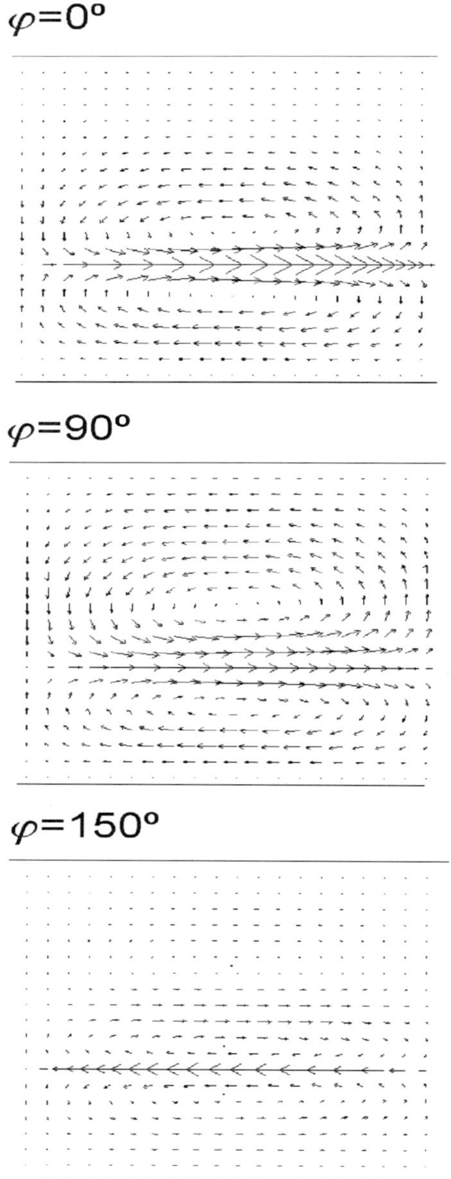

FIG. 3 Thermocapillary instability of a liquid–liquid interface in a two-layer system. Flow structure at different phases of a standing wave. The size of all arrows is proportional to the actual value of the velocity at a given point. (From Ref. 4.)

At phase $\varphi = 130^\circ$ the inertial flow has generated a high enough temperature gradient in a counterflow and the thermocapillary force stops the liquid motion at the interface. There then comes a point of flow rearrangement within the convective cell. The newly born structures are seen in the figure for $\varphi = 150^\circ$. From the very beginning the reversed flow at the liquid–liquid interface gives rise to new vortices, and for a rather long period of time the thicker layer exhibits a two-layer structure that is incapable of providing an effective heat transfer.

Variations of the flow intensity in the layers and the interfacial temperature gradient are shown in Fig. 4. As already mentioned, the zero phase corresponds to a maximum horizontal temperature gradient at the interface. For positive Marangoni numbers a thermocapillary flow in the first layer sustains the existing temperature gradient while the flow in the second layer causes its decrease. Because after $\varphi = 0^\circ$ the flow in the second thicker layer becomes stronger, the interfacial temperature gradient decreases and vanishes at $\varphi = 90^\circ$. The inertial motion going on under its own momentum generates at the interface a temperature gradient with the opposite sign.

At $\varphi = 130^\circ$ the negative temperature gradient becomes so high, in absolute value, that thermocapillary forces suppress interfacial motion and a number of countervortices are generated there. In the thinner layer the direction of motion is reversed faster and the arising flow sustains the temperature gradient at the interface. In the second (thicker) layer the earlier motion responsible for a negative temperature gradient exists until $\varphi = 160^\circ$. Between the phases 130° and 160° a two-vortex structure is generated and heat transfer through the layer is again ineffective. Only at $\varphi = 180^\circ$ is the thermal effect of each layer equilibrated and

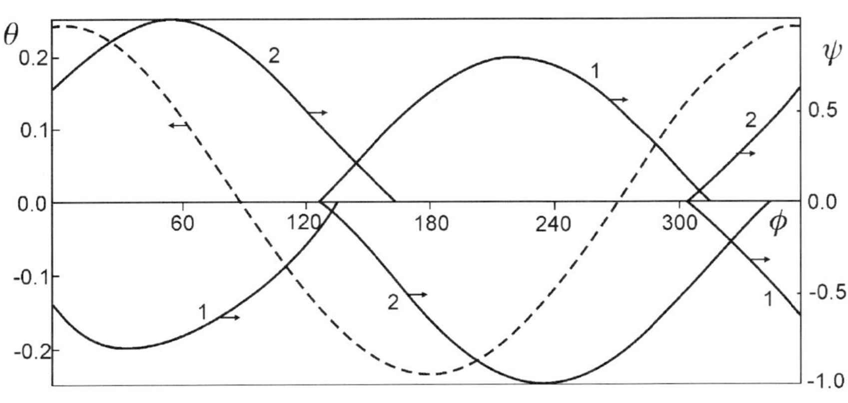

FIG. 4 Thermocapillary instability of a liquid–liquid interface in a two-layer system. Variation of the flow intensity (solid lines) in the layers (ψ stream function, ordinate on the right) and temperature gradient (broken line) on the interface. (From Ref. 5.)

the temperature gradient does not grow any longer. Then a decisive part is again played by the flow in the thicker layer, which leads to a decrease of the temperature gradient at the interface.

Thus, the role of the thinner layer in the generation of undamped oscillations is to sustain temperature disturbances at the interface in such phases when the motion in the thicker layer reverses its direction and there exists a two-vortex structure.

II. THERMOCAPILLARY INSTABILITY IN REAL LIQUIDS SYSTEMS

To illustrate our study, let us consider two particular cases, water–benzene and *n*-heptane–air systems. The first system consists of liquids with slightly different properties. While the second contains fluids with dissimilar properties and approximately corresponds to the single layer model with free surface. The monotonic thermocapillary instability in the water–benzene system was studied by Zeren and Reynolds [7]. This system is interesting in the sense that for some layer thickness ratios monotonic thermocapillary instability is expected when heating on either of the two external boundaries of the system.

A. Water–Benzene System

Let us consider the neutral curves and flow motion modes for the water–benzene system at mean temperature 20°C. The corresponding dimensionless parameters characterizing the liquid properties have the following values: $\nu_1 = 0.574$, $\eta_1 = 0.605$, $\chi_1 = 0.594$, $\kappa_1 = 0.803$, and $\alpha_1 = 0.128$ (α denotes the thermal expansion coefficient). Here and in the following the index 1 refers to the water layer, and the index 2 to the benzene layer. For the system under consideration the Prandtl number is equal to 7.29.

In Fig. 5 the lowest critical Marangoni number, Ma_*, is plotted vs. the layer thickness, h_1 (solid and broken lines denote monotonic and oscillatory instability, respectively), for the corresponding wavenumber, k_*, and the oscillatory disturbance frequency, ω_*. At small h_1 an additional convective heat transfer to the interface is mainly a result of motion in the thicker benzene layer. Therefore, monotonic instability is observed when the system is heated on the side of the second layer ($Ma < 0$, curve 1) whereas cooling on the side of the same layer ($Ma > 0$, curve 2) gives rise to oscillatory instability.

When the value of h_1 increases the layers exchange the roles they play for the onset of instability (curves 3, 4). Unlike the model-problem described in the preceding chapter a change in the instability modes does not occur at $h_1 = 0.5$, but rather it is shifted towards the now thicker water layer ($h_1^* \approx 0.55$), as a result of the higher thermal conductivity of water. However, there is a region of short-

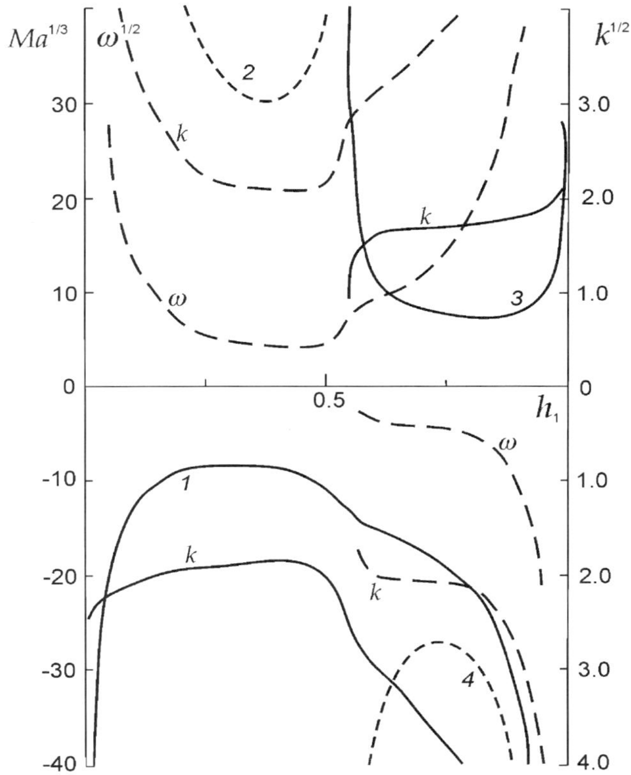

FIG. 5 Thermocapillary instability of a liquid–liquid interface in a two-layer system. Critical Marangoni numbers for monotonic (solid lines) and oscillatory (broken lines) instability and the corresponding wavenumbers, k, (ordinate on the right), and frequencies, ω vs. thickness, h_1, of the bottom water layer. (From Ref. 6.)

wavelength disturbances, with the wavelength comparable to the thickness of the benzene layer, where the mechanism of instability is still related to the motion in this layer (curves 1, 2). This peculiarity also distinguishes the system under consideration from the earlier studied model-problem. For monotonic instability the structure of motions at threshold for large values of h_1 is shown in Fig. 6.

The distributions of longitudinal velocity component u (solid lines) and temperature θ (broken lines) are given for $h_1 = 0.7$. At $k = 3$ ($Ma = 544$, curves 1) the flow intensity and the amplitude of temperature disturbances are higher in the water layer and the motion in this layer is defined by the sign of the interface temperature. For large values of the wavenumber ($k = 10$, $Ma = -6745$, curve 2) the leading role of the benzene layer in the instability mechanism is revealed.

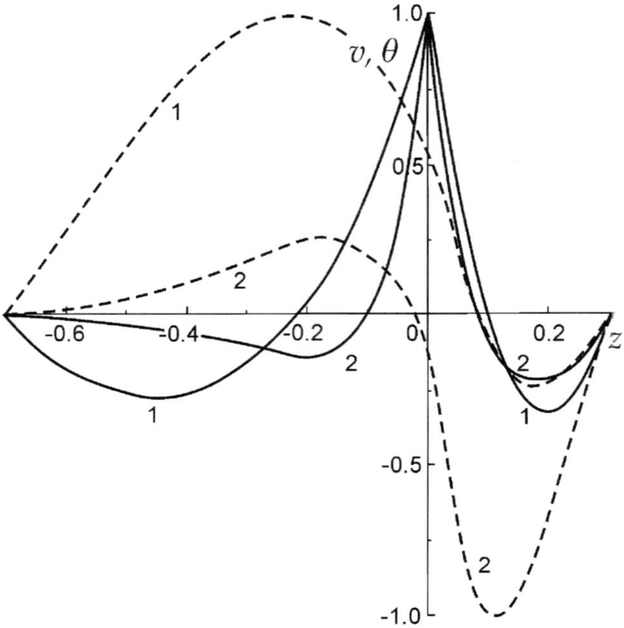

FIG. 6 Thermocapillary instability of a liquid–liquid interface in a two-layer system. Profiles of longitudinal velocity component (solid lines) and temperature (broken lines) plotted along the transverse coordinate for $h_1 = 0.7$: Curve 1 for $k = 3$, $Ma = 544$; curve 2 for $k = 10$, $Ma = -6745$. (From Ref. 6.)

There is higher amplitude of the temperature disturbances in this layer, the flow intensity in the layers being practically the same.

In the region of small h_1, there is no short-wavelength monotonic instability caused by the motion in the thinner water layer, which may be caused by the fact that the dissipative properties of water are less favorable for the growth of disturbances.

The analysis of dispersion relations, $\omega(k)$, shows that the oscillatory instability is related to the elasticlike action of the Marangoni effect. A distinguishing feature of these oscillations is the independence of frequency on the wavenumber in the region of small k, a dispersionless character.

The intensity of motion at different phases of neutral oscillatory disturbances for $h_1 = 0.7$ is illustrated in Fig. 7. This disturbance has the wavenumber $k = 4$ and is excited at $Ma = -21{,}392$ (heating on the benzene side). The solid lines show the time variation of the stream function maximum in the water layer (curve 1) and in the benzene layer (curve 2). The broken line describes the variation in the temperature gradient at the liquid–liquid interface.

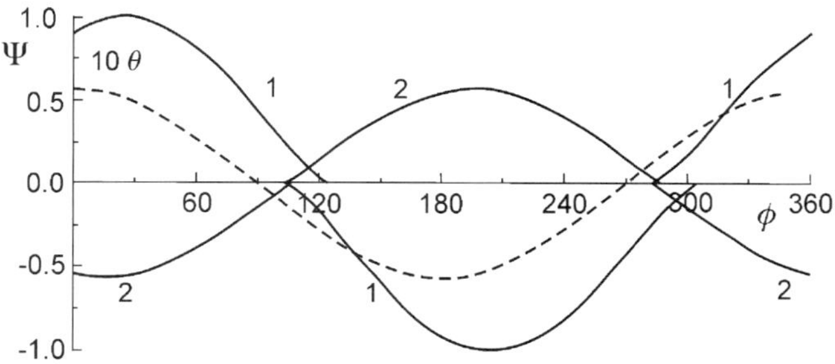

FIG. 7 Thermocapillary instability of liquid–liquid interface in a two-layer system. Variation of the flow intensity (solid lines) using the stream function and temperature gradient at the interface (broken line) in a standing wave for the water–benzene system. (From Ref. 6.)

The phase $\varphi = 0^\circ$ has been chosen to ensure that the temperature gradient along the boundary is maximum when the thermocapillary force acting on the surface is also maximum and there exists a well-developed flow motion in the form of two vortices aligned with the force.

Further, as the flow intensity in the thicker water layer increases the temperature gradient drops. Because of viscous dissipation, the intensity of flow motions in the layers is reduced with the increase of the temperature gradient. At phase $\varphi = 90^\circ$ because of a stronger flow in the water layer the sign of the temperature gradient and the direction of the interfacial force change. At phase $\varphi \approx 102^\circ$ a new pair of vortices appears at the interface. The flow in the thicker water layer changes, becoming a double-vortex structure with lowered heat transfer, while the flow in the thinner benzene layer involves a single rapidly twisting vortex, which sustains the temperature growth at the interface. In the vicinity of $\varphi = 180^\circ$ the thicker water layer recovers its leading role in the formation of the interface temperature, and the temperature gradient falls.

Figure 7 illustrates the contribution of each layer to the onset of undamped oscillations. To support their existence the system must be cooled on the side of the layer capable of providing the highest convective heat transfer.

B. *n*-Heptane–Air System

The *n*-heptane–air system involves fluids with dissimilar properties and it is usually treated as a single-layer system (*n*-heptane) with free surface as air is expected to play no active mechanical role. The choice of *n*-heptane is motivated by considerations of its suitability for experimental study, although the results are expected to be the same for other organic liquids (toluene, acetone, benzene, etc.) and water. The dimensionless parameters of *n*-heptane in the two-layer system at temperature 20°C and normal pressure have the following values: $\nu_1 = 0{,}040$, $\eta_1 = 0{,}958$, $\chi_1 = 0{,}004$, $\kappa_1 = 0{,}840$.

Monotonic instability arises only if the system is heated on the side of the n-heptane ($Ma > 0$). The minimum critical Marangoni number for the n-heptane layer of thickness $h_1 = 0.4$ is equal to 8.4 and reaches this value when the wavenumber is $k = 4.9$ (in terms of the characteristics of the liquid layer only, we obtain $Ma_* = 98.9$ and $k_* = 1.96$). This corresponds to the earlier recalled result (Chapter 5) found by Pearson for a single-layer system [9]. The overestimated value of the Marangoni number may be a result of the influence of the convective heat transfer in the layer of air.

Compared to a single-layer system, the n-heptane–air system, when heated on the side of air ($Ma < 0$), might give rise to oscillatory instability. The minimum critical Marangoni number is $|Ma| = 7900$. The wavenumber and the frequency of critical disturbances are 3.7 and 8.8, respectively. For a liquid depth of about 5 mm in the n-heptane the dimensional frequency of oscillatory disturbances is about 1 Hz. As shown in the first section of this chapter, undamped oscillations in a two-layer system with a flat interface occur as a result of the existence of the oscillatory phase, in which the motion in the heated layer defines the interface temperature. Because the heat diffusivity of air is relatively higher than that of n-heptane, thermal disturbances in the air layer decay faster, and the Marangoni number should be fairly high (in absolute value) to sustain the undamped oscillations.

III. COMBINED SURFACE TENSION GRADIENT-DRIVEN AND BUOYANCY-DRIVEN INSTABILITY OF A TWO-LAYER SYSTEM

As discussed earlier, in experiments on Earth the mechanical equilibrium of a nonuniformly heated fluid in a two-layer system can be destroyed for two reasons: the dependence of density of fluids on temperature and the dependence of the interfacial tension on temperature. When both agents act together, to obtain the instability threshold we must solve a multiparameter boundary-value problem. Let us simply, using heuristic arguments only, try to understand what mechanism generates flow motion, in what areas it is generated, and also what areas are passive. The mechanism of excitation of fluid flow depends on the ratios of fluid parameters and the layer depth ratio. For example, when the latter changes, the driving force for convection, and the type of convection, can change. As noted earlier for low layer depths, surface tension gradient (Marangoni) driven convection dominates while for deeper fluid layers, buoyancy (Rayleigh) driven convection dominates. Narayanan and colleagues have carefully analyzed these problems [10–12].

As equilibrium is lost when the driving force overcomes dissipative processes, the thinner layer in this case appears as a passive medium merely augmenting the dissipation in the system. When the fluid layers have a large depth and close fluid properties then the bulk flow in the layers arises practically simultaneously and can be supported by either viscous or thermal coupling between the layers. Finally, for thinner layers with asymmetry in fluid properties or difference in layer depths the surface tension gradient drives the flow.

A clear indication of what is occurring in a two-layer system can be inferred from the behavior at the fluid–fluid interface disregarding the bulk. Four different interfacial structures have been identified for any given convecting two-layer system with a deformable interface [10]. These interfacial structures are shown in Fig. 8. Each of these structures depends on whether fluid was flowing into or away from the trough or the crest, and whether the fluid was hotter or cooler at the trough or crest of the interface.

Because, generally, interfacial tension decreases with increasing temperature, at cooler interface regions the interfacial tension will be higher and stretching the interface. Where the interface is hotter, the interfacial tension will be lower causing the fluid to move away from warmer regions. One reason the interface deforms is because of bulk convection caused by buoyancy pushing against the interface. Consider two fluids whose dynamic (shear) viscosities are close. If buoyancy-driven convection occurs mostly in the lower layer, then the liquid will flow up from the lower layer into a crest, while if it flows down from the top layer it moves into a trough, and hence buoyancy-driven convection is expected to occur mostly in the upper fluid. An interface also deforms as a result of variations in the interfacial tension.

In each of the four cases, the interfacial structure can be used to indicate the driving force of the convection. In the first interfacial structure, the dominating driving force is the surface tension gradient. Indeed, the cold fluid with the higher interfacial tension pulls the interfacial fluid into the crest. In the second interfacial structure, buoyancy drives convection in the lower phase. The hot, rising fluid pushes the interface upwards. As the fluid moves along the interface, it cools and eventually sinks back down. The third interfacial structure is dominated by

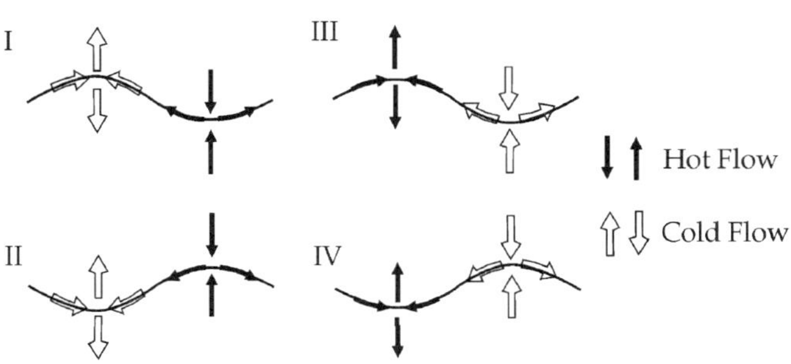

FIG. 8 Combined surface tension gradient (Marangoni) driven and buoyancy (Rayleigh) driven instability of a two-layer system. The four possible interfacial structures at a fluid–fluid interface. (From Ref. 10.)

buoyancy-driven convection in the upper phase or by surface tension gradient-driven convection as the interfacial tension increases with increasing temperature (the nonstandard case). The fourth interfacial structure occurs in the rare instance where the lower fluid has density increasing with an increase in the temperature (e.g., water below 4°C), causing the cooler, lower fluid to flow up into a crest.

Let us consider the evolution of small neutral perturbations on the mechanical equilibrium of two plane horizontal fluid layers with a common interface and linear distribution of temperature in the transverse direction to gravity. In the equations for the amplitudes (2) and (3) we shall add buoyancy and we shall consider only monotonic disturbances ($\lambda = 0$). Using the notation of the first section of Chapter 7, and $\alpha = \alpha_1 + \alpha_2$, $\beta_s = \alpha_s/\alpha$, these equations become

$$\nu_s\left(\frac{\mathrm{d}^4\mathrm{v}}{\mathrm{d}z^4} - 2k^2\frac{\mathrm{d}^2\mathrm{v}}{\mathrm{d}z^2} + +k^4\mathrm{v}\right) - Ra\ k^2 Pr^{-1}\beta_s\theta = 0 \tag{7}$$

$$Pr^{-1}\chi_s\left(\frac{\mathrm{d}^2\theta}{\mathrm{d}z^2} - k^2\theta\right) + a_s\mathrm{v} = 0 \tag{8}$$

$$Ra = g\alpha\Theta d^3/\nu\chi, \qquad k^2 = k_x^2 + k_y^2, \qquad s = 1, 2 \tag{9}$$

where Ra is the Rayleigh number, and α is the thermal expansion coefficient. The dimensionless temperature gradients are related: $a_1 = \kappa_2/(h_1\kappa_2 + h_2\kappa_1)$, $a_2 = \kappa_1/(h_1\kappa_2 + h_2\kappa_1)$.

Boundary conditions are taken as in the preceding case but accounting for a deformable interface. For monotonic perturbations we obtain

$$z = -h_1, h_2 : \mathrm{v} = \frac{\mathrm{d}\mathrm{v}}{\mathrm{d}z} = \theta = 0 \tag{10}$$

$$z = 0 : \mathrm{v}_1 = \mathrm{v}_2 = 0, \qquad \frac{\mathrm{d}\mathrm{v}_1}{\mathrm{d}z} = \frac{\mathrm{d}\mathrm{v}_2}{\mathrm{d}z} \tag{11}$$

$$\theta_1 - a_1\xi = \theta_2 - a_2\xi, \qquad \kappa_1\frac{\mathrm{d}\theta_1}{\mathrm{d}z} = \kappa_2\frac{\mathrm{d}\theta_2}{\mathrm{d}z} \tag{12}$$

$$\eta_1\left(\frac{\mathrm{d}^2\mathrm{v}_1}{\mathrm{d}z^2} + k^2\mathrm{v}_1\right) - \eta_2\left(\frac{\mathrm{d}^2\mathrm{v}_2}{\mathrm{d}z^2} + k^2\mathrm{v}_2\right) = -k^2 Pr^{-1} Ma(\theta_1 - a_1\xi) \tag{13}$$

$$\eta_1\left(\frac{\mathrm{d}^3\mathrm{v}_1}{\mathrm{d}z^3} - 3k^2\frac{\mathrm{d}\mathrm{v}_1}{\mathrm{d}z}\right) - \eta_2\left(\frac{\mathrm{d}^3\mathrm{v}_2}{\mathrm{d}z^3} - 3k^2\frac{\mathrm{d}\mathrm{v}_2}{\mathrm{d}z}\right)$$
$$= (Cak^4 + k^2 Ga\ Pr^{-1})\xi \tag{14}$$

where ξ denotes the dimensionless deformation amplitude of the interface.

The problem of Eqs. (7)–(14) contains three independent dimensionless parameters: Ra, Ma, Ca, $Ga = (\rho_1 - \rho_2)gd^3/\eta\chi$, and Pr, relative parameters of the fluid layers and wavenumber. A significant dimensionless parameter of the problem is once more the *dynamic* Bond number, $Bd \equiv Ra/Ma = \rho g\alpha d^2/\sigma_1$. The controllable quantity in the dynamic Bond number is the total width, d, of the system.

A. Marangoni–Rayleigh Instability of the Water–Benzene System

Let us discuss the stability results obtained for the two-layer system benzene over water [7,13]. Using the low layer fluid characteristics, the Rayleigh and Marangoni numbers are now

$$Ra_1 = \frac{g\alpha_1 A_1 d_1^4}{\nu_1 \chi_1}, \qquad Ma_1 = \frac{\alpha_1 A_1 d_1^2}{\eta_1 \chi_1}$$

where, as earlier, quantities concerning the lower fluid are marked by the index "1." The new parameters are related to the earlier defined ones as

$$Ra_1 = -Ra\left[\frac{\kappa^*(1+\nu^*)(1+\chi^*)}{h_2 + \kappa^* h_1}\right],$$

$$Ma_1 = -Ma\left[\frac{\kappa^*(1+\eta^*)(1+\chi^*)}{h_2 + \kappa^* h_1}\right]$$

where the asterisk, *, denotes physical property ratios of the upper to the lower fluid.

Calculations were made for layers of benzene over water with a total depth $d = 2\,\text{mm}$ and various water depth fractions h_1. The Marangoni number as a function of wavenumber when the heating is done from above is shown in Fig. 9.

The system appears to respond in modes depending on the water depth fraction h_1. As the depth fraction increases from 0.1, the system is most unstable to disturbances of moderate wavenumber, corresponding to disturbance scales of the order of the total depth. As h_1 increases to about 0.4 the configuration becomes stable for all but a narrow wavenumber range. For h_1 exceeding about 0.48, the minimum Marangoni number for moderate wavenumbers is greater than 10^5, and the least stable state then corresponds to large wavenumber, or small-scale, disturbances. As h_1 increases beyond about 0.65, the critical Marangoni number is associated with $k = 0$, a long-wavelength disturbance. Asymptotic values of the critical Marangoni number for $k \to 0$ are shown as vertical marks along the $(k = 0.1)$ abscissa.

Comparison of the critical Marangoni numbers obtained for purely thermocapillary flow [14] and Fig. 9 shows that the addition of a stable density

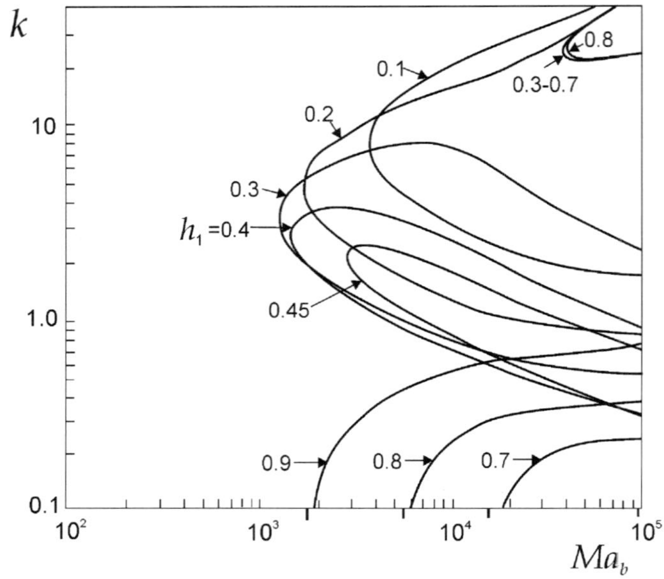

FIG. 9 Combined surface tension gradient (Marangoni) driven and buoyancy (Rayleigh) driven instability of a two-layer system. Monotonic instability neutral curves for various water depth fractions when heating from above: The benzene is over the water; total depth is $d = 2$ mm. (From Ref. 7.)

gradient increases the critical Marangoni number for moderate and large wavenumber disturbances. There density has little effect on the value of the critical Marangoni number at $k = 0$.

The variation of the critical Marangoni number with depth fraction and eigenfunctions $\hat{w}(z) \sim$ v, $\hat{u}(z) \sim$ v′, and $\hat{\theta}(z) \sim \theta$ at the neutral states (except for the low wavenumber cases where they are calculated for $k = 0.1$), and the energy integral terms, E_b (total buoyancy integral) and E_s (surface power term), are shown in Fig. 10.

Each of these eigenfunctions has been suitably rescaled to fill the diagrams. The total buoyancy integral, E_b, and the surface power term, E_s, are given as percentages of the total dissipation integral, usually rounded off to the nearest integer value. The surface deformation is qualitatively indicated by S. For cases where surface deformation occurs, $S \neq 0$, the perturbation temperature eigenfunctions are not continuous across $z = 0$. However, the temperature is continuous across the interface.

The eigenfunction curves show surface deformation for neutral states with small wavenumber while no surface deformation occurs for moderate wavenumber states. The small wavenumber, large-scale disturbances act like tipping on the

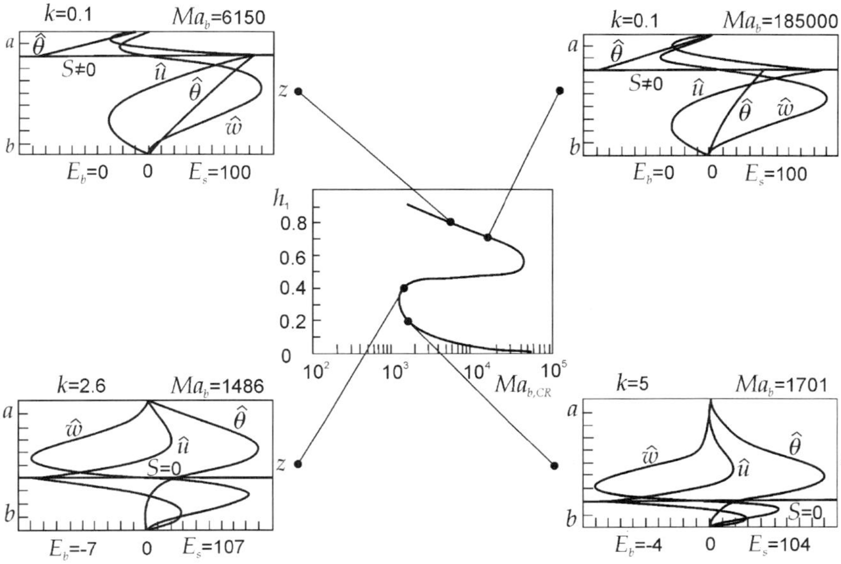

FIG. 10 Combined surface tension gradient (Marangoni) driven and buoyancy (Rayleigh) driven instability of a two-layer system. Variation of critical Marangoni number, eigenfunctions, and energy integrals with depth fraction when heating from above: Benzene over water; total depth is $d = 2\,\text{mm}$. E_b = buoyancy energy integral, E_s = surface power term, as percentage of dissipation integral. Here S qualitatively indicates surface deformation. (From Ref. 7.)

surface. The surface tension gradient brings hot fluid into a crest and, consequently, cold flow into a trough. In this case we have the second interface structure, although buoyancy does not drive convection in the lower phase. The temperature eigenfunctions for these disturbances are nearly linear, indicating a minimum response to surface motion. As shown by the energy terms, there is very little energy drain from the surface power input to the stirring of the stable density stratification for these disturbances. This may justify the slight differences between the critical Marangoni numbers obtained with and without the density gradient. On the other hand, there is a noticeable energy drain to the buoyancy field for the neutral states corresponding to moderate wavenumbers and this increases as h_1 increases.

The sensitivity of the moderate wavenumber disturbances to depth fraction around $h_1 = 0.5$ is presumably a result of the increased energy demand indicated in Fig. 10. The surface mechanism is unable to supply enough energy to drive these relatively large-scale motions against the stabilizing density gradient and hence small-scale large wavenumber motions with no surface deformation are excited.

Let us now consider the benzene–water system when the heating is done from below, hence from the bottom of the water layer. For this case the Marangoni number is shown as a function of wavenumber in Fig. 11. When heating from below the Marangoni number, Ma_b, is negative. Again there seem to be three response modes. Steady patterned convection curves for h_1 increasing from 0.1 up and for moderate wavenumbers is shown with solid lines. In the same depth fraction range a second mode has a minimum Marangoni number at $k = 0$, which is always higher than the minimum Marangoni number for the first mode. A third mode occurs for h_1 greater than about 0.6, with the minimum Marangoni number corresponding to moderate wavenumber. The most striking difference relative to the pure (buoyancy-free) thermocapillary case is that the first mode does not appear there, hence indicating that buoyancy brings instability. The second and third modes apparently correspond to thermocapillary convection. The minimum Marangoni numbers are only slightly displaced for these modes. In fact the stability curves for $h_1 = 0.7$, 0.8, and 0.9, with and without the buoyancy, fall on

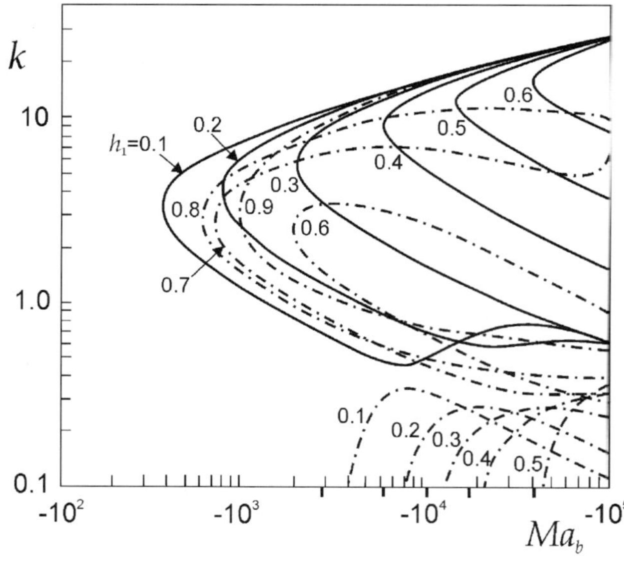

FIG. 11 Combined surface tension gradient (Marangoni) driven and buoyancy (Rayleigh) driven instability of a two-layer system. Neutral curves for monotonic instability with various water depth fractions when heating from below: Benzene over water; total depth is $d = 2$ mm. Solid lines, buoyancy mode; dot–dashed line, surface deformation mode for $h_1 = 0.1$–0.5, Marangoni effect mode for $h_1 = 0.6$–0.9. (From Ref. 7.)

top of one another except at wavenumbers greater than about three times the critical wavenumber. The lowering of the critical Marangoni number for low wavenumber disturbances, at small h_1, apparently results from the additional driving mechanism.

The variation of the critical Marangoni number with depth fraction, the eigenfunctions, and energy integrals are shown in Fig. 12. The energy input to the perturbation field for the neutral state corresponding to $h_1 = 0.8$ comes almost entirely from the surface work terms. For $h_1 = 0.2$ and 0.5, the buoyancy term makes a significant contribution to the energy input. The low wavenumber mode, which again involves interfacial deformation, is driven primarily by the Marangoni effect with a small contribution from buoyancy.

Critical Rayleigh and Marangoni numbers for the case of heating from below and based on the properties, temperature gradients, and depths of each phase are shown in Table 1 [7]. The Rayleigh numbers for the water phase remain substantially below 669, the critical value for fluid resting on an isothermal solid surface with a constant heat-flux-free upper surface. This implies that the water is

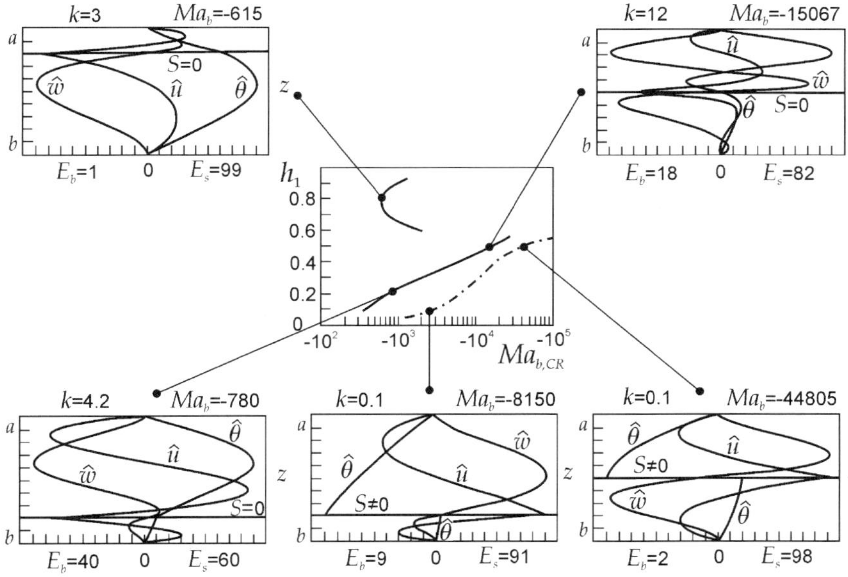

FIG. 12 Combined surface tension gradient (Marangoni) driven and buoyancy (Rayleigh) driven instability of a two-layer system. Variation of critical Marangoni number, eigenfunctions, and energy integral with depth fraction when heating from below: Benzene over water, total depth is $d = 2\,\text{mm}$. E_b = buoyancy energy integral, E_s = surface work term, as percentage of dissipation integral, and S = surface deformation. (From Ref. 7.)

TABLE 1 Combined Surface Tension Gradient (Marangoni) Driven and Buoyancy (Rayleigh) Driven Instability of a Two-Layer System. Critical Rayleigh and Marangoni Numbers for Layers of Benzene Over Water with Total Depth $d = 2$ mm, Heating from Below [7]

		Marangoni number		Rayleigh number	
h_1	k_{cr}	Benzene	Water	Benzene	Water
0.1	3.5	2473	3.7	1780	0.005
0.2	4.2	4108	31.2	2336	0.18
0.3	5.8	7968	178	3468	2.3
0.4	9	17,980	971	5749	22
0.45	—	—	—	—	—
0.5	12	31,000	3770	6880	134
0.6	2.5	2570	703	366	36.0
0.7	2.8	539	357	43.1	24.8
0.8	3	202	394	7.2	35.8
0.9	3.2	79.3	0.780	0.7	89.7

essentially passive to buoyancy-driven motions. The benzene layer is also apparently passive to the action of buoyancy when $h_1 = 0.6$, 0.7, 0.8, and 0.9, which supports the conclusion that the instability is driven by the Marangoni effect. In this case we have the first interfacial structure.

The benzene-layer Rayleigh number for the buoyancy-driven modes ($h_1 = 0.1$–0.5) exceeds 1708, the critical Rayleigh number for a single fluid confined between horizontal isothermal surfaces [14]. This suggests that if buoyancy is driving the instability this is coming from the benzene layer and then the Marangoni effect tends to retard the onset of motion. This corresponds to the third interfacial structure.

Examination of the eigenfuctions for the buoyancy-driven mode neutral states tends to substantiate the passive role of the water layer. The perturbation temperature profiles in the water layer are nearly linear, giving the minimum response necessary to maintain temperature and heat flux continuity at the interface. The velocity component response also appears to be minimal. The eigenfunctions for the surface tension gradient-driven mode show more involvement of the water in the flow motion.

The velocity patterns can also be inferred from direct consideration of the driving mechanisms. The water heated from below is unstably stratified but the Rayleigh number is always far too small for natural convection to occur. The benzene, therefore, sees the water interface as a nearly isothermal, practically rigid boundary; so once the benzene layer Rayleigh number approaches 1708 there is a tendency for buoyancy-driven convection to begin. In the upper fluid, however, buoyancy tends to cause fluid to rise at relatively hot spots and thus draws surface fluid to these spots. The surface tension gradient, on the other hand, tends to pull the surface and adjacent fluid away from these hot spots. Buoyancy in the water layer also tends to support motion in the direction of the thermocapillary-induced surface motion. Two opposing mechanisms thus compete to control the surface flow, so that motion occurs only after one mechanism dominates in the surface flow. Here surface tension gradients are large enough to overcome the counterflow provided by the density gradient, at least for the fluid adjacent to the surface. A single-cell response in each fluid is expected, as it occurs for the thermocapillary modes of the system, so that the two-cell response of the benzene layer at the critical state is unusual. Two cells certainly will result in more viscous dissipation than would occur in a single cell. However, this fact alone does not preclude a response of this type. Figure 13 shows the Rayleigh number as a function of wavenumber for $h_1 = 0.4$, with and without the Marangoni effect.

The critical Rayleigh number is substantially reduced when surface tension gradients are not allowed. The eigenfuctions for the pure buoyancy-driven flow show only one cell in each layer. Instability also occurs for $k \to 0$ for the purely buoyancy-driven case. The perturbation temperature profiles in both layers are nearly linear for $k = 0.1$, again showing a minimum response to surface tipping.

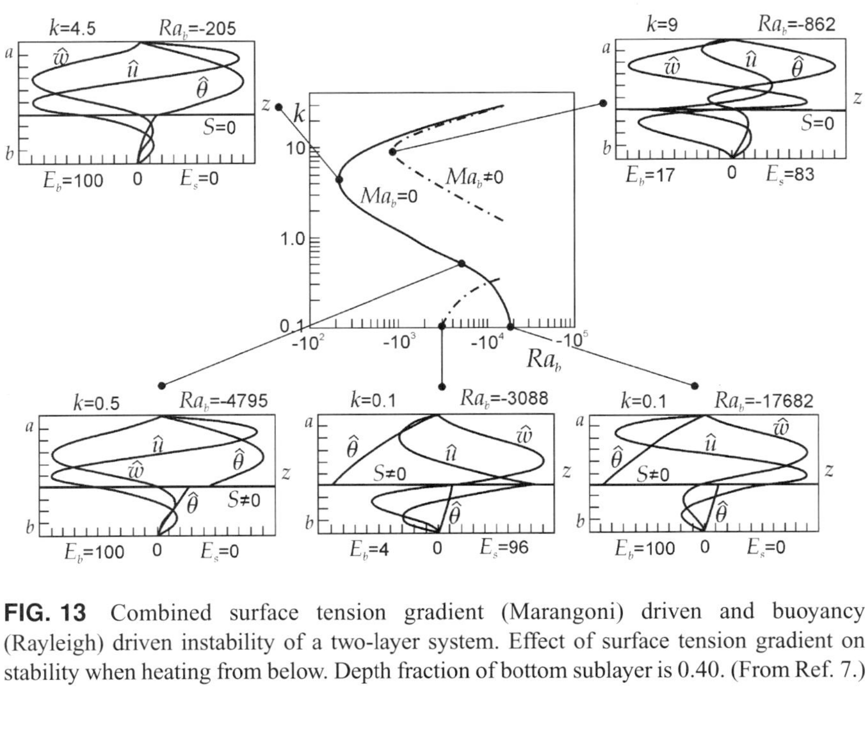

FIG. 13 Combined surface tension gradient (Marangoni) driven and buoyancy (Rayleigh) driven instability of a two-layer system. Effect of surface tension gradient on stability when heating from below. Depth fraction of bottom sublayer is 0.40. (From Ref. 7.)

Single-cell response occurs in both layers for the surface tension gradient-driven modes ($h_1 = 0.6-0.9$). The Rayleigh numbers in both liquids remain well below the corresponding critical values, so the two mechanisms are not in actual conflict. The Marangoni effect must overcome the tendency for the benzene to circulate in the direction opposite to the surface motion, but may not be able to do it.

The effects of the total depth and the liquid depth ratio ($r = h_2/h_1$) on the onset of Marangoni–Rayleigh convection in a horizontal two-layer system with nondeformable interface have also been studied [13]. In Fig. 14 we provide the stability diagram for the benzene–water system with $r = 1$. In this case, the parameters, Ra_1 and Ma_1, were defined with the benzene layer parameters. The diagram indicates the condition for the onset of convection in a two-layer system with arbitrary liquid depths. A thick solid line depicts the neutral line for monotonic instability and the thick dotted line for oscillatory convection, respectively. In order to assign a critical condition for a liquid layer with depth h_1, a straight line was drawn through the origin with a slope $\phi = Ra_1/Ma_1$. A point on this line always expresses the status of the liquid layer. The point where the straight line crosses the stability line indicates the corresponding critical condition of the layer.

Figures 15(a)–(h) shows the distributions of eigenfunctions in terms of the stream function, ψ, and isotherms of temperature perturbation, T', at the conditions

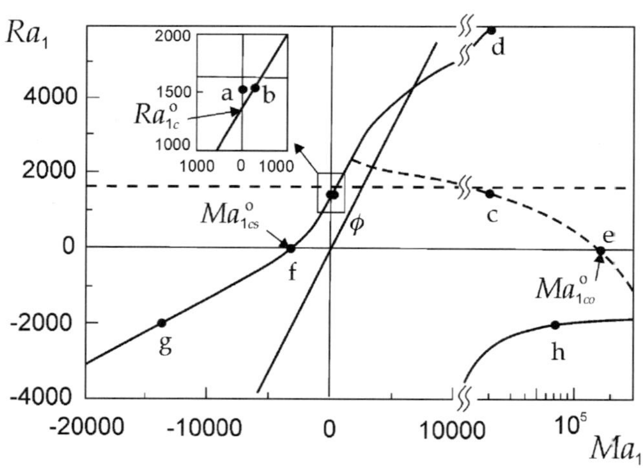

FIG. 14 Combined surface tension gradient (Marangoni) driven and buoyancy (Rayleigh) driven instability of a two-layer system. Stability diagram for the benzene–water system with $h_2/h_1 = 1.0$. Solid lines denote neutral lines for monotonic instability of the system with clean interface, while solid lines with dots account for the oscillatory neutral line, and the dot–dashed line is the neutral line for monotonic instability of the system with contaminated interface. For continuation and details about points (a–h) see Fig. 15. (From Ref. 13.)

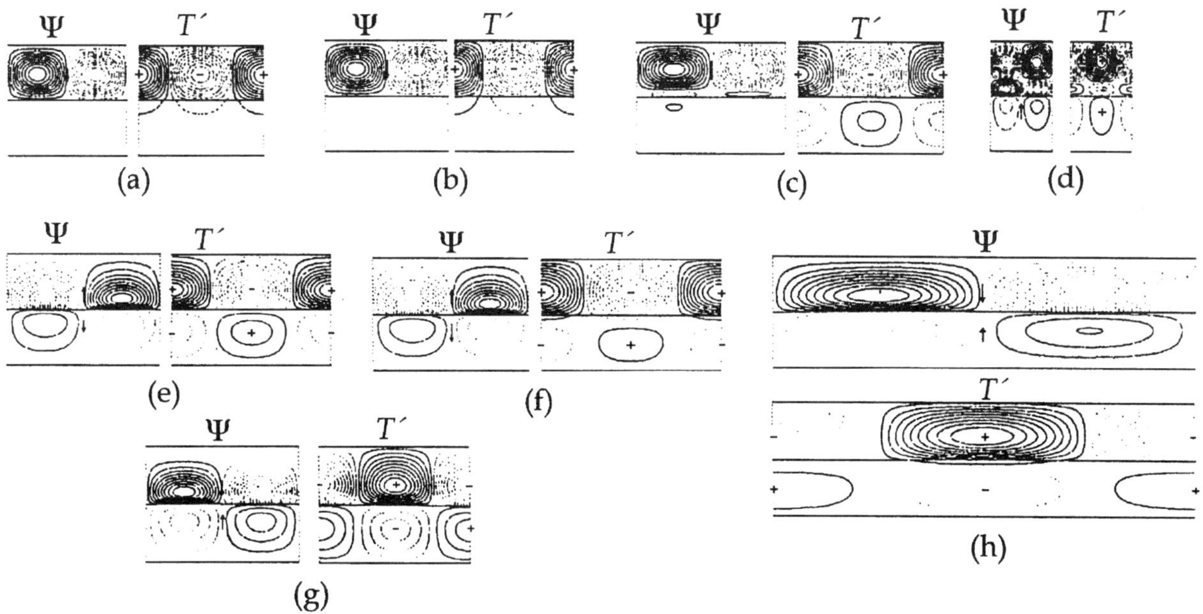

FIG. 15 Combined surface tension gradient (Marangoni) driven and buoyancy (Rayleigh) driven instability of a two-layer system. Streamfunction, ψ, and isotherm, T', of temperature perturbation for the benzene–water system with $h_2/h_1 = 1.0$ according to the (Ra, Ma) plot given in Fig. 14: (a) $Ra_1 = 1500$, $Ma_1 = 0$, $k_1 = 2.9$, $\lambda_r = 0.178$, $\lambda_i = 0.0$ (monotonic instability); (b) $Ra_1 = 1500$, $Ma_1 = 245$, $k_1 = 2.92$, $\lambda_r = 0.0$, $\lambda_i = 0.0$; (c) $Ra_1 = 1500$, $Ma_1 = 19{,}195$, $k_1 = 2.43$, $\lambda_r = 0.0$, $\lambda_i = 3.62$ (overstability); (d) $Ra_1 = 5890$, $Ma_1 = 19{,}195$, $k_1 = 5.79$, $\lambda_r = 0.0$, $\lambda_i = 0.0$; (e) $Ra_1 = 0$, $Ma_1 = 157{,}450$, $k_1 = 2.26$, $\lambda_r = 0.0$, $\lambda_i = 8.98$; (f) $Ra_1 = 0$, $Ma_1 = -3175.6$, $k_1 = 1.99$, $\lambda_r = 0.0$, $\lambda_i = 0.0$; (g) $Ra_1 = -2000$, $Ma_1 = -13{,}693$, $k_1 = 0.84$, $\lambda_r = 0.0$, $\lambda_i = 0.0$; (h) $Ra_1 = -2000$, $Ma_1 = 63{,}075$, $K_1 = 2.43$, $\lambda_r = 0.0$, $\lambda_i = 0.0$. (From Ref. 13.)

indicated by points (a)–(h) in Fig. 14. The distributions of the eigenfunctions for the monotonic and oscillatory thermocapillary convection are compared in Fig. 15(e) and (f). The purely thermocapillary convection when heating from above [Fig. 15(f)] exhibits a slightly longer wavelength than the oscillatory convection due to the heating from below [Fig. 15(e)].

Other distributions in Fig. 15 indicate how buoyancy and the Marangoni effect interact. For the cases heated from below with $Ra_1 = 1500$, Figs 15(a)–(c) indicate the following. If $Ma_1 = 0$, the purely buoyancy-driven perturbations dominate only in the benzene layer. If the Marangoni effect is weak [Fig. 15(b), $Ma_1 = 245$], it slightly retards the fluid motion near the interface. By increasing Ma_1, the buoyancy-driven convection is reduced and the liquid layer is stabilized by the Marangoni effect. But when the Marangoni number increases further up to $Ma_1 = 19{,}195$, it eventually drives oscillatory convection. The double roll cell convection [Fig. 15(c)] looks similar to, but has much longer wavelength, than the steady convective structure [Fig. 15(d)].

On Earth, when the layer is heated from the top the liquid layers are stably stratified due to (negative) buoyancy and if monotonic instability occurs it is brought about by the Marangoni effect. The wavelength becomes longer as Ra_1 decreases but again becomes shorter at $Ra_1 < -5890$, as shown in Fig. 15(g) and (h).

The onset of the combined thermocapillary- and buoyancy (Marangoni–Rayleigh) driven convection has been experimentally observed [7]. Measurements of critical Marangoni numbers for layers of total depth $d = 2\,\text{mm}$ with benzene over water were made for both ways of heating, from above and from below. Because of the difficulty of establishing relatively thin layers of benzene over water, experiments were performed only at water depth fractions of 0.40. The onset of convection was determined by detecting the corresponding increase of the heat transfer through the layers above the value for pure heat diffusion, hence the Nusselt number. Data from amplitude motions can be extrapolated downwards to determine the (linear) critical Marangoni number.

The apparatus used was primarily a device for determining the overall heat transfer of horizontal liquid layers. The test liquids were contained inside a glass cylinder between $\frac{3}{4}$-in thick, 12-in diameter copper plates spaced by quartz guards. One plate was electrically heated and the other water-cooled. Heat transfer through the layers was deduced from an energy balance on the heated plate. Both plates were carefully heated or cooled as necessary to minimize energy losses to the environment. The primary plate temperature difference was measured with a four-couple thermopile; additional thermocouples allowed monitoring of the plate temperature level and plate-to-guard temperature difference. Each plate assembly was constructed so that either the heated or cooled plate can be the upper or lower face of the test section. The lower plate assembly was supported on jacks to allow leveling. Adequate peripheral equipmentwas used to stabilize, control, and measure the input electrical power, to regulate, control, and measure the energy flux of the cooling water, and to monitor all temperatures. The apparatus and the associated data reduction program reproduced the thermal conductivity of water to within 2% of known tabulated values and yielded a critical Rayleigh number of 1700 ± 150 for a single liquid layer heated from below, in agreement with theory and earlier known data [14].

Results corresponding to the case of heating from above for two water depth fractions appear in Fig. 16. No increase in effective heat transfer of the layers was measurable for Marangoni numbers up to five times the predicted critical value. This may be a result of one of two causes. The presence of surface pollutants, even in trace amounts, can suppress the Marangoni effect. Any thermally induced convective motion tends to sweep a contaminant away from the relatively low surface tension, warmer regions of the surface. Since surface tension is normally reduced by contamination, the resulting surface concentration profile causes a surface tension gradient, which opposes the thermally induced gradient. Surface tension is much more strongly dependent on surfactant concentration than on temperature, so that extremely small concentrations of contaminants may drastically block the action of the thermal Marangoni effect. It is also possible that, while this system is unstable to infinitesimal disturbances when heated from

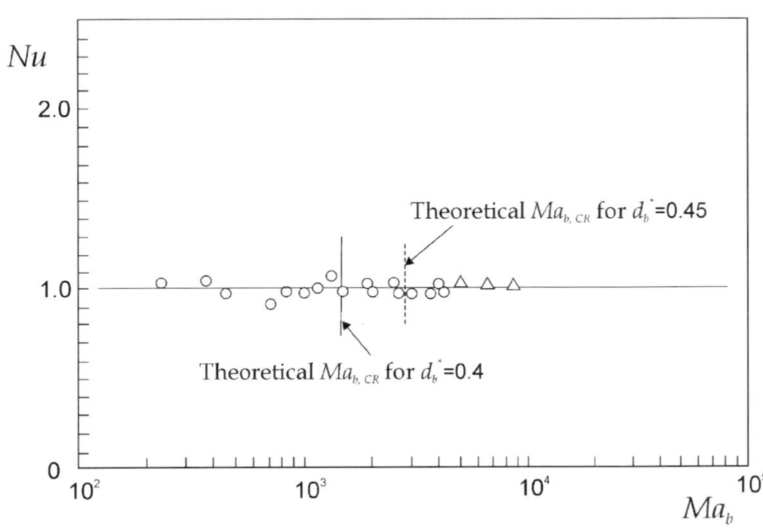

FIG. 16 Combined surface tension gradient (Marangoni) driven and buoyancy (Rayleigh) driven instability of a two-layer system. Nusselt number vs. Marangoni number when heating from above: Benzene over water, total depth, $d = 2$ mm. Water depth fraction: △, 0.39; ○, 0.46. (From Ref. 7.)

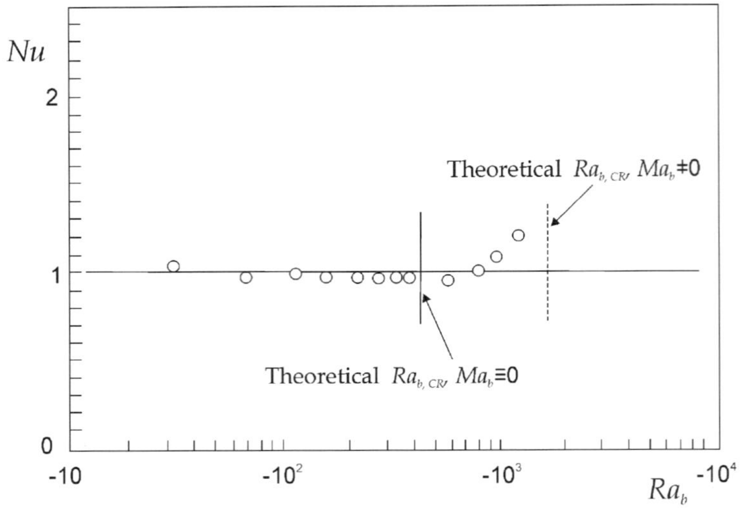

FIG. 17 Combined surface tension gradient (Marangoni) driven and buoyancy (Rayleigh) driven instability of a two-layer system. Nusselt number vs. Rayleigh number when heating from below: Benzene over water, total depth $d = 2$ mm, water fraction depth 0.40. (From Ref. 7.)

above, finite disturbances are damped or amplitude-limited by interactions with the stable density gradient field and no detectable convection actually occurs.

Results when the water depth fraction is 0.4 and the heating is done from below are shown in terms of the Rayleigh number in Fig. 17. A clear break in the data occurs, corresponding to a Rayleigh number of 700 or a Marangoni number of about 2800. Values of the critical Rayleigh number, with and without the Marangoni effect, are also indicated in the figure. Figure 13 shows the earlier noted fact that purely buoyancy-driven flow occurs at a critical Rayleigh number below the critical Rayleigh number for the case when the Marangoni effect is also present. The fact that the measured critical Rayleigh number falls between these limits suggests that surface contamination has retarded, but not completely blocked, the Marangoni effect in these experiments.

B. Marangoni–Rayleigh Instability in the Glycerin–Silicone Oil System

Let us now consider the case of a two-layer system of silicone oil over glycerol [13]. The system is bounded on the top of the upper liquid and at the bottom of the lower one by solid, conducting plates. The inner interface between the upper and lower liquids is expected to deform. The thermophysical properties of the system

TABLE 2 Thermophysical Properties of the Glycerin/Glycerol–Silicone Oil System

Property (units)	ρ (g/cm^3)	$\alpha \times 10^4$ (1/K)	κ (W/mK)	$\chi \times 10^6$ (m^2/s)	$\nu \times 10^4$ (m^2/s)	σ_0 (mN/m)	σ_1 (mN/mK)
Glycerin/Glycerol	1.26	4.9	0.294	0.89	7.45	25	−0.13
Silicone oil	0.97	9.45	0.16	1.16	4.99	—	—

Source: From Ref. 13.

are given in Table 2. In this case the Rayleigh and Marangoni numbers are defined by the parameters of the lower liquid (glycerol), its layer depth and the temperature difference across it.

The Rayleigh number as a function of wavenumber, when heating from below the glycerol–silicone oil system, is shown in Fig. 18. Glycerol, which has the higher density, lies below the layer of silicone oil. The depth of the glycerol is 4.15 cm, and the depth of the silicone oil is 3.0 cm. In this system switching between different instability mechanisms and three different interfacial structures can be seen.

Let us focus on results found for monotonic instability only. The Rayleigh number was calculated, and when the imaginary part of the Rayleigh number became nonzero, it was inferred that the onset of convection was oscillatory. While for real values of the Rayleigh number the numbers make sense, for complex values, the numbers obtained are meaningless. Recall that, as earlier mentioned, when the Rayleigh number is negative, as in the case of negative buoyancy in a

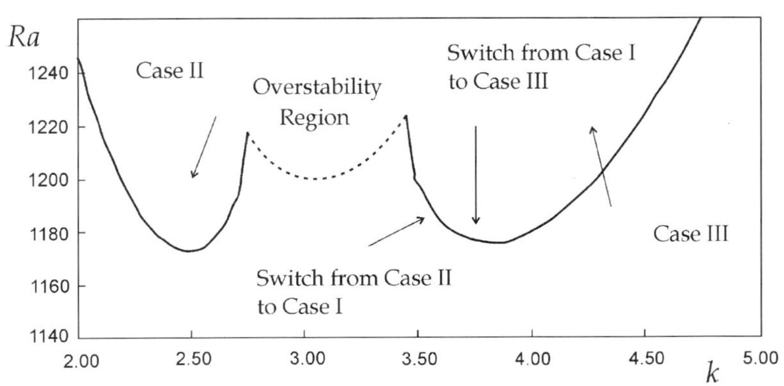

FIG. 18 Combined surface tension gradient (Marangoni) driven buoyancy (Rayleigh) driven instability of a two-layer system. Plot of the Rayleigh number of the glycerol vs. the wavenumber. Glycerol is below with a layer depth $h_1 = 4.15$ cm and silicone oil is above with corresponding layer depth $h_2 = 3.0$ cm. (From Ref. 11.)

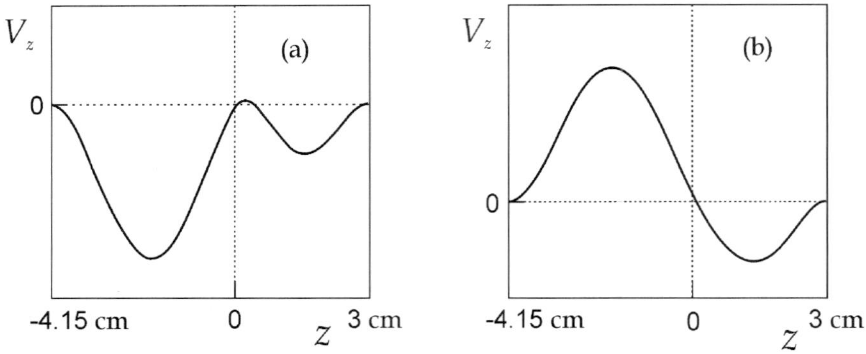

FIG. 19 Combined surface tension gradient (Marangoni) driven and buoyancy (Rayleigh) driven instability of a two-layer system. Plot of the vertical velocity component for $h_1 = 4.15$ cm of glycerol and $h_2 = 3.0$ cm of silicone oil. In panel (a), the two fluid flows are thermally coupled, with a small counter roll in the upper layer, $k = 2.7$. In panel (b), the two fluid flows are (mechanically) viscously coupled, $k = 3.5$. The vertical dotted line depicts the unperturbed interface. (From Ref. 11.)

stably stratified fluid layer heated from above, its square root is proportional to the Brunt–Väisälä frequency of internal wave motion. The dotted line in Fig. 18 is the region where oscillations occur. We shall not consider this case here.

A plot of the vertical velocity component for $k = 2.7$ and 3.5 is given in Fig. 19. For $k = 2.7$, the flows are nearly thermally coupled (both upper and lower rolls turn in the same direction), with a small counter-roll in the upper fluid. For $k = 3.5$, the convection is (mechanically) viscously coupled (with opposite roll rotation as in two wheels in contact). The oscillations in Fig. 18 are a result of a competition between the thermal and viscous coupling.

The points marked with arrows in Fig. 18 indicate wavenumber values at which the interfacial structure changes. For a wavenumber below 3.8, the interfacial structure is Case II of Fig. 8. Between $k = 3.8$ and 3.9, the interfacial structure is Case I. For wavenumbers greater than $k = 3.9$, the interfacial structure is Case III. Here also the interfacial structure changes near the values of the wavenumber where overstability is expected.

The coupling between interfacial structures as well as the flow (thermal or viscous) can be used to describe what is occurring in the two liquid layers. For wavenumbers smaller than in the oscillatory region, the two fluids are nearly thermally coupled and the interfacial structure is a Case II. As earlier described, a Case II interfacial structure denotes buoyancy-driven convection occurring in the lower layer. The hot plumes of the lower layer rise up and drive convection in the upper liquid, creating a thermally coupled two-layer, with a small counter-roll. For wavenumbers much larger than in the oscillatory region ($k = 3.9$), the flows are viscously coupled, and the interfacial structure is a Case III. For a Case III interfacial structure, buoyancy-driven convection occurs mostly in the upper layer. The cold, sinking fluid in the upper layer pushes down and depresses the interface. The interfacial structure seems to indicate that viscous coupling occurs when the upper fluid "initiates" convection, causing the lower fluid to flow and to connect in a gearlike manner (mechanical coupling).

What seems somewhat peculiar is the Case II interfacial structure at wavenumbers between $k = 3.4$ and 3.8, where the flows in the two liquid layers are viscously coupled. This may be a result of the difference in their dynamic (shear) viscosities. The dynamic (shear) viscosity of glycerol is about twice that of silicone oil. Even though convection may start in the silicone oil layer, it has to overcome the higher dynamic viscosity of the glycerol. Only when convection in the silicone oil is really vigorous does it begin to deform the interface to give a Case III interfacial structure. The Case I interfacial structure is a transient one between Cases II and III.

> High viscosity liquids are lazy. Low viscosity liquids are the victims of the laziness of high viscosity liquids because they are easy to push. (Joseph et al. 1984, as highlighted in Ref. 1, p. I, 27).

REFERENCES

1. Joseph, D.D., Renardy, Y.Y. Fundamentals of two-fluid dynamics (2 vols.); Springer-Verlag: Berlin, 1993.
2. Simanovskii, I.B., Nepomnyashchy, A.A. *Convective Instabilities in Systems with Interface*; Gordon & Breach: Amsterdam, 1993.
3. Colinet, P., Legros, J.C., Velarde, M.G. *Nonlinear Dynamics of Surface-Tension-Driven Instabilities*; Wiley-VCH: Berlin, 2001.
4. Birikh, R.V., Rudakov, R.N. Mechanism of a thermocapillary oscillation in a system with interface. In *Vibrational Effects in Hydrodynamics*; PGU: Perm, 1998; 38–48 (in Russian).
5. Birikh, R.V., Briskman, V.A., Rudakov, R.N., Velarde, M.G. Interface Marangoni instability in the high-frequency vibration field. Proceedings of the Joint Xth European and Vith Russian Symposium on Physical Sciences in Microgravity. St. Petersburg, Russia, June 15–21 June, 1997; V. 1, pp 205–212.
6. Birikh, R.V., Briskman, V.A., Rudakov, R.N., Velarde, M.G. Thermocapillary instability of real liquids interfaces. In *Convection in Immiscible Liquid–Liquid Systems*; UB RAS: Ekaterinburg, 1999; 25–44 (in Russian).
7. Zeren, R.W., Reynolds, W.C. Thermal instabilities in two-fluid horizontal layers. J. Fluid Mech. **1972**, *53*, 305–327.
8. Koschmieder, E.L., Prahl, S.A. Surface tension driven Benard convection in small containers. J. Fluid Mech. **1990**, *215*, 571–583.

9. Pearson, J.R.A. On convection cells induced by surface tension. J. Fluid Mech. **1958**, *4*, 489–500.
10. Zhao, A.X., Wagner, C., Narayanan, R., Friedrich, R. Bilayer Rayleigh–Marangoni convection: transitions in flow structures at the interface. Proc. Royal Soc. London A **1995**, *451*, 487–502.
11. Johnson, D., Narayanan, R. Geometric effects on convective coupling and interfacial structures in bilayer convection. Phys. Rev. E **1997**, *56*, 5462–5472.
12. Johnson, D., Narayanan, R. Marangoni convection in multiple boundded fluid layers and its application to materials processing. Phil. Trans. Royal Soc. London A **1998**, *356*, 885–898.
13. Jing, Ch., Sato, T., Imaishi, N. Rayleigh–Marangoni thermal instability in two-liquid layer systems. Microgravity Sci. Technol. **1997**, *X/1*, 21–28.
14. Smith, K.A. On convective instability induced by surface tension gradients. J. Fluid Mech. **1966**, *24*, 401–414.

8

Thermocapillary Instability in Multilayer Systems

A multilayer system allows the study of flows generated at more than one surface or inner interface with eventual interaction through the bulk phases. Because of the diversity of multilayer systems and complexity in their mathematical analyses, in this chapter we shall restrict attention to three-layer symmetric plane systems with a transverse temperature gradient. Floating liquid layers and films earlier discussed are particularly simple examples of such systems. We shall further restrict consideration to the case of two outer layers of finite width. The hottest and coldest layers will be labeled 1 and 3, respectively. In the middle layer, labeled 2, there is a different immiscible liquid. In this particular system, the 1–2 and 1–3 interfaces are identical. In microgravity, there is no Rayleigh–Taylor instability in such a symmetric situation.

I. LINEAR STABILITY ANALYSIS FOR AN INFINITELY EXTENDED SYSTEM ALONG THE HORIZONTAL

Let us first consider the case of a three-layer system confined between two outer isothermal solid boundaries of infinite extension in the plane of the interfaces [1]. The geometry and coordinate system are shown in Fig. 1. For small perturbations, using as earlier suitable Fourier normal modes with generic wavenumber, k, and growth rate, λ, the amplitudes of transverse velocity component, $v_l(y)$, and

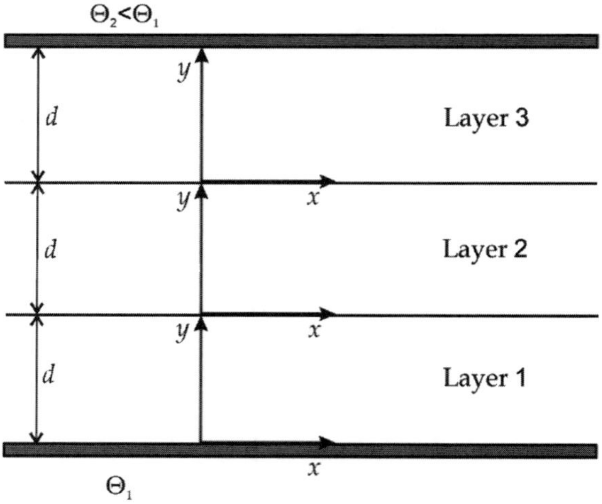

FIG. 1 Thermocapillary instability in multilayer systems: Geometry of the three-layer structure, with two inner interfaces, and the coordinate system.

temperature disturbances, $\theta_l(y)$, obey the system of equations

$$-\lambda\left(\frac{d^2 v_l}{dz^2} - k^2 v_l\right) = \frac{d^4 v_l}{dz^4} - 2k^2\frac{d^2 v_l}{dz^2} + k^4 v_l, \qquad l = 1, 3 \tag{1}$$

$$-\lambda\nu\left(\frac{d^2 v_2}{dz^2} - k^2 v_2\right) = \frac{d^4 v_2}{dz^4} - 2k^2\frac{d^2 v_2}{dz^2} + k^4 v_2 \tag{2}$$

$$-\lambda Pr\theta_l = \left(\frac{d^2\theta_l}{dz^2} - k^2\theta_l\right) + v_l, \qquad l = 1, 3 \tag{3}$$

$$-\lambda Pr\theta_2 = \left(\frac{d^2\theta_2}{dz^2} - k^2\theta_2\right) + \kappa\chi v_2 \tag{4}$$

where all quantities have been made dimensionless by choosing d (thickness in every layer) as unit of length, d^2/ν_1 (ν_1 is the kinematic viscosity of layer 1) as unit of time, and $\Delta T_1 Pr$ (ΔT_1 is the temperature difference in layer 1; $Pr = \nu_1/\chi_1$ is the Prandtl number) as unit of temperature. $\nu = \nu_1/\nu_2$ is the kinematic viscosity ratio, $\chi = \chi_1/\chi_2$ is the thermal diffusivity ratio, and $\kappa = \kappa_1/\kappa_2$ is the thermal conductivity ratio. The above-given balance laws in each layer are written with their own y-coordinate, which varies between zero (at the hot boundary) and unity (at the cold boundary).

At the boundaries the following conditions must be satisfied

$$\mathrm{v}_l(0) = \mathrm{v}_l(1) = 0, \qquad l = 1,\ 2,\ 3 \tag{5}$$

$$\frac{\mathrm{d}\mathrm{v}_1}{\mathrm{d}z}(0) = \frac{\mathrm{d}\mathrm{v}_3}{\mathrm{d}z}(1) = 0, \qquad \frac{\mathrm{d}\mathrm{v}_l}{\mathrm{d}z}(1) = \frac{\mathrm{d}\mathrm{v}_{l+1}}{\mathrm{d}z}(0), \qquad l = 1,\ 2 \tag{6}$$

$$\theta_1(0) = \theta_3(1) = 0 \tag{7}$$

$$\theta_1(1) = \theta_2(0), \qquad \theta_2(1) = \theta_3(0) \tag{8}$$

$$\kappa\frac{\mathrm{d}\theta_1}{\mathrm{d}z}(1) = \frac{\mathrm{d}\theta_2}{\mathrm{d}z}(0), \qquad \frac{\mathrm{d}\theta_2}{\mathrm{d}z}(1) = \kappa\frac{\mathrm{d}\theta_3}{\mathrm{d}z}(0) \tag{9}$$

$$\frac{\mathrm{d}^2\mathrm{v}_1}{\mathrm{d}z^2}(1) - \frac{1}{\eta}\frac{\mathrm{d}^2\mathrm{v}_2}{\mathrm{d}z^2}(0) = Mak^2\theta_1(1) \tag{10}$$

$$\frac{1}{\eta}\frac{\mathrm{d}^2\mathrm{v}_2}{\mathrm{d}z^2}(1) - \frac{\mathrm{d}^2\mathrm{v}_3}{\mathrm{d}z^2}(0) = Mak^2\theta_2(1) \tag{11}$$

where the Marangoni number is defined as $Ma = \sigma_1 \Delta T_1 d/\eta_1\chi_1$, and $\eta = \eta_1/\eta_2$ is the dynamical (shear) viscosity ratio.

The Conditions of Eqs. (5) and (6) describe the continuity of all the velocity components along solid walls and interfaces. Equation (8) expresses that the solid walls are perfect thermal conductors. Equations (8) and (9) arise from the continuity of the temperature and of the heat flux, respectively, along each liquid–liquid interface. Equations (10) and (11) account for the longitudinal shear stress balance along the free interfaces.

A. Monotonic Instability

As in earlier cases for the monotonic neutral case ($\lambda = 0$) it is easy to write the general solution of the system of Eqs. (1)–(4). Using here the temperature difference between the solid boundaries the compatibility of the boundary equations (5)–(11) determines the critical Marangoni number [2]:

$$Ma = \pm\frac{8k^2}{\eta\kappa}(A/B)^{1/2} \tag{12}$$

where

$$A = (2+\kappa)^2[(\kappa+1)^2C^2 - 1][(1+\eta)^2(SC-k)^2 - (S-kC)^2]$$
$$B = (1-\chi)^2(S^3 - k^3C)^2 - \chi^2k^2(kSC + k^2 - 2S^2)$$
$$C = \cosh k, \qquad S = \sinh k$$

The lowest Marangoni number, Ma_c, and corresponding wavenumber, k_c, calculated from Eq. (12) for the system air–water–air, are equal to 947 and 1.91, respectively. Changing scales and using the water layer properties we obtain

$Ma_c = 94.1$. This value is higher than the 87.5 obtained by Funada [3]. The difference is a result of the fact that here we consider conducting interfaces rather than insulating interfaces defined by a very low Biot number. We have also taken into account the viscous shear stresses of all three layers.

For the system Fluorinert FC70 – Silicone Oil 10 cS – Fluorinert FC70, Ma_c and k_c are equal to 500.78 and 1.965, respectively; and for Fluorinert FC70 – Silicone Oil 50 cS – Fluorinert FC70, Ma_c and k_c are equal to 1008 and 1.92, respectively.

The horizontal component of the velocity at the neutral state and the corresponding two-dimensional flow pattern are shown in Fig. 2(a) and (b), respectively. The flow is driven by the hotter interface. The flow in the middle layer will drag the colder interface, creating a small vortex in the top layer. The colder interface is merely passive and only contributes with shear.

When a positive thermal disturbance in layer 1 reaches the 1–2 interface, it creates a hot spot and the interfacial tension gradient that results from it induces a flow that will amplify the initial disturbance. In layer 2, because of the continuity of velocity across the interface, the induced flow will bring colder fluid, which damps the temperature gradient along the interface. It thus exerts a stabilizing effect on the system. When there is silicone oil in the middle layer the thermal diffusivity is about three times larger than that of the outer layers (Fluorinert FC70). Heat is transported by the flow more effectively in layer 1 (thermal disturbances decay slowly due to the low thermal diffusivity in this layer), so the destabilizing effect predominates making the 1–2 interface active. The same argument applies verbatim to the 2–3 interface, which should be passive. When a hot spot reaches the 2–3 interface, the induced thermal gradient is damped by the corresponding downward flow in layer 3 because of its lower thermal diffusivity.

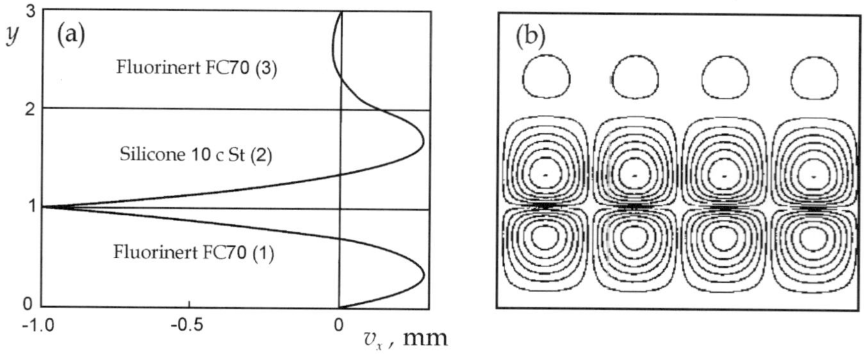

FIG. 2 Thermocapillary instability in multilayer systems. Linear stability results: (a) Normalized longitudinal velocity profile at the neutral state ($Ma = Ma_c$ and $k = k_c$), (b) Streamlines at the neutral state. (From Ref. 5.)

If the thermal diffusivity of the middle layer is lower than that of the outer layers ($\chi > 1$) a different convective flow structure is expected. Now heat is transported by the flow more effectively in layer 2 and the 2–3 interface will be active. However, in the middle layer slowly decaying temperature disturbances bring cold liquid to the 1–2 interface and there a weak counterflow is generated. This motion at the interface is also supported by the bulk flow in layer 1. The cell structure of steady convection for various values of the χ ratio is shown in insets of Fig. 4.

B. Overstability

The argument of the square root in Eq. (12) is always negative when $\chi \approx 1$, whatever the value of the other parameters. For layers of equal thicknesses this case is completely symmetric for stationary heat transfer. It explains the disappearance of monotonic instability. As shown earlier, the onset of oscillatory instability for a two-layer system with a flat interface originates in the phase shift of fluctuations in the layers. In the three-layer system a gearlike driven motion in the two neighboring layers has a characteristic time different from that of the motion in the third layer. Therefore in the three-layer system with close properties of all liquids one can expect overstability at rather large gradients of temperature. Indeed, a neutral oscillatory stability curve exists when both heat diffusivities are equal. One thus expects an oscillatory region around that peculiar value of χ.

When the imaginary part of the growth rate, $\mathrm{Im}\lambda = \omega$, i.e., the disturbance frequency was different from zero, the boundary-value problem (1)–(11) was

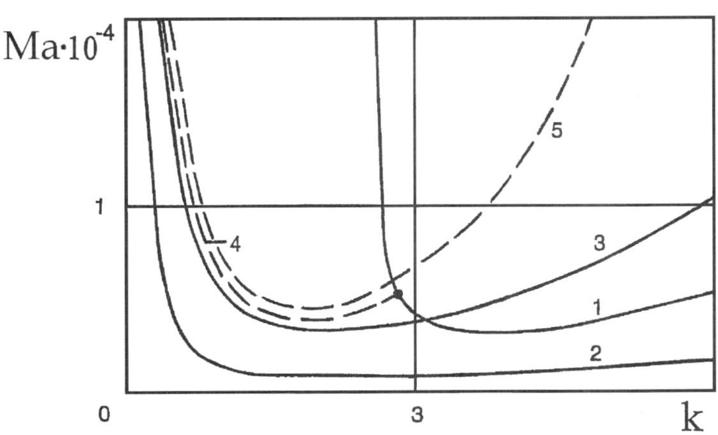

FIG. 3 Thermocapillary instability in multilayer systems. Monotonic neutral curves (solid lines) and oscillatory neutral curves (dashed lines) for the model system: $\chi = 2$, lines 1 and 4; $\chi = 4$, line 2; $\chi = 0.5$, line 3; $\chi = 1$, line 5. The Marangoni number, *Ma*, is defined with the temperature difference between the solid boundaries. (From Ref. 2.)

solved by a numerical method. A rather simple technique was used to convert the boundary-value problem into an initial-value one [2]. Using various initial conditions the required number of linearly independent solutions of the differential system was constructed by a fourth-order Runge–Kutta method in each layer. From the boundary conditions, for a linear combination of the solutions the critical Marangoni number and the frequency of oscillatory disturbances were determined.

One particular case studied corresponds to the following parameter values: $Pr = 1$, $\kappa = 1$, $\eta = \nu = 0.5$, and χ a free parameter. Neutral curves for various thermal diffusivity ratios are shown in Fig. 3 (solid lines, monotonic instability; dashed lines, oscillatory instability). To compare theory with experiment in Fig. 3 the Marangoni number, *Ma*, contains the temperature difference between the solid boundaries.

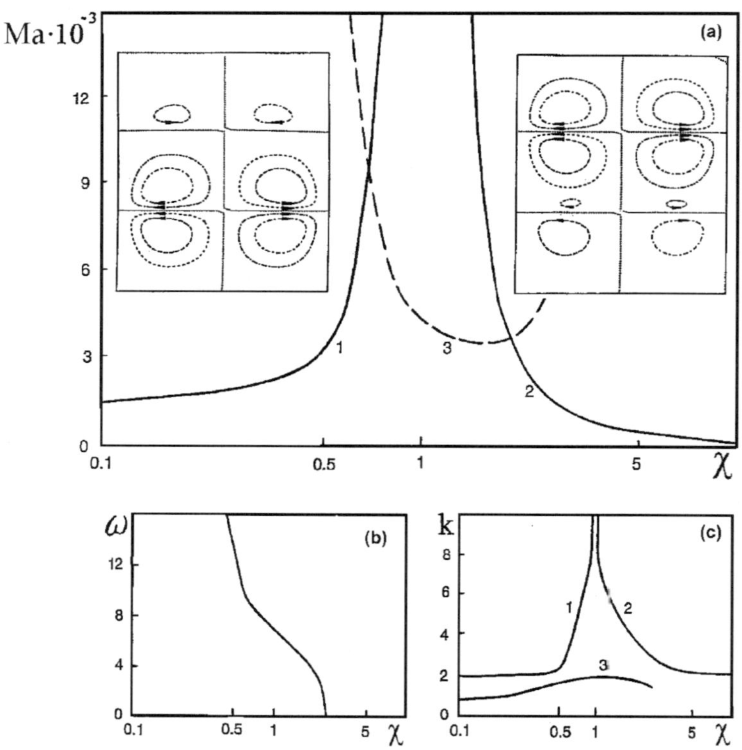

FIG. 4 Thermocapillary instability in multilayer systems. Influence of the thermal diffusivity ratio, χ, on the type of unstable mode: (a) critical Marangoni number, (b) critical frequency, and (c) critical wavenumber as function of χ. The Marangoni number, *Ma*, is defined with the temperature difference between the solid boundaries. (From Ref. 2.)

The disappearance of the monotonic instability ($\chi = 4$, curve 2; $\chi = 0.5$, curve 3) and the onset of oscillatory instability are illustrated with curves 1 and 4 ($\chi = 2$). Overstability is the only possible neutral mode when $\chi = 1$ (curve 5) and it is the most dangerous type of instability for χ equal to around 1 ($0.66 < \chi < 1.92$ for the model system). The critical Marangoni numbers, critical frequency, and critical wavenumbers for monotonic and oscillatory instability are given in Fig. 4. The critical wavenumber, k_0, of the oscillatory instability is always smaller than the corresponding value, k_m, for monotonic instability.

Note that the neutral oscillatory curve exists for arbitrary low χ, but the corresponding critical Marangoni number is very high. The dependence of critical frequency on χ is shown in Fig. 4(b). In the region $\chi > 2.55$, the minimum of the neutral oscillatory curve coincides with the point where the frequency of oscillations vanishes.

II. THE CASE OF A SYSTEM OF FINITE LATERAL EXTENT

It is now shown that for thermocapillary convection in a single layer the confinement provided by rigid walls is slightly stabilizing for not too small aspect ratios [4]. For instance, in a $L/d \times W/d = 6 \times 4$ rectangular box, $Ma_c = 84.4$ (instead of 79.6 for an infinitely extended layer).

The thermal characteristics of the lateral walls introduce a new feature, which is more important than the pure geometrical aspect from confinement. Indeed, heat is transferred across the lateral walls and the large difference in thermal conductivity between the Fluorinert and the silicone oil acts together with the highly conductive walls of, say, quartz, to induce a lateral gradient. Thus, in a finite container, a steady motionless state cannot actually exist in the vicinity of the lateral walls. The effect of conducting lateral walls has to be suitably incorporated in the numerical analysis. Let us consider various results found amenable to experimental test in such a three-layer system (Fig. 11).

A. Numerical Solution of the Problem

The continuity, the nonlinear Navier–Stokes and the energy equations have been solved numerically in two dimensions (x, y) for a symmetric three-layer system of finite lateral extent, all layers of equal depth, and with flat interfaces under zero effective gravity conditions [5]. Using the stream function, $\varphi;(x, y, t)$, and the vorticity function, $\varphi(x, y, t)$, in each layer, we have

$$\frac{\partial \varphi}{\partial t} + \frac{\partial \psi}{\partial y}\frac{\partial \varphi}{\partial x} - \frac{\partial \psi}{\partial x}\frac{\partial \varphi}{\partial y} = \frac{\nu_l}{\nu_1}\Delta\varphi \tag{13}$$

$$Pr\left(\frac{\partial T}{\partial t} + \frac{\partial \psi}{\partial y}\frac{\partial T}{\partial x} - \frac{\partial \psi}{\partial x}\frac{\partial T}{\partial y}\right) = \frac{\chi_l}{\chi_1}\Delta T \tag{14}$$

$$\Delta\psi = \varphi, \qquad \mathrm{v}_x = \frac{\partial \psi}{\partial y}, \qquad \mathrm{v}_y = -\frac{\partial \psi}{\partial x} \tag{15}$$

The equations were discretized on a uniform rectangular grid using the centered-difference technique and solved using the ADI scheme. The Poisson Eq. (3) was solved using the over-relaxation technique.

The computational domain was divided into five parts: three for the liquids ($NX \times NY$) and two for the lateral quartz walls ($WX \times WY$) (Fig. 5). In each of the five domains, nodes were identified by two coordinates (i, j). The index i changed along the x-axis from 1 to NX in the liquids and from 1 to WX in the lateral quartz walls. The index j changed along the y-axis from 1 to NY in the liquids and from 1 to WY in the lateral quartz walls. Highly conductive (horizontal) sapphire plates are represented by a single line of nodes kept at constant temperature. The knife edge shown in Figs. 5 and 11) was modeled using a single row (KN) of nodes where rigid boundary conditions apply. The Navier–Stokes, energy, and Poisson equations are then solved for the internal nodes in each layer independently and the boundary conditions were applied to compute the values of vorticity and temperature on the boundary nodes.

The boundary conditions were formulated in the following form. For simplicity, the labels of the liquid domain will be either dropped or replaced by a and b when referring to coupling conditions across the interface a–b that separates the liquids a and b (this choice is made to avoid confusion with other uses of

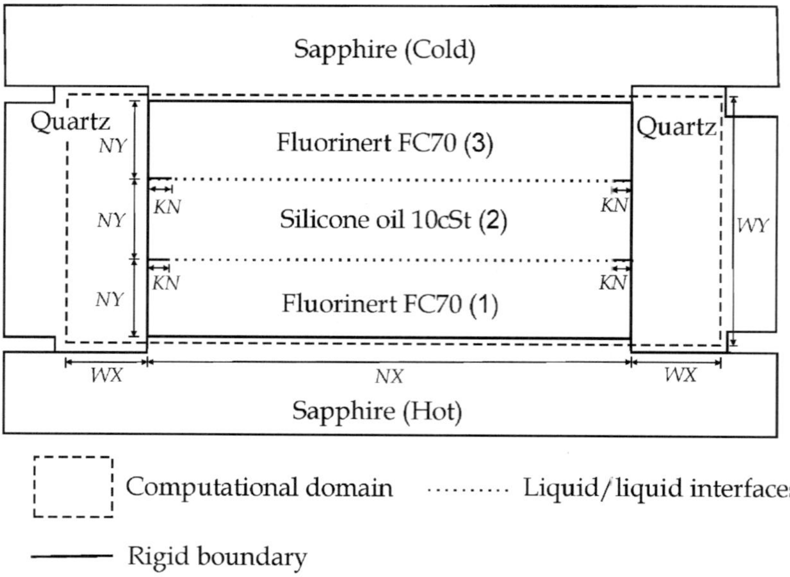

FIG. 5 Thermocapillary instability in multilayer systems. Grid organization for two-dimensional numerical computations. (From Ref. 5.)

subscripts 1, 2, 3). Along the rigid walls, the stream function, ψ, is constant and its first-order derivatives are zero. From the definitions of Eq. (15), developing ψ near the walls in a Taylor series, one finds along the left vertical wall

$$\varphi_{(1)(j)} = \frac{8\psi_{(2)(j)} - \psi_{(3)(j)}}{2\Delta x^2}$$

where

$$\Delta x = \frac{L/h}{NX - 1}$$

in each liquid layer.

At the interface a–b, the temperature field is continuous and the heat flux is held fixed. To second-order accuracy the expression for the temperature of the nodes located on the interface is

$$T_{a/b(i)} = \frac{(\kappa_a/\kappa_b)(4T_{a(i)(NY_a-1)} - T_{a(i)(NY_a-2)}) + 4T_{b(i)(2)} - T_{b(i)(3)}}{3(1 + \kappa_a/\kappa_b)}$$

The interfaces are assumed to remain flat. The expressions for the vorticity along interfaces are found from the equality of the horizontal velocities and the expression of the Marangoni effect:

$$\varphi_a - \frac{\eta_b}{\eta_a}\varphi_b = \frac{Ma}{Pr}\left(\frac{\partial T}{\partial x}\right)_{a/b} \tag{16a}$$

$$\varphi_{a(i)(NY_a)} = -\frac{Ma}{Pr}\frac{\eta_1/\eta_a}{1 + \eta_b/\eta_a}\frac{T_{a(i+1)(NY_a)} - T_{a(i-1)(NY_a)}}{2\Delta x} \tag{16b}$$

with the Marangoni and the Prandtl numbers being defined, as earlier, by the properties of the first liquid ("1") layer.

In the experiment to be discussed later, the ambient temperature around the fluid cell was colder than the temperature of the cold flange. Although the fluid cell was carefully designed, when the temperature of the hot flange increased, a small part of the heat flux unavoidably escaped through the lateral quartz walls. These lateral heat losses have been estimated to be small and about $(6\ \mathrm{W/m^2\,K})\Delta T$ and we have to account for them in the numerics. The boundary condition for the temperature on the external faces of the quartz walls was

$$T_{(WX)(j)} = (4T_{(WX-1)(j)} - T_{(WX-2)(j)} - 2\Delta x\phi/\kappa_{\mathrm{wall}})/3$$

where ϕ is the dimensionless heat loss and κ_{wall} denotes the heat conductivity of the quartz wall.

The steady-state flow patterns were computed in each layer using a 120×30 grid (Fig. 5). The temperature field in the lateral walls was computed using a 24×88 grid. The convergence of the code was tested with different grids by increasing $NX \times NY$ to 500×70 for $Ma = 977.4$, corresponding to a temperature difference between the hot and the cold flange of $\Delta T^{\text{num}} = 10°\text{C}$. Because the uncertainty in the values of the parameters of the liquids was at least a few percent, the convergence obtained was considered satisfactory. There was no need to include an artificial disturbance at the beginning of the computation to induce a steady flow due to the influence of the lateral walls. Steady convective patterns were computed for discrete increments of the temperature of the hot side corresponding to increasing values of the Marangoni number. Then, under such conditions, the previously obtained steady-state situation served as an initial condition for the temperature increments.

B. Numerical Results for Stationary Convection

Three steady flow regimes were numerically observed (Fig. 6). Regime 1 [Fig. 6(a)] reflects the wall-induced flow due to the lateral temperature gradients. Regime 2 [Fig. 6(b)] corresponds to the onset of thermocapillary instability influenced, however, by the wall effect. Beyond an unsteady transition zone [Fig. 6(c)], one can see regime 3 [Fig. 6(d)] where the rotation of the vortices is reversed and is no longer determined by the wall.

1. Regime 1: Wall-Induced Flow [Fig. 6(a)]

The difference between the heat conductivities of the liquids and the quartz wall imposes temperature gradients along the free surfaces. The horizontal temperature gradient, A_x, along the 1–2 interface scaled by the vertical temperature gradient, $\Delta T^{\text{num}}/3d$, computed in the conductive state ($Ma = 0$) is given in Fig. 7. Although, close to the wall, the lateral gradient is large (up to 70% of the vertically imposed gradient) yet it does not penetrate in the liquids beyond a few millimeters away from the walls. It is mainly located in the knife-edge region ($10\,\text{mm} < x < 12\,\text{mm}$) where the cell design (Fig. 11) rules out thermocapillary convection.

The effect of the lateral gradient is clearly visible in Fig. 6 (a). Owing to the symmetry of the set-up, the temperature gradients along the free interfaces created by the walls have opposite signs. Consequently, near the walls, the 1–2 interface induces flow towards the center of the cell whereas the 2–3 interface drives flow towards the walls. The flow is very slow and no significant distortion of the isotherms from the state of rest was observed.

2. Regime 2 [Fig. 6(b)]

The existence of wall-induced flow introduces a major difficulty in the analysis: How to find the appropriate criterion defining the critical temperature difference, $\Delta T_{\text{c}}^{\text{num}}$, for the onset of convection. In an ideal situation where the lateral walls are

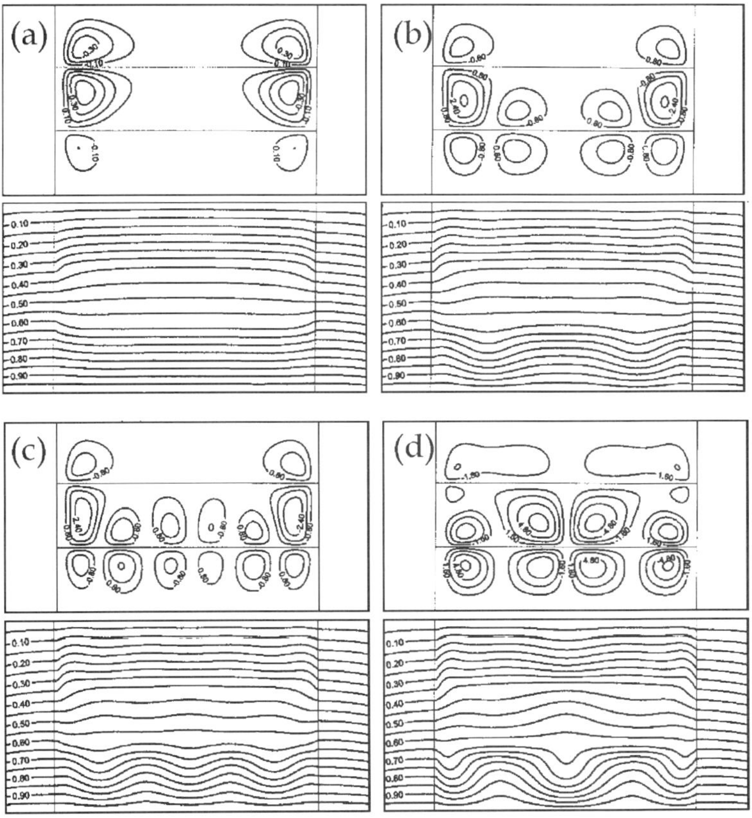

FIG. 6 Thermocapillary instability in multilayer systems. Streamlines ($\times 10^3$) and isotherms ($T/\Delta T^{\text{num}}$) obtained with a two-dimensional numerical computation: (a) $\Delta T^{\text{num}} = 1.0°\text{C}$; wall-induced flow or regime 1 ($\psi_{\text{max}} = 0.464 \times 10^{-3}$, $\psi_{\text{min}} = -0.464 \times 10^{-3}$); (b) $\Delta T^{\text{num}} = 6.0°\text{C}$; regime 2 ($\chi_{\text{max}} = 3.27 \times 10^{-3}$, $\chi_{\text{max}} \times -3.27 \times 10^{-3}$); (c) $\Delta T^{\text{num}} = 6.9°\text{C}$; transition between regimes 2 and 3 (unsteady) ($\chi_{\text{max}} = 2.94 \times 10^{-3}$, $\chi_{\text{min}} = -3.08 \times 10^{-3}$); (d) $\Delta T^{\text{num}} = 14.0°\text{C}$; regime 3 ($\chi_{\text{max}} = 7.03 \times 10^{-3}$, $\psi_{\text{min}} = -7.03 \times 10^{-3}$). Positive values of the stream function correspond to anticlockwise rolls. (From Ref. 5.)

perfectly insulating, ΔT_c^{num} is by definition the temperature difference at which departure from the motionless conductive state is observed, and the dynamic description of that case is clearly a perfect pitchfork bifurcation [4]. However, in experiment, the weak wall-induced flow is superimposed and acts as a bias in the transition that becomes imperfect. A simple, heuristic description of such a case is obtained using the following Ginsburg–Landau equation for a single amplitude A

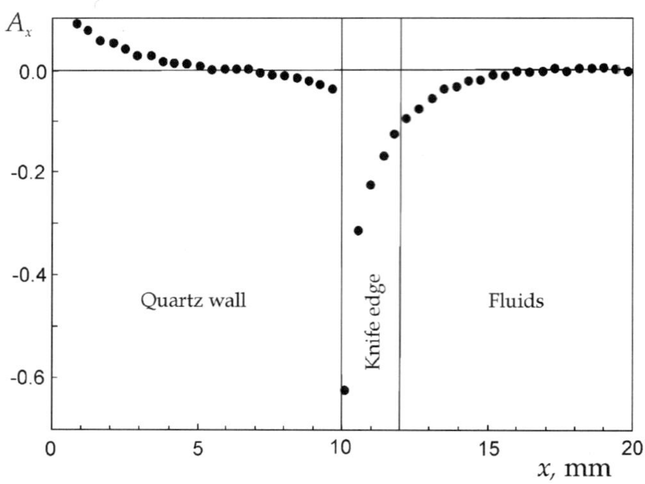

FIG. 7 Thermocapillary instability in multilayer systems. Horizontal temperature gradient, A_x, scaled by the vertical temperature gradient across the quartz wall ($0 < x < 10\,\text{mm}$), along the knife edge ($10\,\text{mm} < x < 12\,\text{mm}$), and along the 1–2 interface ($x > 12\,\text{mm}$). (From Ref. 5.)

(for illustration, take A as velocity amplitude) [6,7]:

$$\frac{\mathrm{d}A}{\mathrm{d}t} = \gamma + \varepsilon A - \beta A^3, \qquad \varepsilon = \Delta T - \Delta T_c \tag{17}$$

The parameter γ here is a measure of the strength of the imposed imperfection, assumed to be small. The strength of the lateral-wall-induced flow is proportional to ΔT in regime 1, so that we can take $\gamma = \zeta\Delta T$ with ζ a constant. In Eq. (17) β is a saturation parameter, whose explicit form is not needed here. At the steady state Eq. (17) reduces to

$$0 = \gamma + \varepsilon A - \beta A^3 \tag{18}$$

The steady amplitude, A_s, which corresponds to the imperfect bifurcation is the single root of the cubic equation, Eq. (18), while the two other roots are complex conjugates when $\varepsilon < 0$. We have

$$A_s = \frac{F}{3\sqrt[3]{2}\beta} - \frac{\sqrt[3]{2}\varepsilon}{F} \tag{19}$$

where

$$F = [27\beta^2\gamma + (729\beta^4\gamma^2 + 108\beta^2\varepsilon^3)^{1/2}]^{1/3}$$

In Fig. 8, values of the vertical component of velocity taken at $x = L/2$, $y = \frac{2}{3}d$, in layer 1 are plotted (open circles) as a function of ΔT. At steady state, they define the amplitudes A. The unknown parameters β, ζ, and $\Delta T_c^{\mathrm{num}}$ can be determined numerically by fitting Eq. (18) with the curve in Fig. 8. The prediction is $\Delta T_c^{\mathrm{num}} = 4.75°\mathrm{C}$. This value is in good agreement with the result obtained from the linear stability analysis (5.13°C).

As $\Delta T_c^{\mathrm{num}}$ is exceeded, the largely quiescent middle zone starts to move and two pairs of rolls appear in layers 1 and 2 [Fig. 6(b)]. This pattern is reminiscent of the previous linear analysis results [Fig. 2(b)]: the flow is driven by the 1–2 interface and layer 3 remains practically motionless in the central region of the container where the wall-induced flow does not penetrate.

3. Transition Between Regimes 2 and 3 [Fig. 6(c)]

At around $\Delta T^{\mathrm{num}} = 6.9°\mathrm{C}$, the flow pattern undergoes a complex transformation. Two rolls appear in the central region of layers 1 and 2. This flow pattern, shown in Fig. 6(c), is unsteady and its stability range has not been investigated.

4. Regime 3 [Fig. 6(d)]

The central rolls that appeared during the stage just described, grow and push the lateral rolls towards the walls. Finally, in layer 2, the lateral rolls are reduced to two very small vortices located close to the walls near the 2–3 interface. In layer 3, the

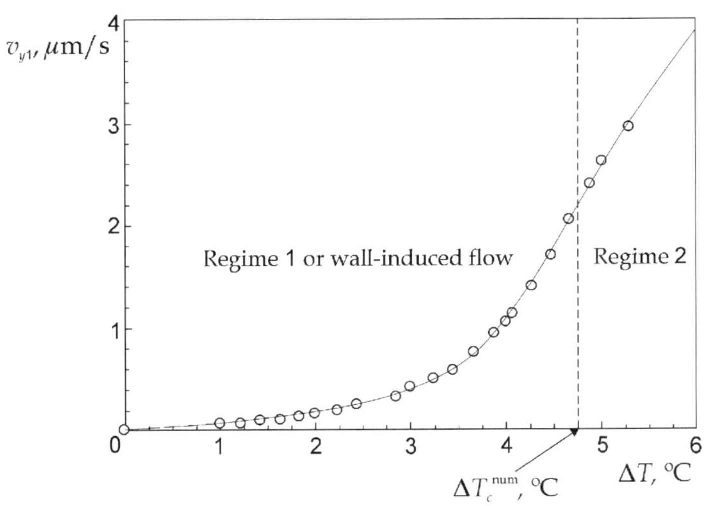

FIG. 8 Thermocapillary instability in multilayer systems: v_y in layer 1 taken at ($x = L/2$, $y = \frac{2}{3}d$) (○) plotted as a function of ΔT fitted with the real root of Eq. (18) (solid line); $\zeta = 0.23\,\mu\mathrm{m/(s°C)}$, $\beta = 0.11°\mathrm{C}^{-1}\,(\mu\mathrm{m/s})^{-2}$ and $\Delta T_c^{\mathrm{num}} = 47.5°\mathrm{C}$. (From Ref. 5.)

flow consists of two wide vortices coupled with the central rolls of layer 2 and with the two small residual vortices. This description of the flow pattern is valid over a wide range of values of the thermal constraints. The structure of the flow remains qualitatively the same for ΔT^{num} up to 20°C.

5. Velocity Profiles

Three horizontal profiles of v_y in layer 1 ($y = \frac{2}{3}d = 5.3$ mm), which account well for the three different regimes, are shown in Fig. 9. In regime 1 ($\Delta T^{\mathrm{num}} = 2.0$°C, ■), the flow is upward near the walls and is not appreciable in the center. Regime 2 ($\Delta T^{\mathrm{num}} = 6.6$°C, ▲) appears essentially as an amplification of the wall-induced flow. Flow is upward near the walls and in the center. In regime 3 ($\Delta T^{\mathrm{num}} = 9.7$°C, ●) the flow is in the opposite direction along the lateral walls. This situation where the flow is going downward along the lateral wall is preferred even though it is not favored by the wall-induced thermal gradient. The reason why the Marangoni effect tends to drive the interface flow towards the rigid conducting boundary rather than the reverse is still an open question. Nevertheless, the existence of a preferred direction is natural because the conditions for the fluid to move away from the lateral walls or towards it are not identical.

C. Numerical Results for Oscillatory Convection

Let us now describe results obtained with the two-dimensional nonlinear numerical computation of the boundary-value problem. They are to be compared

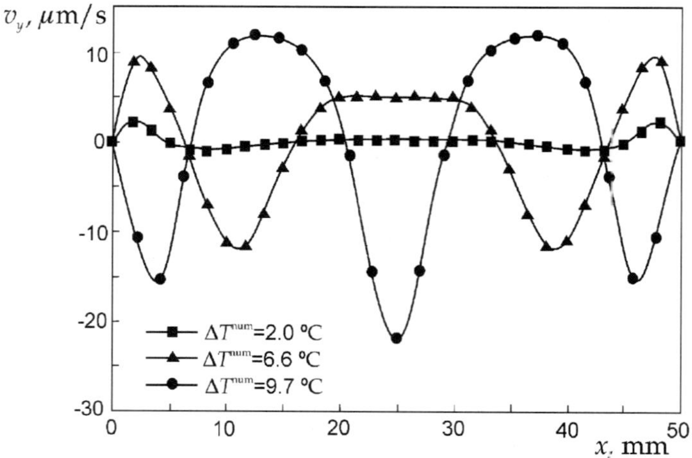

FIG. 9 Thermocapillary instability in multilayer systems: v_y in layer 1 at $y = \frac{2}{3}d$ plotted vs. x for the three different regimes. (From Ref. 5.)

with an experiment [2] realized with fluids of similar heat diffusivity, i.e., *n*-octane–methanol.

The boundary-value problem was solved by the finite-difference method. Equations were approximated on a rectangular uniform 42×84 mesh using a second-order approximation for the spatial coordinates. The Poisson equations were solved by the iterative Liebman successive over-relaxation method on each time step: the accuracy of the solution was 10^{-5}.

The mechanical equilibrium state became unstable with respect to oscillatory disturbances for $Ma \geq Ma_c^n = 3400$. The instability leads to time periodic oscillations, which appear in a supercritical way. The streamlines during one period are shown in Fig. 10. As expected, near threshold the oscillations are quasi-harmonic. One can see that there are parts of the period when the motion takes place mostly near the upper cold interface, and others when the lower hot interface plays an active role in supporting the motion.

Please note that flow transfer from the upper cold interface to the lower hot interface and its reverse occurs under different schemes. Accordingly, even near threshold we can have nonsinusoidal oscillations. After state (a) in Fig. 10 on the lower interface a thermal "spot" generates counter vortices in the middle layer. A thermocapillary elasticlike force arises on this interface. In the middle layer a two-vortex structure is generated and heat transfer through the middle layer becomes ineffective. In these phases of the oscillation the lower layer effectively brings warm liquid to the lower interface, supporting the initiated motion [Fig. 10(b)]. The previous motion in the middle layer disappears and, simultaneously, the flow in the upper layer stops. Subsequently, the flow in the middle layer becomes single-vortex and strong [Fig. 10(c)]. A gearlike interaction together with a thermocapillary force on the upper interface induces new vortices in the upper layer, and the thermocapillary elasticlike force on the lower interface stops nearby motion [Fig. 10(d)]. The second half of the period of oscillation begins with an opposite direction of rotation of whirlwinds. Thus, the role of the two interfaces in creating the oscillation appears different. Owing to the absence of a solid boundary the middle liquid layer allows flow more easily than the external layers. For the structure of the oscillatory movement it appears important from what side the hot interface is relative to the middle layer.

III. EXPERIMENT UNDER ALMOST VANISHING EFFECTIVE GRAVITY (MICROGRAVITY) WITH A SYMMETRIC THREE-LAYER SYSTEM

Two microgravity experiments were performed, the first in June 1994 during the IML2 mission of Spacelab aboard the US Space Shuttle [5] and the second during the LMS mission in June 1996 [2]. The experiments were conducted in the Bubble Drop and Particle Unit (BDPU) as part of the programme sponsored by the

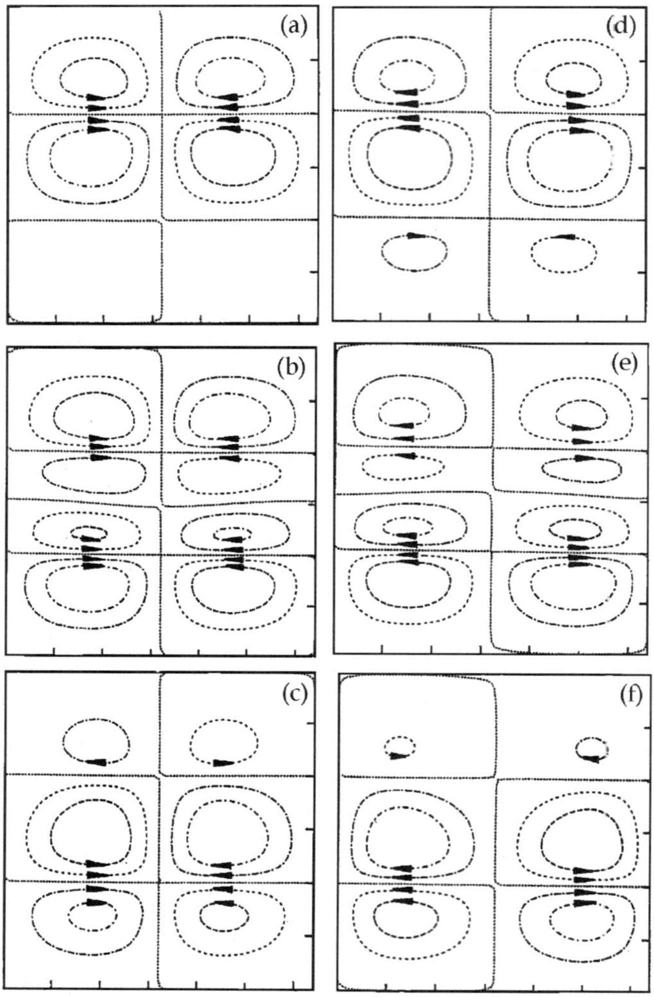

FIG. 10 Thermocapillary instability in multilayer systems. Streamlines for periodic oscillatory motion in the n-octane–methanol system at $Ma = 4500$. (From Ref. 2.)

European Space Agency (ESA). The BDPU is a multi-user facility designed to study fluid physics in variable effective gravity. This instrument provides the thermal control and the flow visualization tools for a dedicated experimental cell. During the IML2 mission, the Fluorinert–Silicone oil–Fluorinert system was investigated and during the LMS mission, the n-octane–methanol–n-octane system was studied.

A. Description of the Experiment

The experimental set-up is depicted in Fig. 11. The container is made of a rectangular quartz frame closed on two sides by two (1 cm thick) sapphire windows. Each liquid layer is 8 mm thick, 50 mm wide, and 35 mm deep. Initially, the layers are separated by two 50-μm thick stainless steel curtains to prevent mixing before the beginning of the experiment. The boundaries between the layers are kept leak-tight by Viton seals moulded on an antiwetting frame forming a knife. Because the knife edge penetrates 2 mm into the fluid cell, the effective lengths of the free interfaces are 46 mm rather than the original 50 mm. At the beginning of the experiment, the curtains are pulled out and gently spooled (0.25 mm/s) on the rollers (Fig. 11), creating two free liquid–liquid interfaces subjected to the Marangoni effect. Heating and cooling are achieved with an electrical resistance and Peltier elements located along the edges of the sapphire windows.

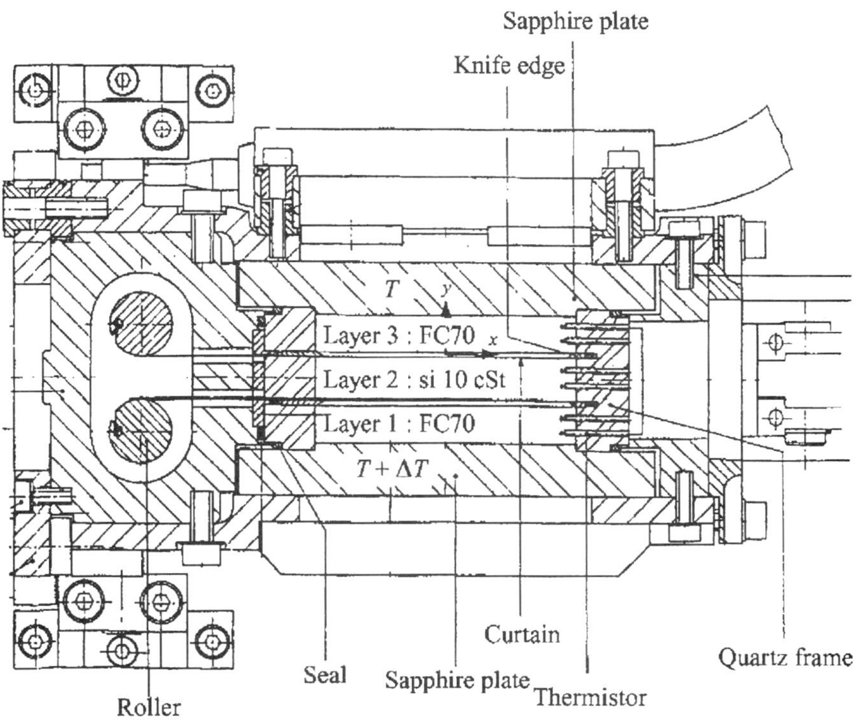

FIG. 11 Thermocapillary instability in multilayer systems. Outline of the experimental set-up. (From Ref. 5.)

TABLE 1 Thermocapillary Instability in Multilayer Systems: Physical Properties of the Liquids for the Monotonic Instability Experiment ($1\,\text{St} = 10^{-4}\,\text{m}^2/\text{s}$) [5]

	Fluorinert (Layers 1 and 3)	Silicone oil 10 cSt (Layer 2)	(1)/(2)
ν (m^2/s)	13.4×10^{-6}	9.08×10^{-6} (35°C)	1.48
η (kg/ms)	2.55×10^{-2}	8.40×10^{-3} (35°C)	3.04
κ (W/mK)	7.00×10^{-2}	1.34×10^{-1}	0.522
χ (m^2/s)	3.44×10^{-8}	9.51×10^{-8}	0.362
ρ (kg/m^3)	1940	934	2.08
Pr	390	95	4.11
σ_0 (mN/m)		6.9	
σ_1 (mN/mK)		-2.7×10^{-2}	

Fluorinert FC70 (3M) for the outer liquid layers and 10 cSt (Dow Corning) silicone oil for the middle layer were used to observe stationary convection. The physical properties of the liquids and their relevant ratios are listed in Table 1.

As noted earlier, in microgravity the static shapes of connected liquid phases are governed by the interfacial forces only. In the absence of walls, a three-layer configuration such as shown in Fig. 1 would be unstable. Indeed, the surface tension forces would act to minimize the surface with respect to the volume of the layers, curving the free interfaces and leading presumably to connected spherical phases (big drops). In a finite container, the situation is different because the layers will remain flat provided that their edges are properly anchored on the walls. Anchorage was achieved using two stainless steel frames coated on one side with a Teflon® film that prevents creeping of the silicone oil along the quartz wall. Indeed, unlike quartz and stainless steel, Teflon is wetted preferentially by Fluorinert rather than by silicone oil. Thus the Teflon-coated sides of the frames are oriented towards the Fluorinert layers.

Besides controlling the wall wetting, the experimental set-up was designed to satisfy another specific requirement: the interfacial tension has to balance effectively the liquid density difference. This condition has to be fulfilled because, as earlier noted, in the Space Shuttle there is a non-negligible low residual effective gravity.

If a Rayleigh–Taylor instability is to occur as a result of the residual gravity being normal to the interfaces, it is expected that the surface tension will stabilize the system with respect to flow motions of short enough wavelengths. The critical wavelength, λ_m, for the onset of the Rayleigh–Taylor instability for two liquids of infinite lateral extent is given by the capillary length here expressed as

$$\lambda_m = 2\pi\left[\frac{\sigma_0}{(\rho_2 - \rho_1)g}\right]^{1/2} \tag{20}$$

Thus maximum tolerable effective gravity level, g^*, can be estimated from Eq. (20) knowing that the 46-mm length of the free available surface has to be smaller than λ_m to exclude the appearance of the Rayleigh–Taylor instability. The prediction is $g^* \approx 1.30 \times 10^{-2}\, g_0$, where $g_0 = 9.81\,\text{m/s}^2$.

As in the Space Shuttle, the mean microgravity level is lower than $10^{-4}\, g_0$, and the Rayleigh–Taylor instability is ruled out for the geometry of the experiment (Fig. 11) and indeed was not observed.

For flow visualization the liquids were seeded with silver-coated ceramic spheres with diameters in the range 80–100 μm. The flow was visualized using a He–Ne laser light sheet oriented parallel to the longest side of the test container and located 10 mm from the front wall. To avoid sedimentation due to residual acceleration, the densities of the particles matched the densities of the liquids to within a few percent.

B. Experimental Results for the Monotonic Instability Experiment

1. Thermal Data

Six thermistors, penetrating 5 mm into the liquid bulk, were inserted along the lateral wall of the quartz frame to record the temperatures near the lateral walls, throughout the experiment. The precise locations of these thermistors and the recording of the signals during the experiment are shown in Fig. 12.

The instrumentation has to obey several constraints. For instance, in order to change the temperature difference across the three-layer system, the reference point on the hot side had to be varied by jumps of one degree. The time span allowed for this experiment was rather short and it did not allow very accurate measurements of the critical temperature difference at the onset of convection.

As noted earlier, for high Prandtl number fluids, beyond the onset of convection, the velocity field is slaved to the thermal field. Accordingly, the intensity of the flow will be reflected by the temperature recordings at the thermistors. From these signals, the numerical results were checked and the three earlier discussed different regimes were identified (wall-induced flow, regime 2 and regime 3) depending on the temperature difference imposed between the two sapphire conductive plates. Because the distance between the thermistors above and beneath any interface was 5.75 mm (Fig. 12), a parameter, Δ_2, was used to compare the temperature gradients across the interfaces:

$$\Delta_2 = \frac{T_4 - T_5}{5.75} - \frac{T_2 - T_3}{5.75} \tag{21}$$

To compare experimental data and numerical predictions, Δ_2 is plotted vs. the temperature difference, $\Delta T(\Delta T^{\text{exp}}$ and $\Delta T^{\text{num}})$, across the three layers in Fig. 13.

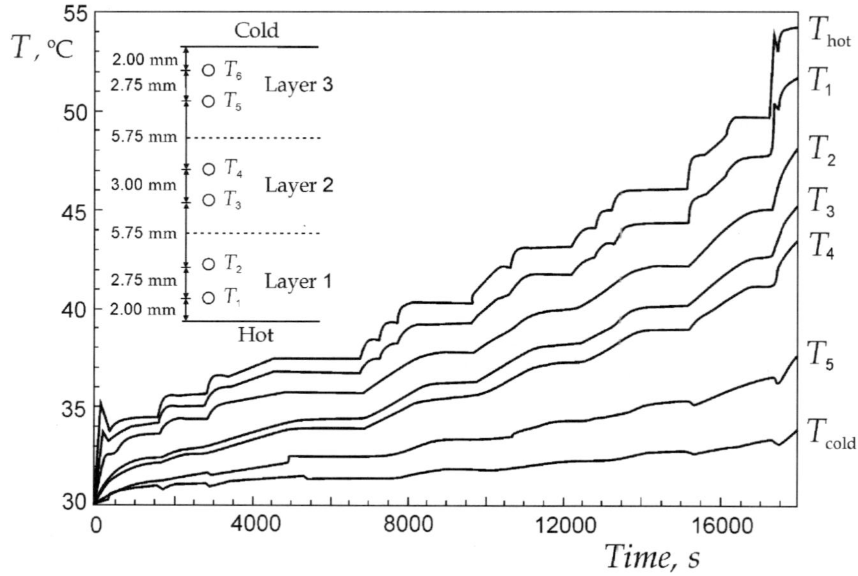

FIG. 12 Thermocapillary instability in multilayer systems. Position of thermistors and recordings during the duration of the experiment. (From Ref. 5.)

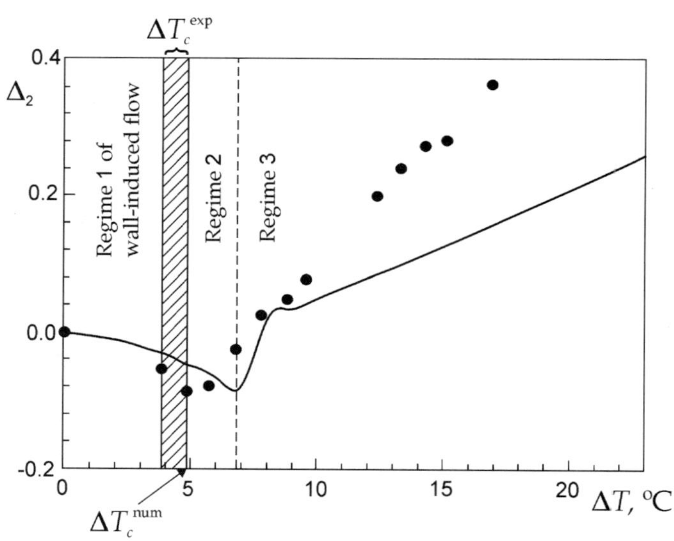

FIG. 13 Thermocapillary instability in multilayer systems. Comparison between experimental (●) and numerical (solid line) values of Δ_2 at quasi-steady state. (From Ref. 5.)

The trend is very similar for both the experimental data and the numerical results. The reason for choosing Eq. (21) is that in regime 1, Δ_2 has to be close to zero because of the symmetry of the configuration and because the isotherms are not perturbed; $T_4 - T_5$ is more or less equal to $T_2 - T_3$ in the range of the imposed ΔT. When ΔT is increased, the hotter 1–2 interface pushes the liquids away from the walls, bringing hot Fluorinert and cold silicone oil to the 1–2 interfaces near the lateral wall. But in layer 3, the flow is very slow and the thermal field is negligibly undistorted from the conductive state. This means that $T_2 - T_3$ increases whereas $T_4 - T_5$ barely changes and hence Δ_2 becomes negative. In the region $\Delta_2 < 0$, the experimental curve reaches its minimum value earlier than the numerical curve, indicating that ΔT^{exp} is smaller than $\Delta T^{\text{num}} = 4.75°\text{C}$.

As ΔT is increased still further, the surface tension gradient (Marangoni) driven convection becomes more and more important, reducing the relative thermal influence of the walls on the flow. This means that the rotation direction of the vortices at the lateral walls is no longer determined by the wall lateral gradient effect. It is rather dictated by the effect of the surface tension gradient. When the 1–2 interface is pushing the liquids towards the wall, the colder Fluorinert and the hotter silicone oil move away from that interface. But, $T_2 - T_3$ then decreases and $T_4 - T_5$ increases, so that Δ_2 increases towards zero and becomes positive. Both the numerical computation and the experiment exhibit such a behavior. Generally, the numerical curve matches well the experimental points for $\Delta T < 10°\text{C}$. Beyond this value of ΔT, the recorded experimental values of Δ_2 are significantly larger than their numerical counterparts, indicating that the real flow strength is stronger than the numerical one (Fig. 13). The transition between regime 2 and regime 3 corresponding to the reverse of the rotation of the lateral vortices occurs at $\Delta T^{\text{exp}} = 6.9°\text{C}$ in experiment whereas it occurs at $\Delta T^{\text{num}} = 8.6°\text{C}$ in the numerical computations. The agreement for this nonlinear effect thus appears acceptable.

2. Flow Pattern Analysis

The flow was visualized using a light sheet parallel to the widest side of the experimental cell; it is 1 mm thick and located 10 mm away from the front window inside the liquids. The images were digitized and the velocity of the tracer particles was measured using a gray level correlation algorithm. The flow was resolved in $1.2\,\text{mm}^2$ windows. To improve the signal-to-noise ratio and to increase the number of velocity vectors computed, the velocity maps were time averaged. The average velocity vector, $\bar{\mathbf{v}}(x, y)$, was computed as follows

$$\bar{\mathbf{v}}(x, y) = \frac{1}{N} \sum_{t=t_0}^{t_1} \mathbf{v}(t)(x, y)$$

where N was the number of samples acquired between t_0 and t_1; t_0 denotes the start of the sampling sequence and t_1 the end of the sampling sequence.

3. Onset of Convection

No flow was observed inside the fluid cell for time $t < 850$ s when the temperature differences imposed across the three layers were below $\Delta T^{\mathrm{exp}} = 3.8°\mathrm{C}$. Flow was observed from $t = 2860$ s ($\Delta T^{\mathrm{exp}} = 4.9°\mathrm{C}$) onwards. Instability appeared for $\Delta T_{\mathrm{c}}^{\mathrm{exp}}$ between 3.8 and 4.9°C.

4. Regime 2

A flow pattern was observed in layers 1 and 2 with two rolls located in the central region of the cell. Layer 3 was at rest [Fig. 14(a)]. In the middle of the cell the flow in layer 1 was found rising towards layer 2, as observed in the numerical computations for regime 2. Near the lateral walls, the flow appeared to be very weak. In fact, convection always starts near the lateral walls [8]. It is very likely that, in that region, the largest components of the velocity vectors were perpendicular to the visualization plane, and thus no measurements were possible.

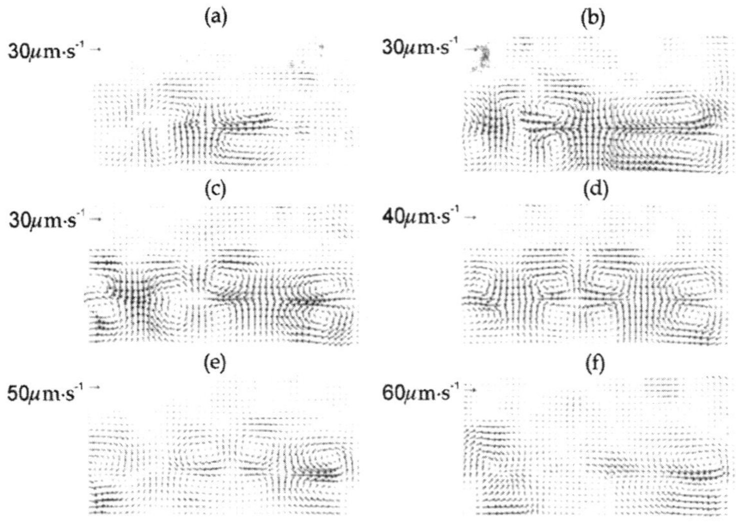

FIG. 14 Thermocapillary instability in multilayer systems. Experimental velocity fields, vector scale is given in upper left corner: (a) $\Delta T^{\mathrm{exp}} = 5.5°\mathrm{C}$, $v_x(\max) = 44\,\mu\mathrm{m/s}$, $v_y(\max) = 28\mu\mathrm{m/s}$; (b) $\Delta T^{\mathrm{exp}} = 6.7°\mathrm{C}$, $v_x(\max) = 38\,\mu\mathrm{m/s}$, $v_y(\max) = 34\mu\mathrm{m/s}$; (c) $\Delta T^{\mathrm{exp}} = 7.8°\mathrm{C}$, $v_x(\max) = 41\,\mu\mathrm{m/s}$, $v_y(\max) = 36\mu\mathrm{m/s}$; (d) $\Delta T^{\mathrm{exp}} = 8.7°\mathrm{C}$, $v_x(\max) = 57\,\mu\mathrm{m/s}$, $v_y(\max) = 37\mu\mathrm{m/s}$; (e) $\Delta T^{\mathrm{exp}} = 9.7°\mathrm{C}$, $v_x(\max) = 72\,\mu\mathrm{m/s}$, $v_y(\max) = 43\mu\mathrm{m/s}$; (f) $\Delta T^{\mathrm{exp}} = 12.4°\mathrm{C}$, $v_x(\max) = 82\,\mu\mathrm{m/s}$, $v_y(\max) = 44\mu\mathrm{m/s}$. (From Ref. 5.)

As ΔT^{exp} was increased, four convective rolls became clearly visible in layers 1 and 2 [Fig. 14(b)].

5. Regime 3

The roll at the right-hand side, initially much larger than the others, progressively shrank [Fig. 14(c),(d)]. The pattern became more regular as different rolls became of equal size [Fig. 14(e)]. The long exposure time picture given in Fig. 15 allowed visualization of the streamlines in regime 3. While convection appeared to be intense in layers 1 and 2, it was very weak in layer 3. At the walls, the direction of rotation was found to be opposite to that of regimes 1 and 2, as predicted in the numerical computation. Near the interface 2–3, close to the lateral walls, the two residual vortices were clearly visible. The characteristic shapes of the central vortices in layer 2 agree with the numerical predictions.

As ΔT^{exp} was further increased, the v_x velocity component perpendicular to the light sheet became dominant in the central region of layers 1 and 2 and the structure of the flow could not be made observable using two-dimensional visualization [Fig. 14(f)]. Vortices were, however, still visible near the lateral walls.

6. Velocity Profiles

Corresponding to the numerical results (compare with Fig. 9), experimental horizontal v_y profiles (taken at $y = \frac{2}{3}d$) are shown in Fig. 16. For the profile that

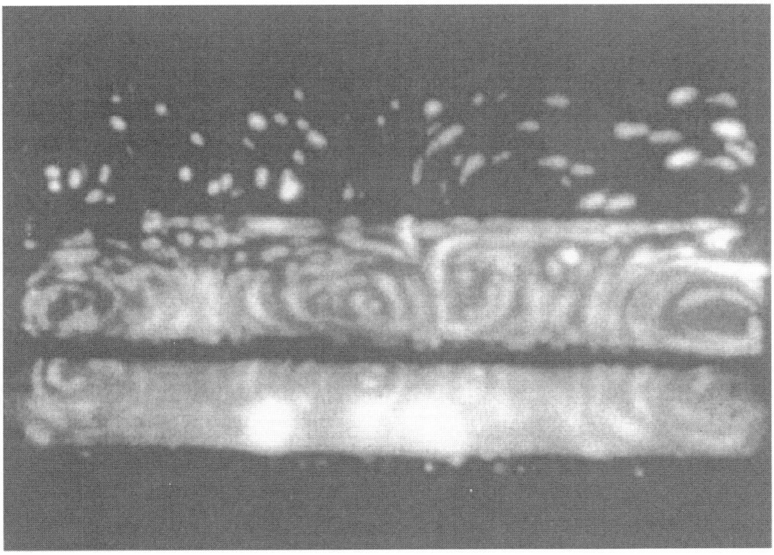

FIG. 15 Thermocapillary instability in multilayer systems. Long exposure time picture of the convective flow, $\Delta T^{exp} = 9.7°C$. (From Ref. 5.)

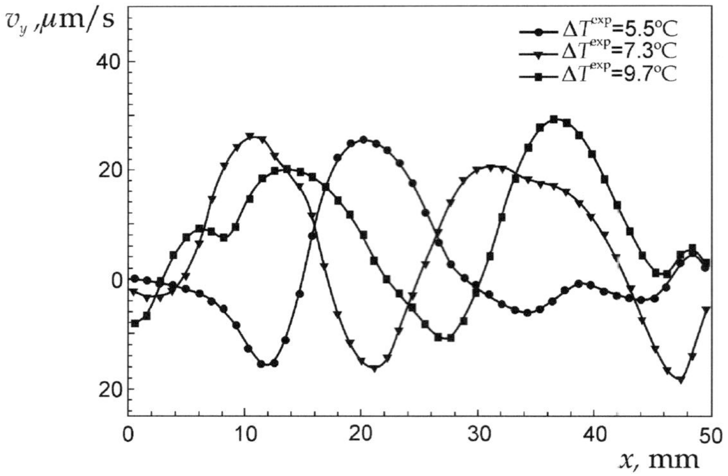

FIG. 16 Thermocapillary instability in multilayer systems. Experimental velocity profile v_y in layer 1 at $y = \frac{2}{3}d$ along the coordinate x. (From Ref. 5.)

corresponds to regime 2 (●), the flow is rising up in the center region and very weak near the walls. The profile is shifted to the left with increasing ΔT^{exp}. The magnitude of the velocity was not found to increase as ΔT^{exp} increased. This is probably an artifact resulting from the three-dimensional nature of the convection cells as the light sheet is not located exactly in a plane where the z velocity component vanishes.

7. Comparison of Experimental Data with the Linear Stability Analysis and the Numerical Results

Visual observations indicated that convection started for a temperature difference between 3.8 and 4.9°C. The comparison of the thermal data using Δ_2 indicates that the effect of the flow on the thermal field is qualitatively identical in the experiment and in the numerical computation (Fig. 13). It was found that $\Delta T_{\mathrm{c}}^{\mathrm{exp}}$ was always below $\Delta T_{\mathrm{c}}^{\mathrm{num}} = 4.75°\mathrm{C}$ but this does not permit the uncertainty in $\Delta T_{\mathrm{c}}^{\mathrm{exp}}$ to be further reduced. Nevertheless, $\Delta T_{\mathrm{c}}^{\mathrm{exp}}$ was found to be within 20% of $\Delta T_{\mathrm{c}}^{\mathrm{num}}$, which is rather satisfactory, because even in an accurately controlled ground-based experiment one expects an error of about 10% on the measured Marangoni number [8].

The experimental results match the results of the two-dimensional numerical computations and the linear stability analysis. The instability is driven by the 1–2 interface in layers 1 and 2 while layer 3 is almost at rest. The number of convection cells (two) is in perfect agreement with the linear stability analysis, which predicts that the critical wavelength, $\lambda_c = 2.56$ cm, is half the length of the container. Both regimes 2 and 3, identified in the numerical study, were observed experimentally through the dependence of Δ_2 on ΔT^{exp} and on the velocity fields. The inversion of the rotation of the vortices in the numerics was visible in the experimental pattern as the comparison of Fig. 14(a) and 14(d) shows. In regime 3, experimental patterns were found to be very similar to the numerical predictions. In particular, specific flow features, related to the effect of the lateral walls, are present in both cases [compare Fig. 6(d) and Fig. 15].

C. Analysis of the Results for the Oscillatory Instability Experiment

Oscillatory instability was observed using *n*-octane for the outer layers and methanol for the middle layer. The physical properties of the liquids and their relevant ratios are listed in Table 2. This system has a thermal diffusivity ratio close to unity and hence one expects only oscillatory instability.

With respect to the above-described previous set-up, three additional thermistors, penetrating 5 mm into the liquid bulk, were inserted along the lateral wall to record the temperature throughout the experiment. The locations of the thermistors are given in Fig. 17 and the recordings of the signals during the experiment are shown in Fig. 18.

As with the previous experiment, the instrumentation had to obey several constraints. In order to change the temperature difference across the three-layer system, the reference point on the hot side had to be varied by jumps of 1°. The time span allowed for this experiment was rather short and, as in the earlier case, it did not allow very accurate measurements of the critical temperature difference at the onset of convection. When the temperature difference across the three layers,

TABLE 2 Thermocapillary Instability in Multilayer Systems. Physical Properties of the Liquids for the Oscillatory Instability Experiment [2]

	n-Octane (Layers 1 and 3)	Methanol (Layer 2)	(1)/(2)
ν (m^2/s)	7.0×10^{-7}	8.0×10^{-7}	0.875
η (kg/ms)	5.51×10^{-4}	5.62×10^{-4}	0.98
κ (W/mK)	2.15×10^{-1}	1.5×10^{-1}	1.43
χ (m^2/s)	1.09×10^{-7}	1.02×10^{-7}	1.07
ρ (kg/m^3)	703	787	0.897
Pr	6.42	7.84	0.82
σ_0 (mN/m)		4	
σ_1 (mN/mK)		1.9×10^{-2}	

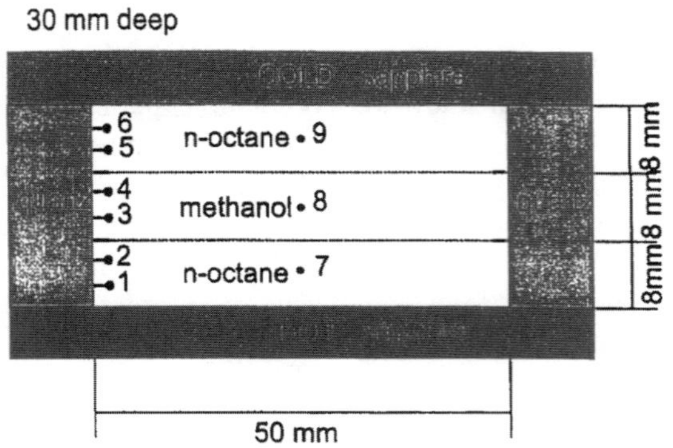

FIG. 17 Thermocapillary instability in multilayer systems. Location of the thermistors in the oscillatory instability experiment. (From Ref. 2.)

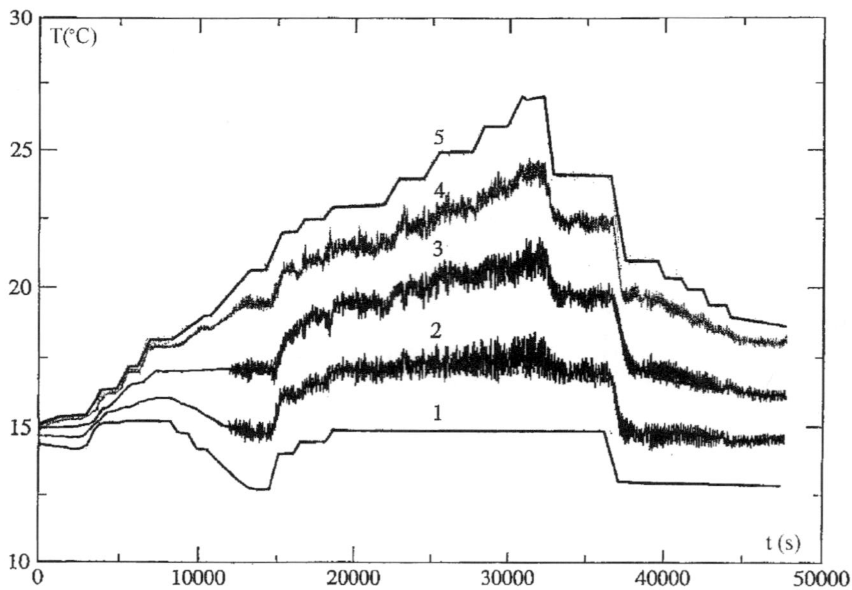

FIG. 18 Thermocapillary instability in multilayer systems. Recording of thermistor signals during the experiment: $T(t): T_{\text{cold}}$ (line 1), T_7 (2), T_8 (3), T_9 (4), T_{hot} (5). (From Ref. 2.)

ΔT, was relatively small, all disturbances decayed, and the system remained at mechanical equilibrium. With the increase of ΔT this mechanical equilibrium became unstable and oscillatory instability developed in the system. Oscillations appeared when the experimental time t reached the value $t_c = 11,700$ s and further during all the experiment ($t = 47,500$ s) oscillations had an irregular character. Note that both interfaces play an active role in the oscillatory process. The critical Marangoni number determined in the experiment was $M_c^{\exp} = 4000$. The critical Marangoni number determined for this system from the linear stability analysis was equal to 3390.6. The agreement thus appears satisfactory.

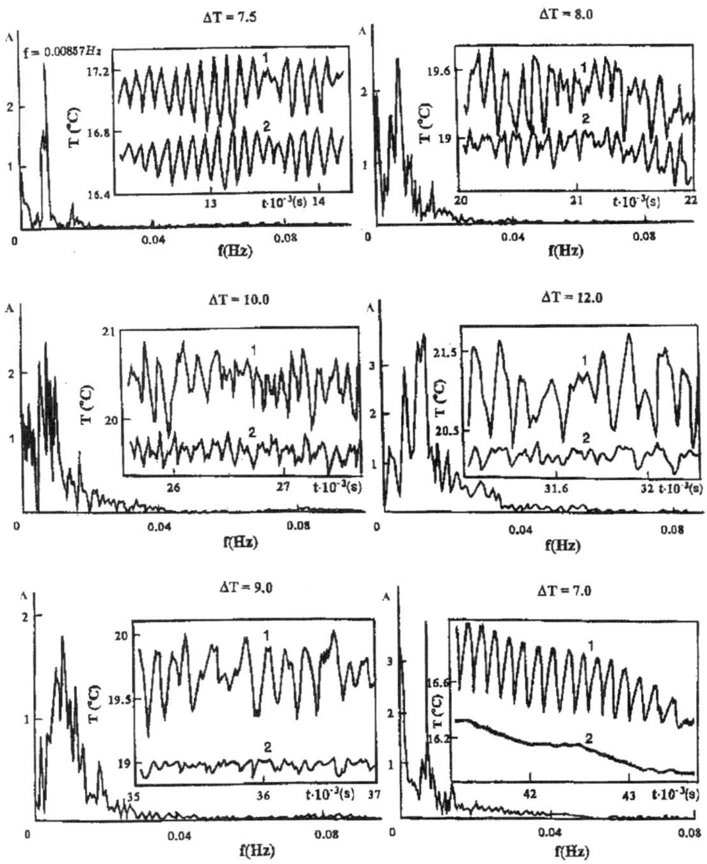

FIG. 19 Thermocapillary instability in multilayer systems. Evolution of temperature recordings inside the methanol layer and the corresponding frequency spectrum. (From Ref. 2.)

The evolution of the temperature recordings inside the methanol layer and the corresponding frequency spectrum are shown in Fig. 19. At the onset of convection, the oscillation pattern appears simple although the amplitude is not regular. The amplitude of the temperature oscillation recorded on the side thermistor and on the one at the center were of the same order of magnitude. The frequency spectrum contained a dominant peak at $f = 0.00857$ Hz. In the corresponding dimensionless form, $\omega_c = 4.92$. There is again good agreement with the linear theory stability analysis.

When ΔT was increased, the oscillation pattern became more and more complex and the spectrum exhibited multiple peaks. Past onset the main frequency tended to increase as ΔT increased. When ΔT was decreased back to the threshold value, the oscillations became regular as at the beginning of the experiment.

REFERENCES

1. Georis, Ph., Hennenberg, M., Simanovskii, I.B., Nepomniashchy, A., Wertgeim, I., Legros, J.C. Thermocapillary convection in a multilayer. Phys. Fluids A **1993**, *5*, 1575–1582.
2. Georis, Ph., Hennenberg, M., Simanovskii, I.B., Nepomniashchy, A., Legros, J.C. Oscillatory instability in multilayer systems (unpublished).
3. Funada, T. Marangoni instability of thin liquid sheets. J. Phys. Soc. Japan **1986**, *55*, 2191–2202.
4. Dauby, P.C., Lebon, G. Benard–Marangoni instability in rigid rectangular containers. J. Fluid Mech. **1996**, *329*, 25–64.
5. Georis, Ph., Hennenberg, M., Lebon, G., Legros, J.C. Investigation of thermocapillary convection in a three-liquid-layer system. J. Fluid Mech. **1999**, *389*, 209–228.
6. Manneville, P. *Dissipative Structures and Weak Turbulence*; Academic Press: New York, 1990.
7. Colinet, P., Legros, J.C., Velarde, M.G. *Nonlinear Dynamics of Surface-Tension-Driven Instabilities*. Wiley-VCH: Berlin, 2001.
8. Koschmieder, E.L. *Benard Cells and Taylor Vortices*; Cambridge University Press: Cambridge, 1992.
9. Funada, T., Kubo, T., Matsuura, S. Marangoni instability of unsteady mode for a liquid sheet. Numazu College Technol. Research Annual **1994**, *28*, 53–59.

9

Thermocapillary Convection of Constrained Interfaces

In the previous chapters we have always considered that interfaces or open surfaces were mechanically free. However, in some experimental conditions the surface of the liquid is constrained by solid guards or structural elements used to hold the liquid. In the present chapter we shall study the stability of the mechanical equilibrium with respect to spatially periodic perturbations in a plane liquid layer whose open surface is covered by a grid or lattice made of parallel solid elements. Both transverse and longitudinal temperature gradients will be considered. First, we shall find the possible instability modes according to linear theory. Then we shall explore the nonlinear growth of convective motions using the finite difference method to integrate the full thermohydrodynamics equations. For simplicity, we shall disregard gravity.

I. LINEAR STABILITY ANALYSIS

As in earlier studied cases, let us consider a plane layer (depth here denoted by h) of a viscous incompressible liquid placed on a solid support held at constant temperature Θ. The upper surface is supposed to be covered with a solid grid consisting of thin flat strips or plates (elements of a lattice) of width a and periodicity d. We shall limit consideration to the case when the heat conductivity of the grid is close to that of the liquid. Hence, a uniform cooling of the upper surface provides equal temperature all over the surface, which we take as a reference temperature. The coordinates, geometrical parameters of the system, and boundary conditions are shown in Fig. 1. Under such conditions mechanical

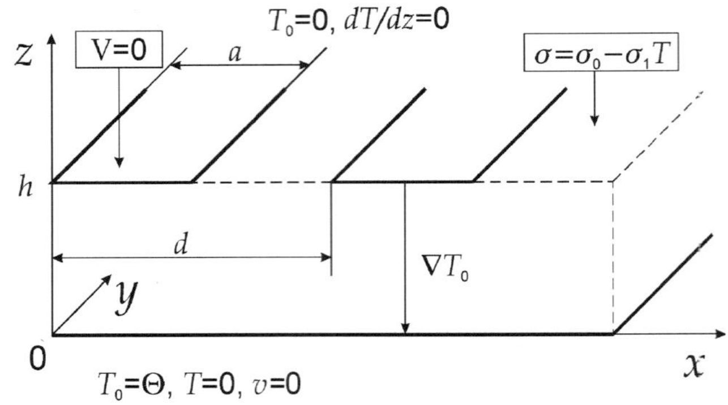

FIG. 1 Thermocapillary convection of a liquid surface with a solid grid. Geometry of the liquid layer and boundary conditions.

equilibrium with linear temperature distribution across the layer is possible. We have

$$\mathbf{v} = 0, \quad T_0 = \Theta\left(1 - \frac{z}{h}\right) \tag{1}$$

We shall study the influence of the Marangoni effect upon its stability.

A. Equations for Small Perturbations

Let us consider the stability of the state of Eq. (1) with respect to small perturbations of velocity, **v**, and temperature, T. We shall restrict consideration to monotonic disturbances. The equations for neutral perturbations can be written as [1]

$$\text{div } \mathbf{v} = 0 \tag{2}$$

$$\Delta \mathrm{v}_x = \Delta \mathrm{v}_y = \Delta \mathrm{v}_z = 0 \tag{3}$$

$$\Delta T = -Pr \mathrm{v}_z \tag{4}$$

where v_x, v_y, and v_z are the velocity projections on the axes x, y, and z, respectively (z is the transverse coordinate in the layer), and Pr is the Prandtl number. Quantities are made dimensionless using the following scales: h, ν/h, and Θ for distance, velocity, and temperature, respectively. On the solid regions of the boundary there is viscous adhesion of the liquid, hence no slip. The open regions of the boundary are assumed flat with the surface tension, as in earlier cases studied, being linearly dependent on temperature, $\sigma = \sigma_0 - \sigma_1 T$ and σ_1 positive. As usual, if flow develops the thermocapillary forces balance the viscous stresses in the open regions.

The boundary conditions for the temperature are defined as follows. As already noted, the lower boundary is assumed to have fixed uniform temperature. At the upper boundary we allow temperature perturbations capable of generating convective instability due to the Marangoni effect. The heat conductivity of the ambient air is assumed small compared to the liquid heat conductivity and hence temperature changes do not affect the heat flux through the open, albeit constrained, liquid surface.

The prescribed boundary conditions can be written in dimensionless form as

$$z = 0\text{: } \mathrm{v}_x = \mathrm{v}_y = \mathrm{v}_z = T = 0 \tag{5}$$

$$z = 1\text{: } \mathrm{v}_z = \frac{\mathrm{d}T}{\mathrm{d}z} = 0 \tag{6}$$

$$F_x(x, y) = 0 \qquad F_y(x, y) = 0 \tag{7}$$

where

$$F_x(x, y) = \begin{cases} \mathrm{v}_x, & nd \leq x < nd + a, \\ \dfrac{\partial \mathrm{v}_x}{\partial z} + \dfrac{Ma}{Pr}\dfrac{\partial T}{\partial x}, & nd + a \leq x < d(n+1) \end{cases}$$

$$F_x(x, y) = \begin{cases} \mathrm{v}_y, & nd \leq x < nd + a, \\ \dfrac{\partial \mathrm{v}_y}{\partial z} + \dfrac{Ma}{Pr}\dfrac{\partial T}{\partial y}, & nd + a \leq x < d(n+1) \end{cases}$$

where $Ma = \sigma_1 \Theta h / \eta\chi, \ldots, n = 0, \pm 1, \pm 2, \ldots, -\infty < y < \infty$, and Ma is the corresponding Marangoni number. The boundary value problem of Eqs. (2)–(7) determines the critical Marangoni number for neutral perturbations.

Let us restrict consideration to perturbations that are periodic along the axes x and y. The distinguishing feature of the boundary-value problem of Eqs. (2)–(7) is that it involves the condition of Eq. (7) periodic along the x-axis. In view of the linearity and the homogeneity of the problem, the solution can be searched once more in the form of Fourier normal modes,

$$(\mathrm{v}_x, \mathrm{v}_y, \mathrm{v}_z, T) = \exp i(k_x x + k_y y) \sum_{n=-\infty}^{\infty} [\Omega_n(z), \Psi_n(z), \Phi_n(z), Pr \times \Theta_n(z)] \exp(ik_d\, nx) \tag{8}$$

where k_x and k_y are the components of the two-dimensional wave vector, $\mathbf{k}$, defining the periodicity along the (x, y) axes. The series must be periodic in x with period d; $\Omega_n(z)$, $\Psi_n(z)$, $\Phi_n(z)$, $\Theta_n(z)$ are the amplitudes of the nth harmonics of velocity components and temperature, and $k_d = 2\pi/d$ is the wave vector of the reciprocal lattice. The boundary value problem determining the instability threshold can be reformulated for the transverse velocity component, v_z, and the temperature, T. Substituting the solution of Eq. (8) into Eqs. (2)–(4) with periodicity d, we obtain the differential equations for the amplitude functions. Thus, Eqs. (2)–(4) reduce to

$$\frac{d^4\Phi_n}{dz^4} - 2\omega_n^2\left(\frac{d^2\Phi_n}{dz^2}\right) + \omega_n^4\Phi_n = 0 \tag{9a}$$

$$\frac{d^2\Theta_n}{dz^2} - \omega_n^2\Theta_n^2 = \Phi_n, \qquad \omega_n^2 = (k_x + k_d n)^2 + k_y^2 \tag{9b}$$

The continuity equation, which can be formulated now as

$$i(k_x + k_d n)\Omega_n(z) + ik_y\Psi_n(z) + \Phi_n'(z) = 0 \tag{10}$$

allows eliminating longitudinal components of velocity from the boundary conditions.

Let us now write the boundary conditions on the grid. For conditions homogeneous all along the boundary surfaces we have

$$z = 0\colon\ \Phi_n = \frac{d\Phi_n}{dz} = \Theta_n = 0 \tag{11}$$

$$z = 1\colon\ \Phi_n = \Theta_n = 0 \tag{12}$$

The boundary condition of Eq. (7) requires that the functions F_x and F_y be equal to zero in each lattice site. Using the continuity equation we have two conditions for the longitudinal components of the velocity on the solid region

$$\sum_{n=-\infty}^{n=\infty} \frac{d\Phi_n/dz}{i(k_x + k_d n)} \exp i[(k_x + k_d n) + k_y y] = 0, \qquad 0 \le x < a \tag{13}$$

Excluding longitudinal components of velocity, on the free area we have

$$\sum_{n=-\infty}^{n=\infty} \frac{d^2\Phi_n/dz^2 + Ma\,\omega_n^2\Theta_n}{k_x + k_d n} \exp i[(k_x + k_d n) + k_y y] = 0, \qquad a \le x < d \tag{14}$$

The solution to the system of Eq. (9) can be written as

$$\Phi_n = C_1 \sinh \omega_n z + C_2 \cosh \omega_n z + C_3 z \sinh \omega_n z + C_4 z \cosh \omega_n z \tag{15a}$$

$$\Theta_n = C_5 \sinh \omega_n z + C_6 \cosh \omega_n z + \left(\frac{C_4}{4\omega^2} - \frac{C_2}{2\omega}\right) z \sinh \omega_n z + \left(\frac{C_3}{4\omega_n^2} - \frac{C_1}{2\omega_n}\right) z \cosh \omega_n - (C_3 \sinh \omega_n z + C_4 \cosh \omega_n z)\frac{z^2}{4\omega_n} \tag{15b}$$

where the constants, C_i, are determined by the boundary conditions of Eqs. (5) and (6),

$$C_1 = -\frac{C_3}{\omega_n}, \qquad C_3 = \frac{C_4 \omega_n \sinh \omega_n}{\sinh \omega_n - \omega_n \cosh \omega_n}, \qquad C_2 = C_6 = 0,$$

$$C_5 = -\frac{C_4}{\omega_n \cosh \omega_n}\left(\frac{\sinh \omega_n}{4\omega_n^2} - \frac{\cosh \omega_n}{4\omega_n} - \frac{\sinh \omega_n}{4}\right)$$
$$+\frac{C_3}{\omega_n \cosh \omega_n}\left(\frac{3\cosh \omega_n}{4\omega_n^2} + \frac{\sinh \omega_n}{4\omega_n} - \frac{\sinh \omega_n}{4}\right)$$

Thus, in the solution for the n-th velocity amplitude and the temperature there is only one undefined constant C_4, which in the following will be denoted as c_n. Then the amplitude functions can be written as

$$\Phi_n = c_n \varphi_n(z), \qquad \Theta_n = c_n \tau_n(z)$$

where φ_n and τ_n are known functions. Substituting the solution of Eq. (15) into the periodic boundary conditions (13)–(14) and applying the collocation method, or any other integral method, to equation $F(x, y) = 0$, with

$$F(x, y) = \begin{cases} \sum\limits_{n=-\infty}^{\infty} c_n \dfrac{\mathrm{d}\varphi_n/\mathrm{d}z}{(k_x + k_d n)} \exp ik_d nx, & 0 \le x < a \\ \sum\limits_{n=-\infty}^{\infty} c_n \dfrac{(\mathrm{d}^2\varphi_n/\mathrm{d}z^2{}_n + \omega_n^2 Ma \tau_n)}{(k_x + k_d n)} \exp ik_d nx, & a \le x < d \end{cases}$$

yields a system of homogeneous algebraic equations for the coefficients c_n. The solvability condition for this system, with the prescribed grid period d, and wave vector, $\mathbf{k}$, governs the spectrum of critical Marangoni numbers. For the numerical implementation of the collocation procedure 60 points were chosen lying on the surface consistent with the grid period, with half the points (from 1 to 30) located on the solid part and the rest (from 31 to 60) on the free part of the liquid surface. An alternative choice was a grid disposition on the computational domain such that the free region (from 16 to 45) occurs between two solid regions (from 1 to 15 and from 46 to 60).

B. Numerical Results

The stability boundary is defined by the neutral surface $Ma(k_x, k_y, d, a)$. Let us analyze the situation when the width of the solid region of the upper boundary is equal to half the grid period $a = d/2$. Of particular interest is the case with $d = \pi/2$, because the perturbations with wavelength $\lambda = 2d$ are the most dangerous in a plane layer whose open surface is unconstrained. The neutral surface $Ma(k_x, k_y)$ for this case is shown in Fig. 2. As k_y increases, the neutral

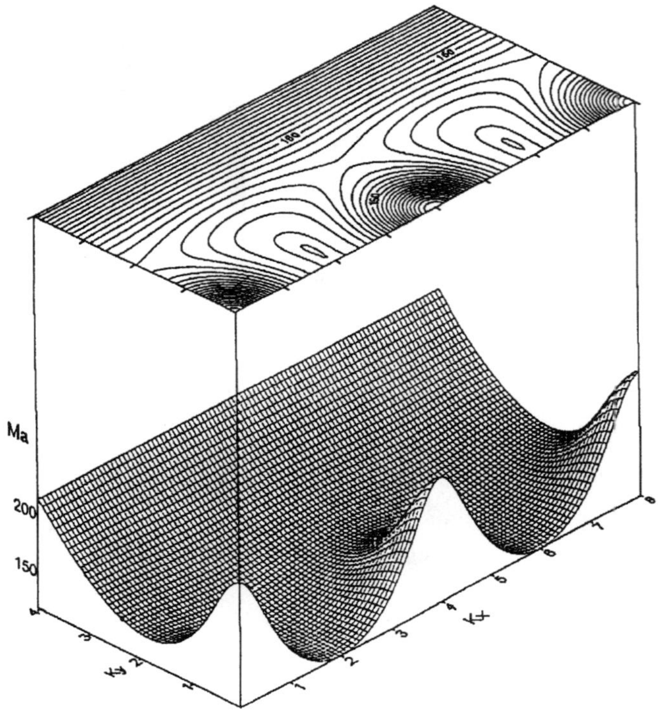

FIG. 2 Thermocapillary convection of a liquid surface with a solid grid. Neutral surface for $a = d/2$ and $d = \pi/2$. (From Ref. 1.)

surface moves rapidly up, but within the region of small k_y the Marangoni number remains finite, changing periodically as k_x changes. The surface is symmetric with respect to the lines $k_y = 0$, $k_x = 0$, and $k_x = nk_d$. The position of these lines reflects the grid periodicity. The global minimum corresponds to motion with $k_x = 2$, $k_y = 0.51$, and differs from the local minimum $k_x = 2$, $k_y = 0$ by less than 1%.

Perturbations in the form of rolls occur at the section of the surface $Ma(k_x, k_y)$ by the plane $k_y = 0$. They are described by $Ma(k_x)$, which is similar to the dispersion law of the electron energy in the periodic field of a crystal lattice. Physically different states exist between the minimum at $k_x = 2$ and the maximum at $k_x = 0$. At equivalent points $k_x = 0$ and $k_x = 4$ (corresponding to an infinite wavelength) with $k_y = 0$ a disturbed through-flow is generated along the whole grid, which is a system of rolls (one at each open region) having the same direction of rotation. This flow provides liquid transfer along the whole surface.

Lowering the grid period, d, the stability limit increases. However, this does not qualitatively affect the shape of the neutral surface. The global minimum tends to the line $k_y = 0$ where the equilibrium state is unstable with respect to rolls parallel to the grid elements. An increase of the grid period, d, lowers the stability threshold and shifts the minimum from the line $k_y = 0$ towards higher values of k_y. Accordingly, the rolls cease to be the most dangerous perturbations. The critical value of the grid period under which the minimum of the neutral surface moves away from the k_x axis is approximately equal to $d_c \approx 1.5$. Further increase of the grid period, d, allows instability in the form of rectangular cells with $k_y > k_x$. For very high values of d the open surface region is so wide that stagnant zones appear under solid areas. In this case the influence of the plates upon the motion under the free area decreases and the global minimum of the neutral surface asymptotically tends to the value obtained by Pearson ($Ma \approx 80$) [2].

Continuing with the case $a = d/2$, for simplicity, let us focus our attention only on the evolution of rolls ($k = k_x$, $k_y = 0$). With the vector of the reciprocal grid fixed ($k_d = 2\pi/d$) we find the dependence of the critical Marangoni number on the wave vector of the perturbation k. The results are given as a state diagram in Fig. 3. It involves three pairs of neutral curves (1, 2), (3, 4), and (5, 6) for $k_d = 2$, 4, and 8 ($d = \pi$, $\pi/2$, and $\pi/4$), respectively. The lower curves 1, 3, and 5 of each pair (solid lines in the figure) correspond to the first critical motion while the upper curves 2, 4, and 6 (dashed lines) correspond to the second critical motion. In these motions the surface forces determining the character of perturbations are distributed differently. In the first critical motion the liquid flow has one direction along the whole open region of the surface whereas in the second case at some point the liquid begins to flow in two opposite directions. Owing to such differences in flow structure we can link the perturbations and their neutral curves to symmetry; asymmetric (curves 1, 3, and 5) and symmetric (curves 2, 4, and 6), respectively.

Different parts of Fig. 3 show the motion for the case $k_d = 4$ when the wavelength is a multiple of the grid period, $\lambda = md$ ($m = 1, 2, 3, 4$): Symmetric motions, (a) $k = 0.5$, (b) $k = 1$, (c) $k = 2$, and (d) $k = 4$, corresponding to curve 4, and asymmetric motions, (h) $k = 0.5$, (e) $k = 1$, and (f) $k = 2$, (g) $k = 4$, corresponding to curve 3. The flow structures for neutral curves 1 and 2 are similar, although their horizontal scale is twice greater while for curves 5 and 6, it is two times smaller. Owing to the linearity of the problem, flows of opposite signs are possible. The relative location of the pair of neutral curves is defined by the period, d, and there exists a region where the symmetric mode lies below the asymmetric one. In the region of short waves ($k > 2.1$ for $k_d = 2\pi/d$ at $k = k_d$) the asymmetric mode is the most dangerous. A distinguishing feature of the mode in this range of wavenumbers is the existence of flow through neighboring cells of the grid.

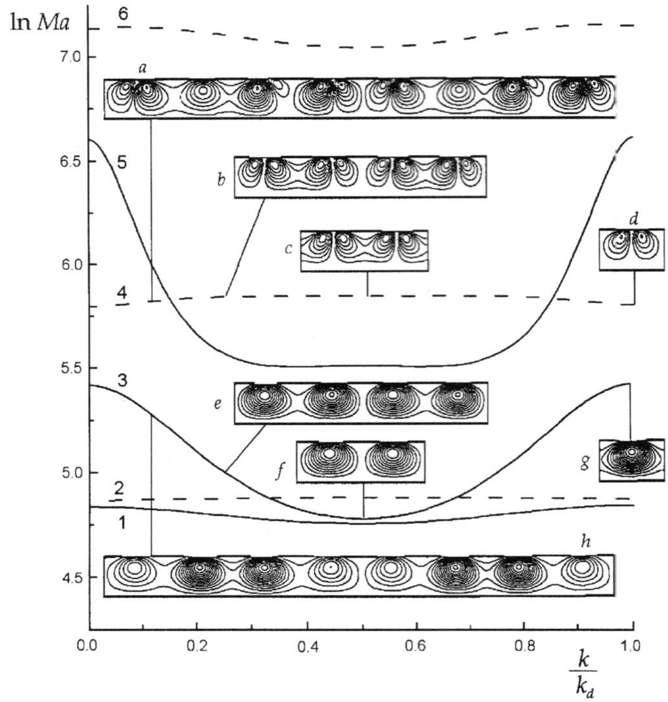

FIG. 3 Thermocapillary convection of a liquid surface with a solid grid. Neutral stability curves for different grid periods. Curves 1, 2: $-d = \pi$ ($k_d = 2$); curves 3, 4: $-d = \pi/2$ ($k_d = 4$); curves 5, 6: $-d = \pi/4$ ($k_d = 8$). Solid lines (1, 3, 5) denote the first critical motions (asymmetric mode) and dashed lines (2, 4, 6) denote the second critical motions (symmetric mode). (From Ref. 1.)

With the increase of cell length (k_d decreases) the harmonic component with lowest period in the solution of Eq. (8) gains greater importance. For long cells several vortices are generated in the vicinity of an open region giving rise to alternation of instability modes. In the region $1.1 < k < 2.1$ the symmetric mode becomes the most dangerous. When the grid period and the wavelength grow very large, there is alternation of dangerous instability modes, the critical Marangoni number tends to the value 80 and a system of vortices with horizontal dimension ~ 1.6 is formed in the open regions.

II. NONLINEAR RESULTS

The nonlinear regime of convection can be investigated by applying the finite differences procedure to the full equations written in terms of the stream function,

ψ, the vorticity, φ, and the temperature, T. We have

$$\frac{\partial \varphi}{\partial t} + \left(\frac{\partial \psi}{\partial z}\frac{\partial \varphi}{\partial x} - \frac{\partial \psi}{\partial x}\frac{\partial \varphi}{\partial z} \right) = \Delta \varphi \tag{16a}$$

$$\Delta \psi = -\varphi \tag{16b}$$

$$\frac{\partial T}{\partial t} + \left(\frac{\partial \psi}{\partial z}\frac{\partial T}{\partial x} - \frac{\partial \psi}{\partial x}\frac{\partial T}{\partial z} \right) = Pr^{-1}\Delta T \tag{17}$$

The boundary conditions on the horizontal planes are

$$z = 0\colon T = 1, \qquad \psi = 0, \qquad \frac{\partial \psi}{\partial z} = 0 \tag{18a}$$

$$z = 1\colon \frac{\partial T}{\partial z} = -1, \qquad \psi = 0, \qquad F(x) = 0 \tag{18b}$$

$$F(x) = \begin{cases} \dfrac{\partial \psi}{\partial z}, & nd \le x < nd + a \\ \varphi + \dfrac{Ma}{Pr}\dfrac{\partial T}{\partial x}, & nd + a \le x < (n+1)d \end{cases}$$

where $n = 0, \pm 1, \pm 2, \ldots$

The calculation was performed on a horizontal computational domain of length L, whose vertical boundaries $x = 0$ and $x = L$ satisfy the periodicity conditions ($L = md$ and $m = 1, 2, 4$). As in the earlier given linear analysis the width of the free region was assumed to be equal to the width of the solid region, $a = d/2$.

Equations (16) and (17) and boundary conditions (18) were written in the finite differences form for which the spatial derivatives are expressed in terms of central differences and time derivatives in terms of the forward difference. The computational mesh was such that each solid region of the upper boundary incorporated 30 nodes. The solution to the system of equations was found by using the two-layer scheme and the transverse–longitudinal sweep method. In both cases the Poisson equation was solved by the over-relaxation method. The values of the vorticity at the boundary of solid regions were calculated using Thom's formula [3]. The desirable motion structure was excited by applying the appropriate temperature perturbation pattern at the upper boundary of the region. The resulting surface forces made it possible to generate a convective cell of required shape and to follow its evolution.

As the most dangerous disturbance in the free liquid layer has wavelength $\lambda = \pi$, the length (period) of the computational domain was set at $L = \pi$. Let us consider changes of the flow shape that occur due to the increase in the number of plates located over that period. Figure 4 depicts structures of stationary convective modes with equal wavelengths and different number of solid areas on the surface.

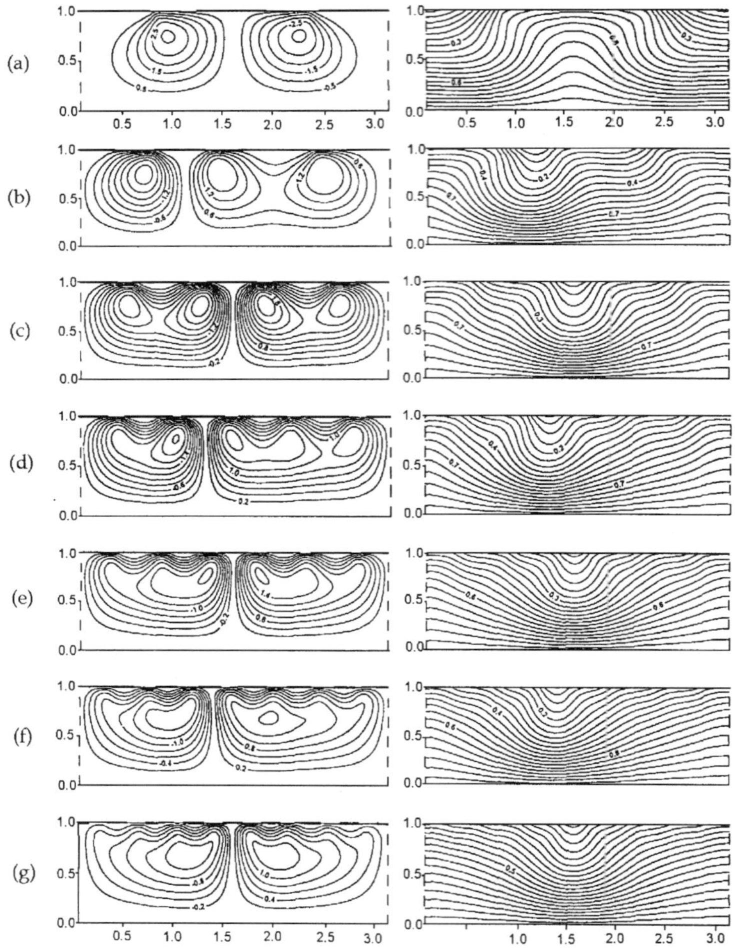

FIG. 4 Thermocapillary convection of a liquid surface with a solid grid. Isolines of stream functions and isotherms of two vortex stationary modes for $Ma = 10^3$ and $L = \lambda = \pi$.

Calculations were carried out for $Ma = 10^3$ and $Pr = 1$. The flow, depicted in Fig. 4(a), has one plate on the calculated area period. The other plates (b)–(g) show flows in the domain with 3, 4, 5, 6, 7, and 8 solid elements on the surface, respectively. The flow covers free surface areas and bends near the plates. Structures containing an even number of plates on the surface have two symmetric vortices. In the case with odd number of plates on the surface, one vortex covers an area that contains one grid period more than the vortex turning opposite. Now let us consider the evolution of the convective modes existing in the domain $L = 2d$.

The results obtained for the nonlinear convective regimes with $d = \pi/2$ are given in Fig. 5. Solid lines depict the motion intensity under steady conditions. The characteristic parameter of the flow is here the half-sum of modules of the stream function extremes in the computational domain, $Q = (|\Psi_{max}| + |\Psi_{min}|)/2$. With the increase of the Marangoni number the first excited mode is an asymmetric flow with $L = 2d$ (curve a), in agreement with the linear theory: each open region contains a single vortex and vortices in adjacent regions rotate in opposite directions. The critical Marangoni number, estimated by extrapolating down to zero the amplitude curve, is $Ma = 116$ (in linear theory $Ma = 119.5$).

At point A the motion became unstable with respect to symmetric disturbances with $\lambda = d$ and reorganized into a symmetric flow (curve s). The symmetric motion with two convective cells in the computational domain $L = 2d$ remained stable for all investigated Marangoni numbers beyond the transition point. However, as the Marangoni number decreased, at point B ($Ma = 392$) there was an inverse transition to the asymmetric mode $\lambda = 2d$. Hence there is hysteresis in an interval of Marangoni numbers where two stable steady convective regimes coexist.

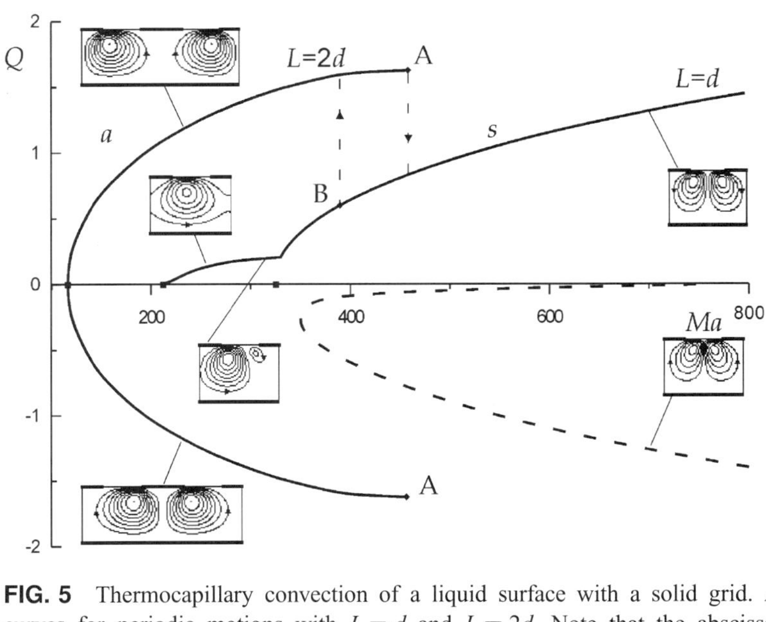

FIG. 5 Thermocapillary convection of a liquid surface with a solid grid. Amplitude curves for periodic motions with $L = d$ and $L = 2d$. Note that the abscissa starts at $Ma = 100$. (From Ref. 1.)

In the numerical integration the flow was also assumed to have periodicity d. In this case with increase of the Marangoni number, at $Ma = 216$ there appears asymmetric motion in the form of through-flow (in the linear theory $Ma = 227$). Further increase of Ma leads to a small vortex of opposite direction being generated below one of the plates and its growth breaks the through-flow. The result is an asymmetric cell with turning vortices of different size. Increasing again the Marangoni number causes equalization in the intensity of vortices, accompanied by a decrease in amplitude of the bigger vortex. Hence, the flow structure becomes closer to the second critical motion. Finally, when complete equalization of both vortices is reached the flow intensity corresponds to the amplitude branch of the symmetric mode. Moreover, the generated symmetric motion has ascending liquid flow in the middle of the open area.

Furthermore, for this periodicity there is one more bifurcation point at $Ma = 323$ (in the linear theory $Ma = 333$), which leads to symmetric motion (curve s). In its nonlinear evolution the asymmetric through-flow is converted into symmetric motion with up-flow in the center of the open regions. At this instant an imperfect bifurcation takes place, which is the reason why the symmetric mode is the extension of the amplitude curve for the asymmetric mode and is stable.

Symmetric flows with opposite directions of rotation are unstable. Using vortices of opposite signs as an initial mode in the calculations causes the flow to return to the mode taking liquid out of an open area. The unstable branches of the imperfect bifurcation are depicted by the dashed line in Fig. 5. Similar behavior has been described in Refs. 5 and 6.

To investigate nonlinear motions covering several grid periods a numerical experiment was carried out for the domain $L = 4d$ with $d = \pi/4$. In this case, depending on the initial perturbation, two asymmetric regimes $\lambda = 2d$ and $\lambda = 4d$ were excited [4]. Amplitude curves for those flows are shown in Fig. 6. Above its critical point the motion with $\lambda = 2d$ appeared to be stable over all the considered range of Marangoni numbers. Another mode, $\lambda = 4d$, is stable above point A, which, however, is located below the crossing point of the amplitude curves. Extrapolation of the amplitude curves down to zero provides a value of the Marangoni number lower than in linear theory: 226 and 248, respectively. The diagram of the Nusselt number (overall heat transfer) dependence on the Marangoni number for the calculated domains $L = d$ and $L = 2d$ with $d = \pi/2$ qualitatively reproduces the diagram for the amplitude curves (Fig. 5). For the case $L = 4d$ and $d = \pi/4$ these diagrams are different, as the curves for the Nusselt number do not cross each other. For the whole region considered the motion with $\lambda = 2d$ provides a greater heat transfer than the motion with $\lambda = 4d$.

In all numerical integrations, extrapolation of the amplitude curves down to zero allowed the determination of the critical Marangoni number that recovers the results of the linear theory with a 10% discrepancy.

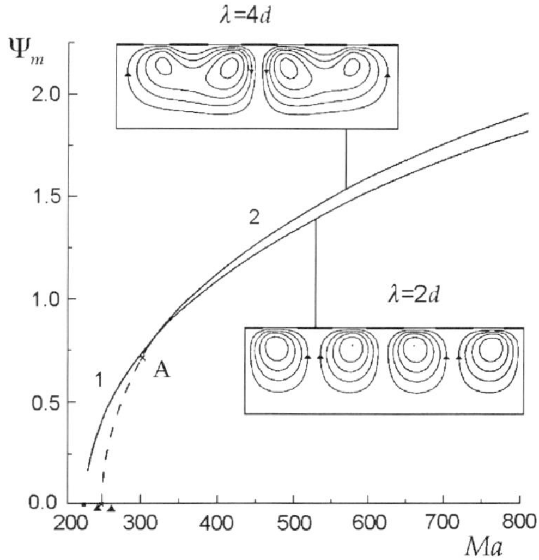

FIG. 6 Thermocapillary convection of a liquid surface with a solid grid. Amplitude curves of nonlinear stationary modes and shape of the convective cell in a layer with $d = \pi/4$ and $L = 4d$. The dashed line indicates an unstable branch. (From Ref. 4.)

III. THERMOCAPILLARY FLOW WITH LONGITUDINAL TEMPERATURE GRADIENT

Let us now consider the case of a constant longitudinal temperature gradient along the surface with the solid grid. We shall consider a plane infinite layer with two open surfaces, hence a film, and we shall study it using the finite differences method. Generally, [7–9] the surface of the liquid was assumed initially homogeneous although it is commonly recognized that, for example, surfaces of molten material are contaminated or covered by scraps of oxide films and the intensity of the thermocapillary convection strongly depends on the surface quality.

A. Formulation of the Problem

Let us consider a film, or an infinite plane liquid layer, of thickness $2h$, in weightlessness. Let two thin solid plates of width a be placed on the layer boundary surfaces to each form a one-dimensional grid along the x-axis with period d (Fig. 7).

A temperature gradient, ∇T_0, is prescribed on the two free boundary surfaces along the grid and maintained constant. Then the free areas of the open surfaces will be under the action of the Marangoni effect, giving rise to flow.

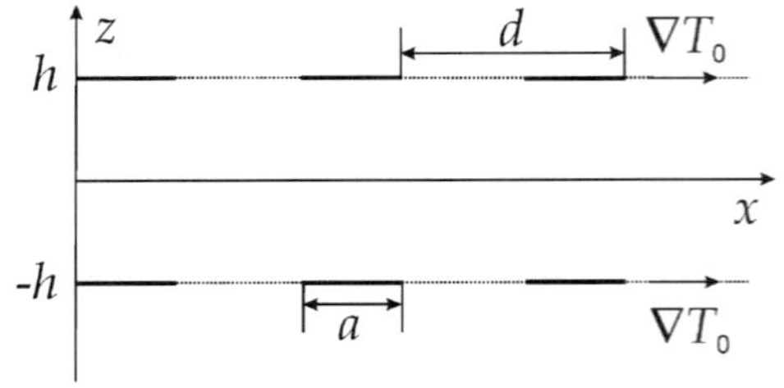

FIG. 7 Thermocapillary flows in systems with longitudinal temperature gradient. Coordinate system and geometry of the problem.

Because the surface temperature distribution is prescribed, the temperature redistribution within the layer due to convection does not affect the structure and intensity of the flow. Therefore the heat balance will be omitted from our consideration and to describe convection under these conditions we need only the continuity and Navier–Stokes equations with corresponding boundary conditions. We have

$$\text{div } \mathbf{v} = 0 \tag{19}$$

$$\frac{\partial \mathbf{v}}{\partial t} + (\mathbf{v} \cdot \nabla)\mathbf{v} = -\frac{1}{\rho}\nabla p + \nu \Delta \mathbf{v} \tag{20}$$

together with the conditions

$$\begin{aligned} &z = \pm h : nd \le x \le nd + a \text{ (on the solid regions)}, && \mathbf{v} = 0 \\ &nd + a \le x \le (n+1)d \text{ (on the free surfaces)}, && \eta \frac{\partial \mathrm{v}_x}{\partial z} = -\frac{\partial \sigma}{\partial T}\frac{\partial T}{\partial x} \end{aligned} \tag{21}$$

We shall also impose the condition that the flow is closed

$$\int_{-h}^{h} \mathrm{v}_x \mathrm{d}z = 0 \tag{22}$$

As earlier, here $\mathbf{v}$ is the liquid velocity, p is the pressure, ρ is the liquid density, ν is the kinematic viscosity, η is the dynamic (shear) viscosity, and σ denotes the surface tension that as usual for standard liquids linearly depends on temperature, $\sigma(T) = \sigma_0 - \sigma_1(T - \Theta)$, where T is the reference temperature; σ_1 is taken constant and positive.

The boundary conditions suggest that the flow is expected to be two-dimensional, (x, z), hence we introduce the stream function, $\psi(x, z)$. Then the problem of Eqs. (19)–(22) becomes

$$\frac{\partial \Delta \psi}{\partial t} + \frac{\partial \psi}{\partial x}\frac{\partial \Delta \psi}{\partial z} - \frac{\partial \psi}{\partial z}\frac{\partial \Delta \psi}{\partial x} = \Delta^2 \psi \tag{23a}$$

with

$$nd \leq x < nd + a : \psi = \frac{\partial \psi}{\partial y} = 0,$$

$$nd + a \leq x \leq (n+1)d : \psi = 0, \qquad \frac{\partial^2 \psi}{\partial y^2} = Re = Ma/Pr \tag{23b}$$

where $Ma = \sigma_1 |\nabla T_0| h^2/\eta\chi$ is the Marangoni number and Re is the associated Reynolds number with the velocity [Eq. (13), Chapter 1] dictated by the Marangoni effect, while d and a are the dimensionless period of the grid and the width of the solid region (note that we are using the same symbols to denote dimensional and dimensionless lengths), and χ is the heat diffusivity. To make quantities dimensionless we have used as scales the half-width of the layer h, h^2/ν, and ν. Thus the parameters determining the intensity and structure of the possible flow are Re, d, and a.

B. Numerical Results

The problem of Eq. (23) was solved for the region corresponding to the period of the grid along the x-axis (Fig. 8) [10]. The limits of the region along the z-axis are 0 and 1 as the boundary conditions are symmetric about the x-axis. Hence the

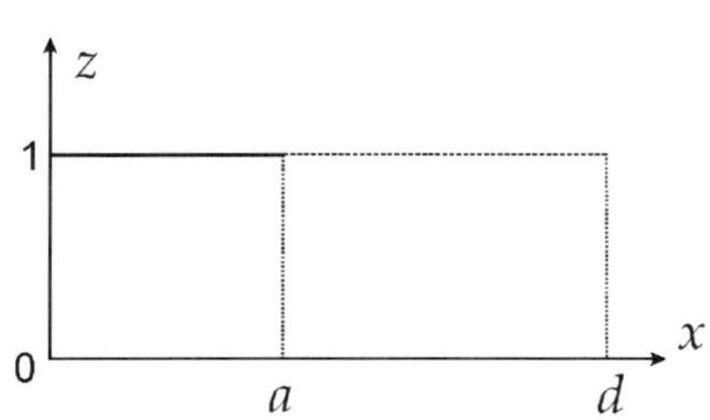

FIG. 8 Thermocapillary flows in systems with longitudinal temperature gradient. Integration domain and coordinate system for numerical integration.

boundary conditions of the integration domain are

$$z = 0 : \psi = \frac{\partial^2 \psi}{\partial z^2} = 0 \quad (0 \le x \le d)$$
$$z = 1 : \psi = \frac{\partial \psi}{\partial z} = 0 \quad (0 \le x \le a), \qquad \psi = 0, \qquad \frac{\partial^2 \psi}{\partial z^2} = Re \quad (a < x < d)$$
$$\psi(0, z) = \psi(d, z)$$

Equation (23a) was integrated by making use of the implicit finite differences scheme with the vorticity introduced as an additional variable. The Poisson equation was solved by Fourier transformation along one of the space variables. The boundary values of vorticity were calculated using Thom's formula [3]. The calculations were carried out on meshes with various steps for different grid spacing.

Let us first consider the dependence of the structure of the flow and its intensity on the Reynolds number for a few values of the grid spacing. Figure 9 shows stream lines for thicknesses with closed and open areas being equal in size over the grid spacing. At the counterflow boundary there exist vortices that are located under the open surface area. Such flow structure is the same for different ratios of closed to open area of the surface over the grid spacing.

The flow intensity is characterized by the maximum of the absolute value of the stream function, Ψ_m, and monotonically increases with increasing Reynolds number (Fig. 10). These features are typical for grids with spacing $d < 1$.

In the case of large grid spacing ($d > 4$) in the examined range of values of the Reynolds number, with corresponding values of a, the flow structure is qualitatively the same, except that it involves a system of vortices circulating in the same direction and separated by stagnant zones (Fig. 11).

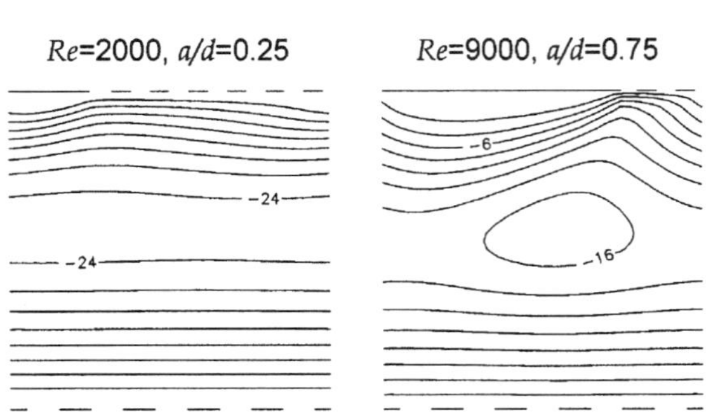

FIG. 9 Thermocapillary flows in systems with longitudinal temperature gradient. Isolines of stream function for the case $d = 1$. (From Ref. 10.)

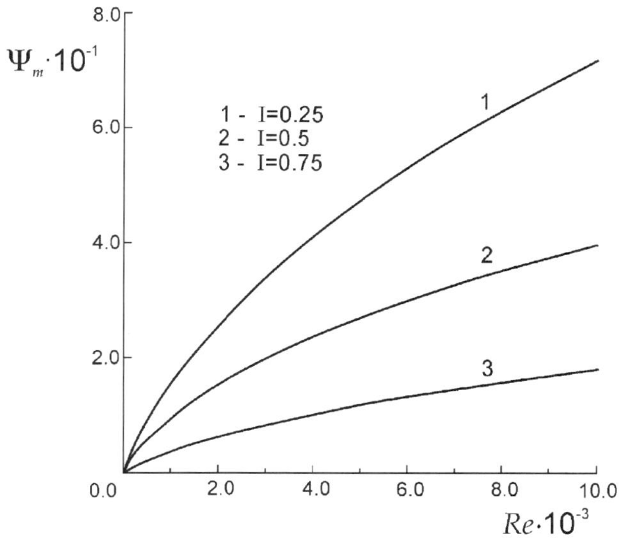

FIG. 10 Thermocapillary flows in systems with longitudinal temperature gradient. Dependence of the maximum value of the stream function on the Reynolds number ($I = a/d$, $d = 6$). (From Ref. 10.)

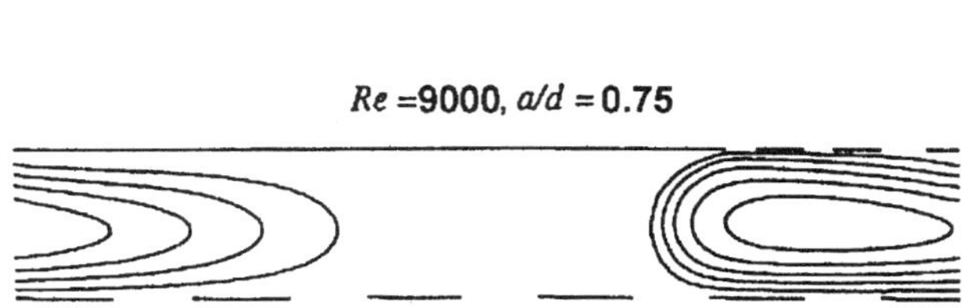

FIG. 11 Thermocapillary flows in systems with longitudinal temperature gradient. Isolines of the stream function for the case $d = 6$. (From Ref. 10.)

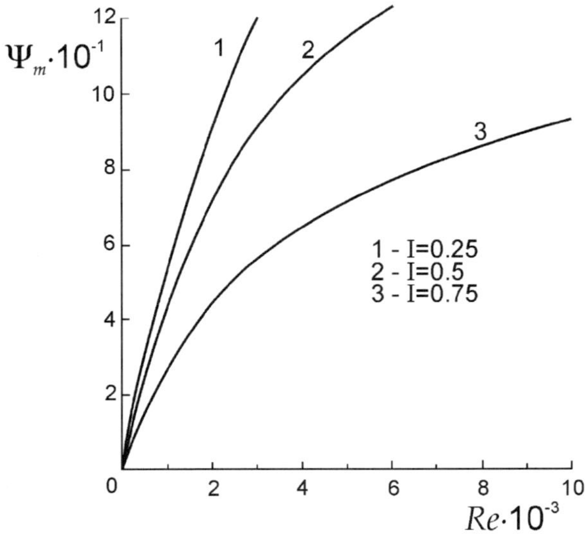

FIG. 12 Thermocapillary flows in systems with longitudinal temperature gradient. Dependence of the maximum value for the stream function on the Reynolds number ($I = a/d$, $d = 6$). (From Ref. 10.)

The vorticity monotonically increases with increasing Reynolds number (Fig. 12). With the grid spacing being approximately equal to the layer thickness (Fig. 13) the motion can change as the Reynolds number increases. If the size of the rigid part of the surface is $a < 0.4d$ (curve 1) the flow structure does not change as shown in Fig. 9. If $a > 0.7d$ (curve 4) the flow structure does not change either as shown in Fig. 11. For the case $0.4d < a < 0.7d$ the flow structure changes as the Reynolds number increases.

The nonmonotonic curves 2 and 3 in Fig. 13 are noteworthy. For $Re < 5 \times 10^3$ the flow structure is similar to the corresponding one for a large grid spacing, i.e., it involves a system of isolated vortices (Fig. 11). As the Reynolds number increases the vortices combine to form a flow almost identical to that appearing for a small grid spacing (Fig. 9). Such flow transformation may occur through self-sustained oscillatory regimes.

Figure 14 shows the time dependence of the maximum value of the stream function and the flow patterns in the time instants which, respectively, correspond to the maximum and minimum values of the $\Psi_m(t)$ for $Re = 9 \times 10^3$, $d = 2$, and $a/d = 0.62$. It appears that the intensity and location of the vortex periodically change with time. These self-excited oscillations are very much like those observed when a weight is placed on a long horizontal moving conveyor fastened by a spring to a stationary vertical support. Figure 15 shows the flow intensity

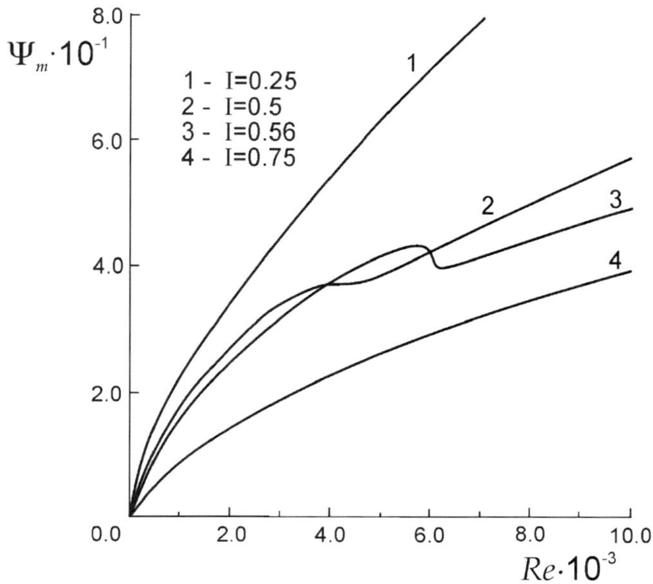

FIG. 13 Thermocapillary flows in systems with longitudinal temperature gradient. Dependence of the maximum value for the stream function on the Reynolds number $d = 2$ ($I = a/d$). (From Ref. 10.)

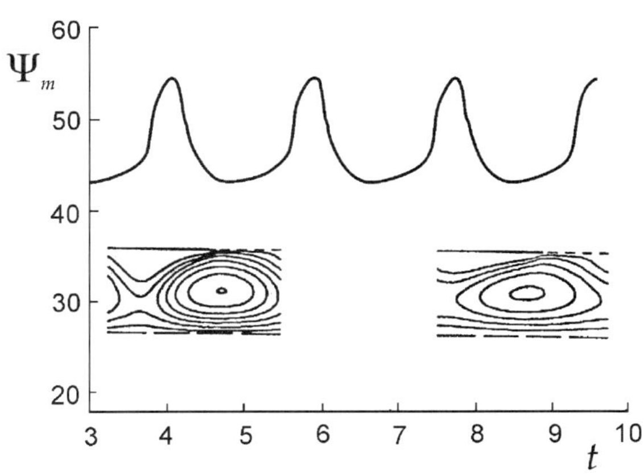

FIG. 14 Thermocapillary flows in systems with longitudinal temperature gradient. Time dependence of maximum value of the stream function for $Re = 9 \times 10^3$, $d = 2$ and $a = 1.24$. (From Ref. 10.)

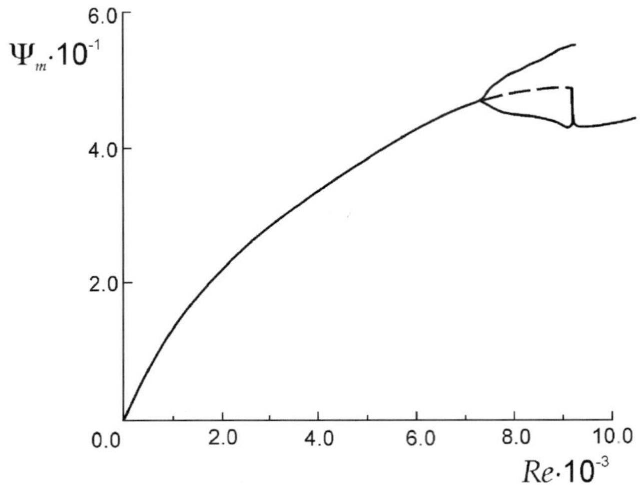

FIG. 15 Thermocapillary flows in systems with longitudinal temperature gradient. Dependence of the maximum value of stream function on *Re* for the case $d = 2$ and $a/d = 0.62$. (From Ref. 10.)

variation with Reynolds number for $a/d = 0.62$. The dashed portion of the curve in the vicinity of local maxima corresponds to oscillatory flow regimes. The variation of the period of the oscillations is shown in Fig. 16.

Oscillations begin with finite period that diverges to infinity if the Reynolds number increases to $Re = 9250$. When comparing the behavior of the $\Psi_m(Re)$ curves for different values of *a* we arrive at the conclusion that upon increasing the latter parameter the region of flow transformation moves in the direction of high Reynolds numbers and disappears as *a* decreases.

Let us discuss the results obtained at a fixed value of the Reynolds number, $Re = 10^3$, different values of the grid spacing, *d*, and different sizes of the closed surface areas over the period, *a*. Figure 17 shows the stream function isolines of the liquid in a computational domain for several values of *a* when $d = 1$. It is readily seen that for small values of *a* the flow remains essentially plane-parallel. As *a* increases the flow structure changes. At the counterflow boundary a vortex appears whose size gradually increases.

The vortex localization is also variable. It constantly moves in the direction of the free surface and finally occupies the position under an open surface area. It should be emphasized that the through-flow component is preserved. The variation of the flow intensity shows that the maximum value of the stream function, Ψ_m, is related to a/d as shown in Fig. 18. As might be expected, with the reduction of the open surface areas the intensity of the flow monotonically decreases and goes to zero at $a/d \rightarrow 1$.

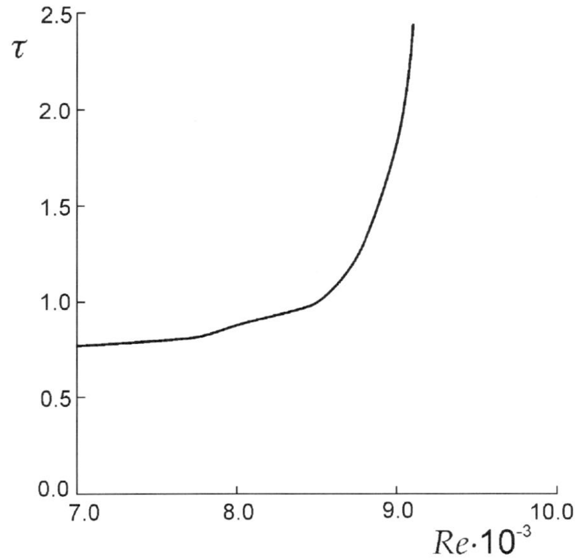

FIG. 16 Thermocapillary flows in systems with longitudinal temperature gradient. Period of convective oscillations vs. Reynolds number. (From Ref. 10.)

At small a, as in the case $d = 1$, the flow appears as a superposition of through-motions and vortices at the boundary of counterflows. A flow in the layer involves vortices localized under the open areas of the surface separated by stagnant parts. To put it differently, a transformation of the flow occurs at $d > 1$. Figure 19 shows the streamlines in the computational domain for $d = 10$. With this particular transformation of the flow the function $\Psi_m(a/d)$ becomes nonmonotonic (Fig. 20). An increase in Ψ_m occurs in the range of a values where the through-flow vanishes. As one would expect, further increasing a results in reduced vortex intensity and at $a/d \to 1$ the flow vanishes.

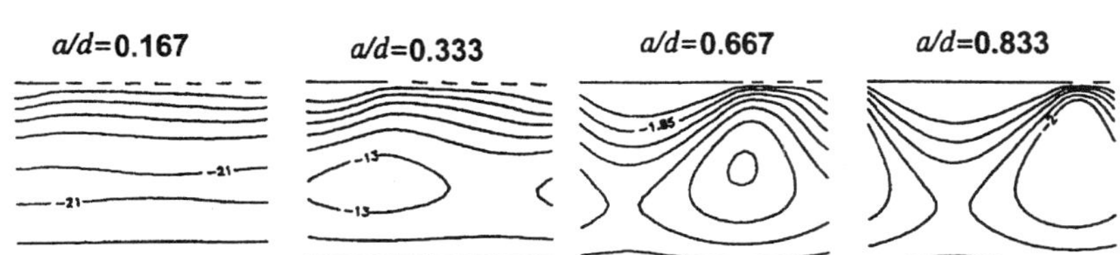

FIG. 17 Thermocapillary flows in systems with longitudinal temperature gradient. Isolines of the stream function for $Re = 10^3$ and $d = 1$. (From Ref. 10.)

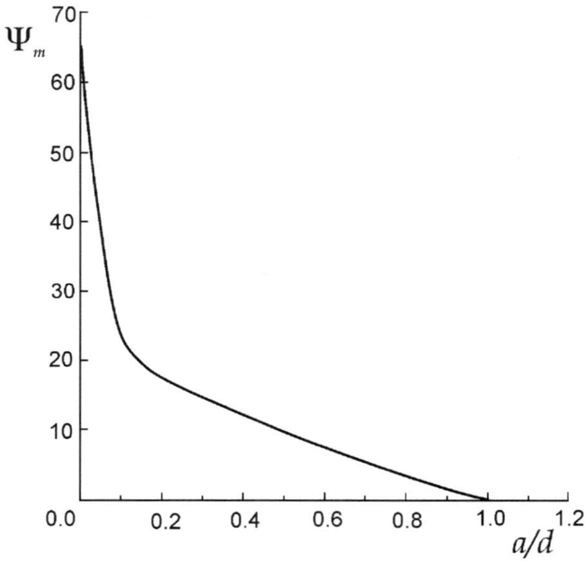

FIG. 18 Thermocapillary flows in systems with longitudinal temperature gradient. Variation of the maximum value of the stream function for $Re = 10^3$ and $d = 1$. (From Ref. 10.)

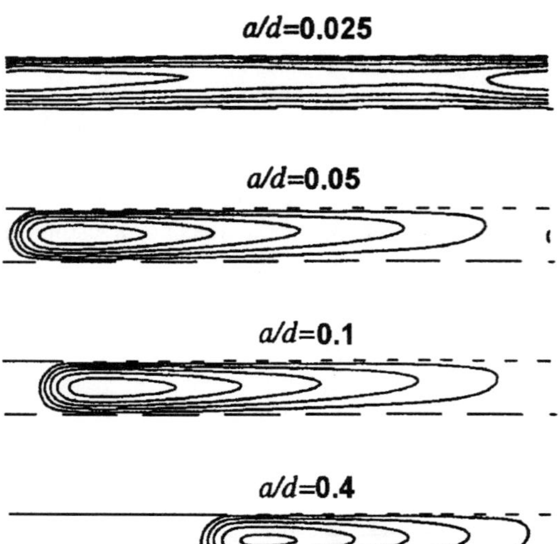

FIG. 19 Thermocapillary flows in systems with longitudinal temperature gradient. Isolines of the stream function for $Re = 10^3$ and $d = 10$. (From Ref. 10.)

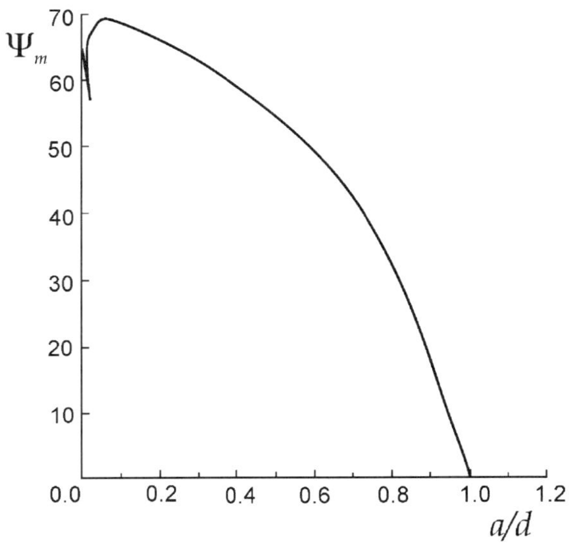

FIG. 20 Thermocapillary flows in systems with longitudinal temperature gradient. Variation of the maximum value of stream function with a/d for $Re = 10^3$ and $d = 10$. (From Ref. 10.)

Let us now consider the flow transformation when $d < 1$. Figure 21 displays the streamlines of the flow regimes at different values of a for $d = 0.5$. Comparison of these flow patterns with the previous ones permits the conclusion that the vortex component of the flow drastically decreases. In fact, for all values of a/d the flow is throughout motion over the whole layer cross-section. The variation of Ψ_m with a/d is monotonically damped at $a/d \rightarrow 1$ as shown in Fig. 18.

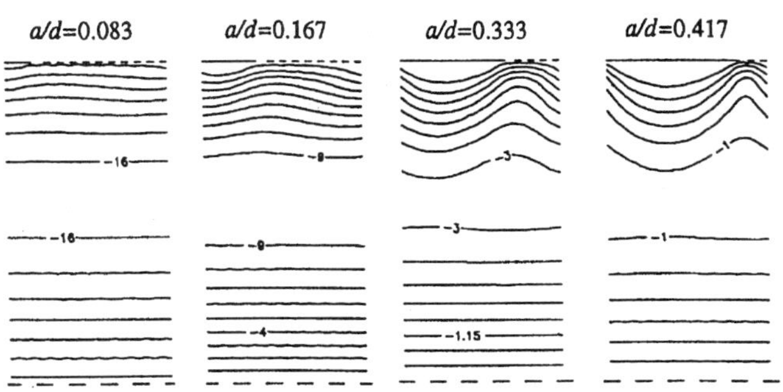

FIG. 21 Thermocapillary flows in systems with longitudinal temperature gradient. Isolines of the stream function for $Re = 10^3$ and $d = 0.5$. (From Ref. 10.)

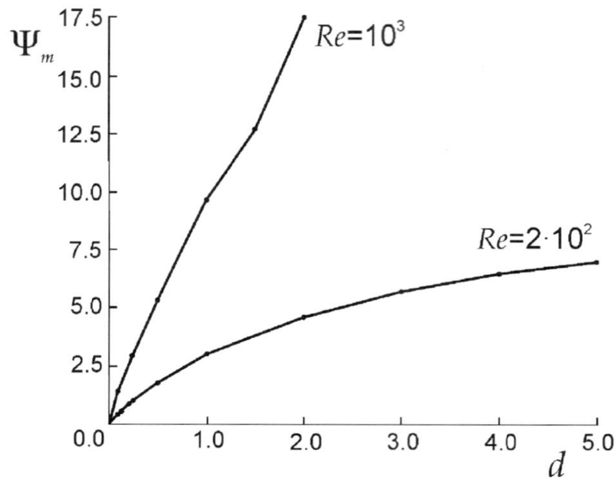

FIG. 22 Thermocapillary flows in systems with longitudinal temperature gradient. Maximum value of stream function vs. grid period. (From Ref. 10.)

Finally, let us see the flow intensity dependence on the grid spacing, *d*, at fixed values of the Reynolds number and the a/d ratio. The latter condition suggests that, with increasing spacing, the grid is reconstructed in such a way that the total area of the open surface remains constant. In Fig. 22 the maximum value of the stream function is plotted against *d* at $Re = 2 \times 10^2$ and 10^3 for $a/d = 0.5$. As shown in the figure $\Psi_m \rightarrow 0$ at $d \rightarrow 0$, i.e., the flow decays with the increase of the grid density regardless of the fact that the area of the open surface remains the same. This conclusion is independent of the particular values of a/d.

IV. THERMOCAPILLARY CONVECTION IN A LIQUID ZONE WITH SOLID ELEMENTS ON ITS OPEN SURFACE

The liquid-zone crystallization method is one of the most efficient procedures to improve quality in a crystal structure. In this technique the liquid region is a liquid cylindrical bridge anchored at two solid surfaces whose temperature is held constant and equal to the melting temperature. Because the temperature distribution along the surface of the liquid phase is axisymmetric and the surface tension coefficient is temperature dependent, thermocapillary convection is expected in the liquid bridge.

Convective mixing adds complexity to the crystallization process, which explains the interest in searching for effective mechanisms to help suppress thermocapillary convection. Authors have considered the influence of magnetic fields on thermocapillary convection in a liquid zone [11] and the effect of high-frequency vibrations [12,13]. Here we shall focus attention on the influence of a grid on the thermocapillary convection in such liquid zones. The study will be carried out for the case of axisymmetric flows with an outer cylindrical, open undeformable surface.

A. Formulation of the Problem

Let us consider a vertical liquid circular cylinder of radius R and height H (Fig. 23), which is confined between two solid surfaces at a common constant temperature T_0. The lateral cylindrical surface is partially covered by thin solid rings of height h_1 periodically distributed along the vertical and separated by regions of free surface of height h_2. The temperature distribution generated near the liquid zone surface by a circular heat source can be approximated by the relation

$$\tilde{T}_a(z) = T_0 + \delta\tilde{T}\exp\left[-\frac{(z - z_0)^2}{d^2}\right] \tag{24}$$

The heat transfer conditions on the lateral surface are given by

$$\frac{\partial\tilde{T}}{\partial r} = b(z)(\tilde{T} - \tilde{T}_a) \tag{25}$$

where $b(z)$ is the corresponding ad hoc heat transfer coefficient (later on we shall use the Biot number). It has different values for solid and free regions of the boundary. The surface tension is assumed to vary with temperature, as in earlier cases studied in this book.

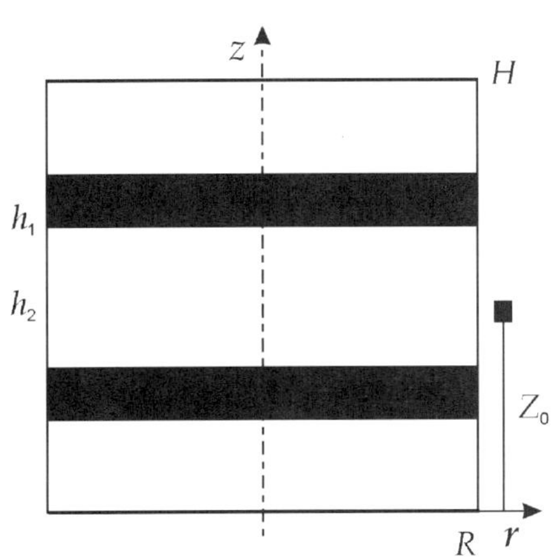

FIG. 23 Thermocapillary convection in a liquid zone with solid elements on its open surface. Schematic representation and coordinate system. (From Ref. 15.)

To describe axisymmetric thermocapillary convection we introduce the stream function, ψ, here related to the velocity components by the equations

$$v_r = \frac{1}{r}\frac{\partial\psi}{\partial z}, \qquad v_z = -\frac{1}{r}\frac{\partial\psi}{\partial r} \tag{26}$$

In terms of the dimensionless stream function, ψ, the vorticity, ζ, and the temperature, T, measured relative to T_0, the evolution equations take the form

$$\frac{\partial\zeta}{\partial t} + \frac{1}{r}\left(\frac{\partial\psi}{\partial z}\frac{\partial\zeta}{\partial r} - \frac{\partial\psi}{\partial r}\frac{\partial\zeta}{\partial z} - \frac{1}{r}\frac{\partial\psi}{\partial z}\zeta\right) = \Delta\zeta - \frac{\zeta}{r^2} \tag{27}$$

$$\frac{\partial T}{\partial t} + \frac{1}{r}\left(\frac{\partial\psi}{\partial z}\frac{\partial T}{\partial r} - \frac{\partial\psi}{\partial r}\frac{\partial T}{\partial z}\right) = Pr^{-1}\Delta T \tag{28}$$

$$r\zeta = \frac{\partial^2\psi}{\partial r^2} - \frac{1}{r}\frac{\partial\psi}{\partial r} + \frac{\partial^2\psi}{\partial z^2} \tag{29}$$

$$\Delta \equiv \frac{\partial^2}{\partial r^2} + \frac{1}{r}\frac{\partial}{\partial r} + \frac{\partial^2}{\partial z^2}$$

The boundary conditions are
at $z = 0$,

$$l \equiv \frac{H}{R} : \psi = \frac{\partial\psi}{\partial z} = T = 0$$

at $r = 0$:

$$\psi = \zeta = \frac{\partial T}{\partial r} = 0$$

at $r = 1$:

$$\psi = 0$$

and

$$\frac{\partial\psi}{\partial r} = 0, \qquad \frac{\partial T}{\partial r} = -Bi(T - T_a) \qquad \text{on the solid regions} \tag{30a}$$

and

$$\zeta = Re\frac{\partial T}{\partial z}, \qquad \frac{\partial T}{\partial r} = -Bi_1(T - T_a) \qquad \text{on the free regions} \tag{30b}$$

The problem involves the Prandtl number, $Pr = \nu/\chi$, the Reynolds number, $Re = \sigma_1\delta\tilde{T}R/\rho\nu^2$ (recall that $Re = Ma/Pr$), and two Biot numbers, Bi and Bi_1 for solid and free regions of the lateral surface, respectively. Other relevant elements are the number of solid rings and the ratio of the free surface height to that of the ring, h_1/h_2. The radius of the cross-section zone, R, R^2/ν, ν, and $\delta\tilde{T}$ are taken as the units of length, time, stream function, and temperature, respectively.

The system of equations (27)–(29) with the boundary conditions of Eq. (30) was solved by the finite differences method. The stationary distributions of vorticity and temperature were obtained by transition to a steady state in the implicit finite differences scheme of the direction variables. At each time step the stream function was calculated by the Chebyshev iterative method [14] with the determined values of vorticity. The vorticity at the solid regions of the boundary was obtained using Thom's formula [3]. The computation was made on uniform or nonuniform grids with respect to the radial coordinate and incorporated 50 nodes. Along the axial coordinate the grid was uniform with a total number of nodes from 200 to 240. The test calculations were made on a 100×200 grid. All numerical computations were carried out for the following fixed values of the parameters: $Pr = 0.023$, $l = 2$, $z_0 = 1$, and $d = 0.2$.

B. Flow Structure and Intensity

Let us first consider the influence of the grid periodicity on the flow structure when liquid sticks to the solid regions of the boundary. Shown in Fig. 24 are the stream function isolines (at the right) and isotherms (at the left) obtained for a fixed value of the Reynolds number, $Re = 400$ ($Bi = Bi_1 = 100$) and different grids on the liquid surface. Figure 24(a) shows the flow pattern in the absence of grid (the number of grid periods $n = 0$). It can be seen from the diagram that owing to the symmetry of the temperature distribution at the boundary, the liquid zone exhibits two symmetric vortices of equal intensity. The temperature distribution is only slightly dependent on the intensity of the motion, because the Prandtl number is small. Figure 24(b) illustrates the flow pattern in the case when the lateral surface

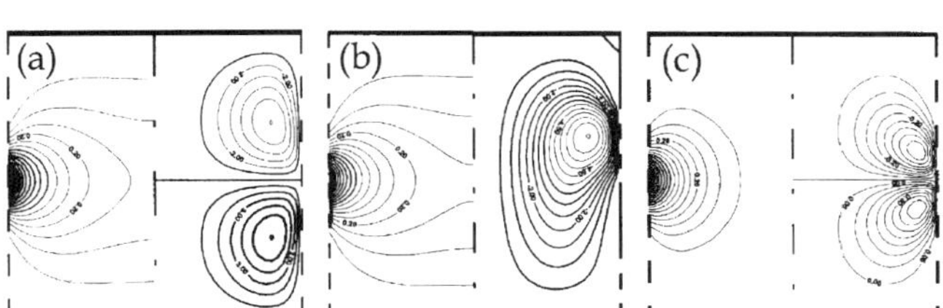

FIG. 24 Thermocapillary convection in a liquid zone with solid elements on its open surface. Stream function isolines (at the right) and isotherms for $Re = 4 \times 10^2$, $Bi = Bi_1 = 10^2$, and different grids on the liquid surface. The number of grid periods is (a) $n = 0$, (b) 2, and (c) 20, respectively. (From Ref. 15.)

of the liquid bridge is covered by the coarse grid generated by two rings whose dimensions are defined by the ratio $h_1/h_2 = 1$. Here only one vortex is observed and circulation throughout the zone occurs in a single direction. Changing the position of the rings along the vertical axis of the zone allows the control of the flow structure by suppressing vortices circulating in one or the other direction.

The flow pattern for a grid with finer grid spacing ($n = 20$) is shown in Fig. 24(c). A comparison with Fig. 24(a) shows that the flow becomes symmetric with a pair of vortices, i.e., similar to the flow pattern occurring in the absence of a grid. The variation of the temperature field is caused by the decrease in the intensity of the motion.

Let the quantity characterizing the flow intensity be the dimensionless flow rate, Q, through a section connecting the extreme points (minimum and maximum) of the stream function

$$Q = \Psi_{max} - \Psi_{min} \tag{31}$$

The effect of the grid on this parameter, Q, is illustrated in Fig. 25 where the flow rate is plotted vs. the grid step, $h = h_1 + h_2$ (note, not be confused with the earlier introduced h for the layer depth). Q asymptotically tends to zero with $h \to 0$. This result seems to be a result of an increase in the hydraulic resistance of the grid. Analogous results are observed when the grid is placed on a liquid–liquid interface [16]. Accordingly, for any value of the Reynolds number we may select the grid that will ensure the creeping flow pattern. In Fig. 26 the flow rate is plotted vs. the Marangoni number, which acts as Reynolds number, for various grid

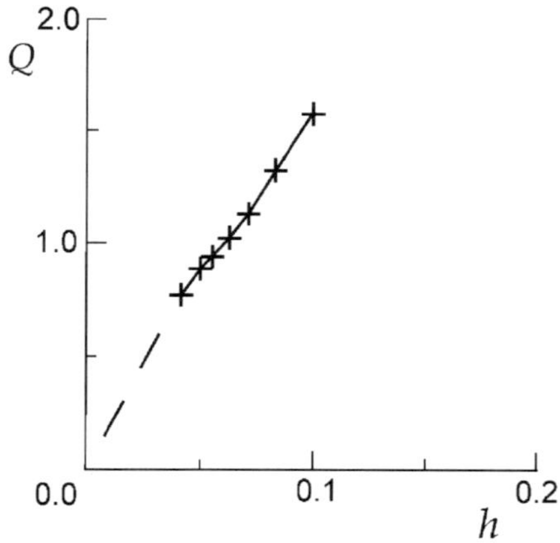

FIG. 25 Thermocapillary convection in a liquid zone with solid elements on its open surface. Flow rate vs. grid spacing, h ($Re = 4 \times 10^2$ and $Bi = Bi_1 = 10^2$). (From Ref. 15.)

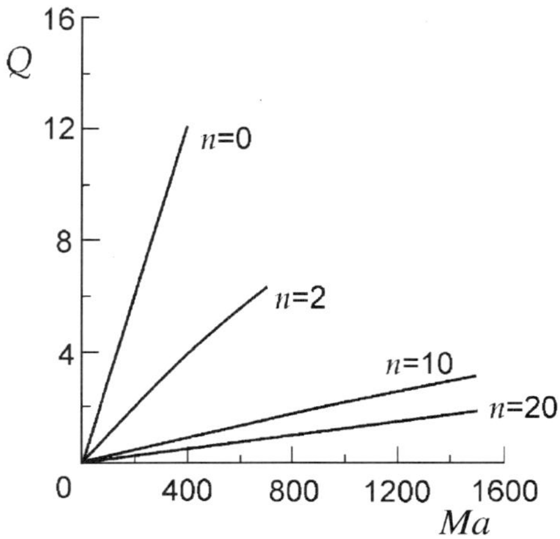

FIG. 26 Thermocapillary convection in a liquid zone with solid elements on its open surface. The flow rate as a function of the Marangoni number for different grid spacings. Recall that here $Ma = Re\,Pr$. (From Ref. 15.)

spacings. The linear dependence suggests that the flow is indeed of a creeping (Stokes) nature. The family of curves shown in Fig. 26 can be analytically described by a simple functional relation: $Q(h, Re) = f(h)Re$. The dependence of the factor f on the number of grid steps is shown in Fig. 27. For the examined range of values of $h = (0.083–0.2)$ we can write $f(h) = 0.025h^{0.81}$.

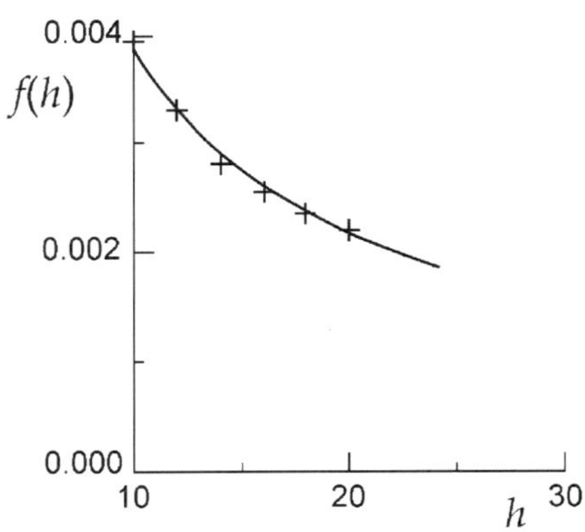

FIG. 27 Thermocapillary convection in a liquid zone with solid elements on its open surface. Proportion coefficient in the flow rate law as a function of the number of grid steps. (From Ref. 15.)

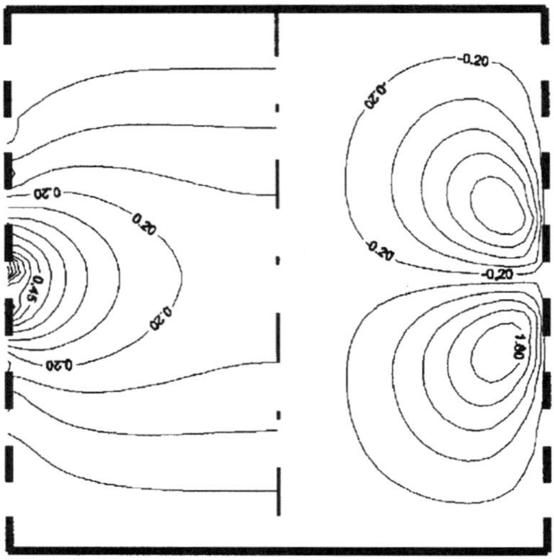

FIG. 28 Thermocapillary convection in a liquid zone with solid elements on its open surface. Stream function isolines (at the right) and isotherms ($Re = 4 \times 10^2$, $n = 12$, $Bi = 20$, and $Bi_1 = 1$). (From Ref. 15.)

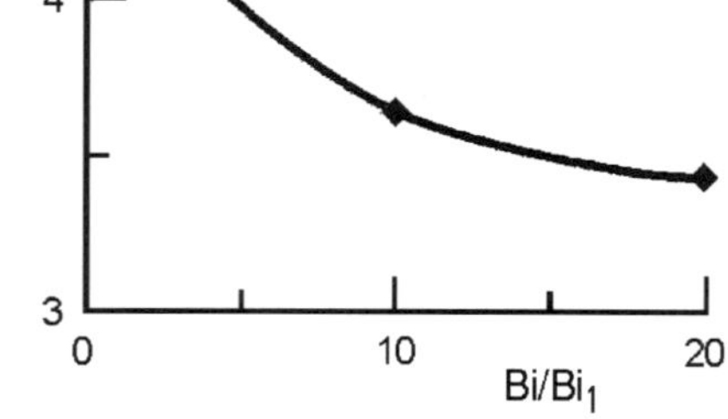

FIG. 29 Thermocapillary convection in a liquid zone with solid elements on its open surface. Flow rate vs. ratio of heat transfer coefficients. (From Ref. 15.)

C. Influence of the Difference Between Liquid and Grid Thermal Properties on the Structure and Intensity of Convection

Since the Marangoni effect depends on the inhomogeneity of the liquid surface temperature, the difference in the thermal properties of the grid and the liquid and, first of all, in the heat transfer coefficients might affect the above-given flow characteristics. This effect was numerically analyzed for a grid with spacing $h = 0.17$ ($n = 12$) and a ratio of grid to liquid heat transfer coefficients ranging from 0.05 to 20.

The stream function isolines and isotherms for the right-hand limiting point of the considered interval are shown in Fig. 28. At different values of heat transfer coefficient ratios the flow has the same global structure. All variations are observed in the boundary layers. In Fig. 29 the flow rate Q is plotted vs. the ratio of Biot numbers defined for the grid, Bi, and the liquid, Bi_l, respectively. The result indicates the nonmonotonic character of the variation in intensity of the flow.

REFERENCES

1. Birikh, R.V., Briskman, V.A., Sklyaev, O.E., Velarde, M.G. Liquid equilibrium instability and supercritical convection caused by the Marangoni effect on a free surface partially covered by solid elements. J. Jpn. Soc. Microg. Appl. **1998**, *15* (suppl II), 388–393.
2. Pearson, J.R.A. On convection cells induced by surface tension. J. Fluid Mech. **1958**, *4*, 489–500.
3. Thom, A., Apelt, C. J. *Field Computations in Engineering and Physics*; Van Nostrand: London, 1961.
4. Birikh, R.V., Briskman, V.A., Sklyaev, O.E., Velarde, M.G. Marangoni instability of liquid with partially free surface. 36th Aerospace Meeting & Exhibit. January 12–15, **1998**, Reno. AIAA-98-0654.
5. Chernatynskiy, V.I., Shliomis, M.I. Convection in the vicinity of the critical Rayleigh numbers at nearly vertical temperature gradient. Izv. Akad. Nauk SSSR, Mech. Zhidk. Gaza **1973**, *1*, 64–70 (in Russian).
6. Dijkstra, H. On the structure of cellular solutions in Rayleigh–Benard–Marangoni flows in small-aspect-ratio containers. J. Fluid Mech. **1992**, *243*, 73–102.
7. Smith, M.K., Davis, S.H. Instabilities of dynamic thermocapillary liquid layers. Part 1. Convective instabilities. J. Fluid Mech. **1983**, *132*, 119–144.
8. Hadid, H.B., Roux, B. Thermocapillary convection in long horizontal layers of low-Prandtl-number melts subject to a horizontal temperature gradient. J. Fluid Mech. **1990**, *221*, 77–103.
9. Polezhaev, V.I. *Hydrodynamics, Heat and Mass Transfer During Crystal Growth*; Springer: Berlin, 1984.

10. Birikh, R.V., Briskman, V.A., Chernatynsky, V.I., Roux, B. Thermocapillary flows due to partially closed nonisothermal liquid surface. 34th Aerospace Meeting & Exhibit. January 15–18, 1996, Reno, AIAA-96-0736.
11. Chen, G., Roux, B. Analytical solution and numerical simulation of thermocapillary convection in floating zones. Adv. Space Res. **1991**, *11*, 151–162.
12. Birikh, R.V., Briskman, V.A., Chernatynsky, V.I., Roux, B. Control of thermocapillary convection in a liquid bridge by high frequency vibrations. Microgravity Q. **1993**, *3*, 23–28.
13. Lyubimov, D.V., Lyubimova, T.P., Roux, B. Mechanisms of vibrational control of heat transfer in a liquid bridge. Int. J. Heat Mass Transfer **1977**, *40*, 4031–4042.
14. Potter, D. *Computational Methods in Physics*; Wiley: New York, 1973.
15. Chernatynsky, V.I., Thermocapillary convection in a liquid zone with solid elements on its free surface. In *Convection in Immiscible Liquid–Liquid Systems*; Briskman, V.A., Birikh, R.V. Eds; UBRAS: Ekaterinburg, 1999; 88–96 (in Russian).
16. Birikh, R.V., Rudakov, R.N. *Convective Instability in a Region with a Thin Permeable Partition* (*Grid*); Perm State Tech. Uni.: Perm, 1996.

10
Thermocapillary Migration of Bubbles and Drops

This chapter is devoted to a description of phenomena once more resulting from nonuniformity in the surface tension. In particular we shall discuss how the Marangoni effect causes displacements of fluid volumes (bubbles, drops). Generally, a liquid drop or an air bubble immersed in a liquid spontaneously migrates in a temperature gradient towards the hotter regions.

I. EXPERIMENTS AND DIFFICULTIES WITH BUBBLE AND DROP MIGRATION

Bubble and drop migration due to the Marangoni effect was first studied by Young et al. [1] (later referred to as YGB). Small bubbles of air were weighed in a drop of silicone oil located between end faces of a machinist's micrometer. The oil was warmed from below, therefore generating a thermocapillary force, also directed downwards against buoyancy. As the latter is proportional to the radius to the third power, a^3, and the surface thermocapillary force to a^2, for very small bubbles ($a < 0.1\,\mathrm{mm}$) and values of the temperature gradient high enough ($\nabla T \sim 10$–$100\,\mathrm{K/cm}$) the thermocapillary force exceeds the buoyancy force and bubbles move downwards, thus solving the "falling bubble paradox" (in Birkhoff's words [2]) due, in Birkhoff's opinion, to the fact that "surface tension (like viscosity) is almost always conventionally assumed to be a constant in fluid mechanics." The selection of the temperature gradient made it possible for the bubble velocity of a given size to become equal to zero, hence levitating in the gravity field. Owing to the smallness of volume size and the large viscosity of the oils, buoyancy-driven convection was negligible. The bubble diameters were measured with a travelling microscope.

In the creeping flow approximation (Hadamard–Rybczynskii–Stokes approximation with flow inside and outside) the vertical migration velocity, u, of a bubble or drop is [1,3–4]

$$u = \frac{2}{3\eta_1}\frac{[3a\sigma_1 A/(2+\kappa_2/\kappa_1) - (\rho_1-\rho_2)ga^2(1+\eta_2/\eta_1)]}{(2+3\eta_2/\eta_1)} \tag{1}$$

where A is the temperature gradient, ρ_i, η_i, and κ_i denote density, (dynamic) shear viscosity, and heat conductivity, respectively; the subscript i refers to the external fluid ($i = 1$) and to the fluid bubble (or drop) ($i = 2$).

In Fig. 1 the temperature gradient, A, necessary to levitate bubbles of various radii are shown. The solid line corresponds to the case $u = 0$ of Eq. (1). The experiment was later repeated by Hardy [5] (black dots in this figure) and other authors. Noteworthy is the experiment conducted by Braun et al. [6] aboard a TEXUS sounding rocket using a nonstandard liquid with surface tension increasing with increasing temperature.

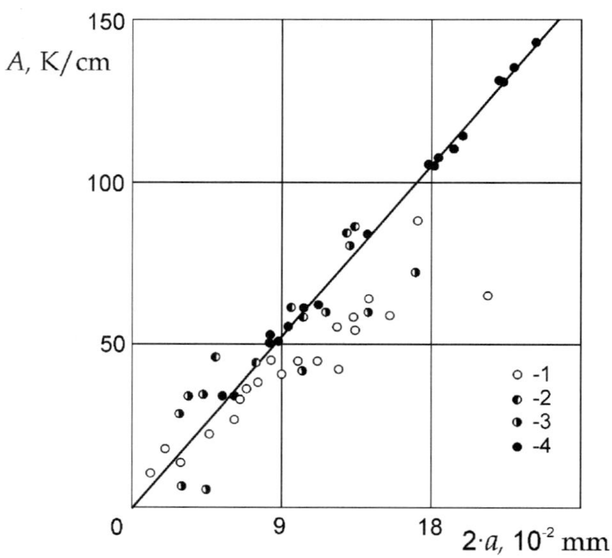

FIG. 1 Thermocapillary migration of bubbles and drops. Temperature gradient necessary to levitate a bubble vs. bubble diameter, $2a$. Points 1, 2, and 3 indicate silicone oil $\eta_1 = 20\,\text{cS}$, $200\,\text{cS}$, and $1000\,\text{cS}$ [1]; points 4 indicate silicone oil $\eta_1 = 200\,\text{cS}$ and $1000\,\text{cS}$; $1\text{S} = 10^{-4}\text{m}^2/\text{s}$ [4]; solid line corresponding to Eq. (1) with $u = 0$. (From Ref. 5.)

These experiments show that the tangential thermocapillary stresses on the spherical liquid–gas bubble interface, induced by the temperature dependence of the liquid surface tension, generates flows in the fluid surrounding the bubble. For standard liquids, as generally referred to in this book, surface tension goes down with increasing temperature, and hence the liquid flows around the bubble in the direction opposite to the surface tension gradient (i.e., from the hot pole to the cold rear) and hence the bubble spontaneously moves towards the warmer regions. For a liquid drop immersed in another immiscible liquid the abovementioned flow is generated on either side of the interface (Fig. 2). The stress transmitted to the fluid within the drop cannot cause any force to be excited on the drop. The stress transmitted to the surrounding liquid will cause a reaction on the drop of the same magnitude, but opposite in direction. If the drop is held fixed by attaching it to a wire, for example, it would experience this reaction force. In the thermocapillary motion of a bubble or a drop, there is no external force.

Refinements of the YGB results [1] were provided by Harper et al. [7], Bratukhin [8], Subramanian [4,9] and others. Ryazantsev, Velarde and their collaborators generalized the theory to account for both *active* and *passive* drops in a thermal gradient [10–16]. Subramanian et al. [4,9,17], Wozniak et al. [18] and Sadhal et al. [19] have provided reviews and monographs on this subject. The book by Subramanian and Balasubramanian [4] is a most valuable addition to the literature.

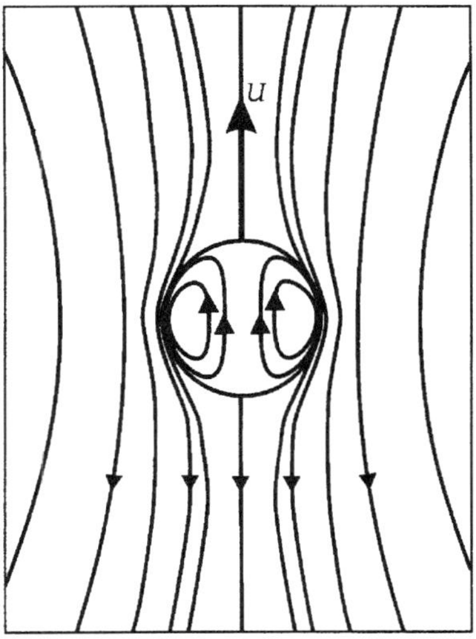

FIG. 2 Thermocapillary migration of bubbles and drops. Thermocapillary flows inside and outside an insoluble drop placed in a liquid with upward-directed uniform temperature gradient.

Thermocapillary migration of drops and bubbles has been the subject of renewed interest largely due to needs posed by material processing methodologies in variable gravity and microgravity conditions. Indeed, as already seen in many cases already discussed here, when buoyancy effects are negligible the Marangoni effect is the major factor determining the behavior of fluids in nonisothermal processes. However, the experimental research on bubble and drop motion did not progress beyond a few laboratory experiments. This can be explained by the fact that on Earth the motion of drops and bubbles is extremely sensitive to buoyancy forces, so that thermocapillary migration can be observed only if the former is either weakened drastically or artificially compensated. To meet this requirement one should use very small bubbles and drops or resort to using a Plateau tank, when the densities of the drop and the continuous phase are nearly the same. It seems pertinent to note that an air bubble of radius 20 μm, placed in a temperature gradient of 1 K/mm in a silicone oil, will move with approximately the same velocity that it would have if rising under the influence of buoyancy. In a space craft with effective gravity about 10^{-6} (microgravity), even under a gentle gradient of 10^{-2} K/mm, a bubble must be large as 0.2 m in radius for the gravitational contribution to its motion to be comparable to that of thermocapillarity.

The approach used to investigate thermocapillary motion of bubbles relies on the fact that the buoyant velocity of a bubble is proportional to the square of its radius, whereas the thermocapillary velocity is proportional to the first power of the radius. Therefore for a small enough bubble in a downward-directed temperature gradient, the thermocapillary force, F_C, balances the buoyancy force, F_A, and hence the bubble will be as if in zero effective gravity [Fig. 3(a)]. This allows the measurement of the temperature gradient needed to hold the bubble in a steady state and to compare it with the predicted value. However these studies [1,5,20,21] are restricted to bubbles of very small diameter (about 100–200 μm) and low

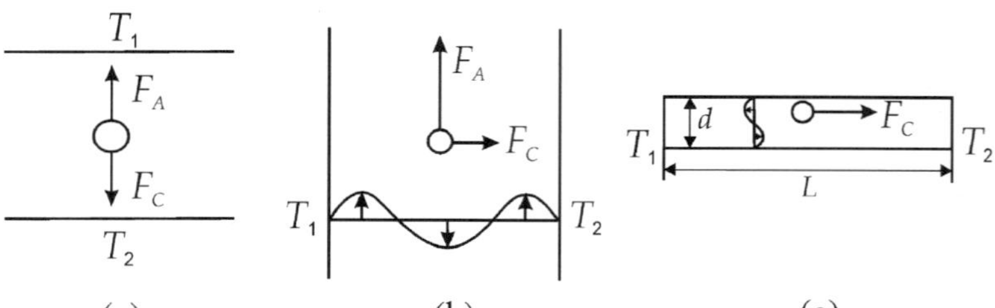

FIG. 3 Thermocapillary migration of bubbles and drops. Schematic representation of various experimental set-ups used to observe thermocapillary bubble migration when gravity points downward.

enough temperature gradients to ensure that the influence of the background buoyancy-driven flow is negligible.

Experiments with liquid drops have been confined to systems with similar densities, using Plateau tanks, in which pairs of immiscible liquids match their densities at a certain temperature (e.g., ricin/castor oil in silicone oil). Upon changing the temperature the drop velocity is determined by following each drop over a short distance to a new equilibrium point and calculating the slope of the plot of the drop position vs. time. However, such systems are not well suited for a detailed investigation of the migration process, since the drops quickly come to rest. Experiments in variable and, eventually, low effective (micro) gravity laboratories offer new possibilities for quantitative investigation of thermocapillary migration of large-scale bubbles and liquid drops at high Marangoni (Peclet) and Reynolds numbers [22–27]. They also provide much wider choice to pick from a variety of liquids. Until recently there have been only few attempts, using various facilities, to carry on short-duration low gravity experiments on the motion of gas bubbles [28,29] and liquid drops [30,31]. Results from experiments on thermocapillary migration in microgravity conditions exist thanks to the D1 Shuttle mission [23,24,32–35]. As there are too many parameters involved most of the available experiments provided qualitative descriptions and quantitative data that were often contradictory.

Two different methods were tested allowing laboratory observations on Earth and velocity measurements of thermocapillary bubble migration. In both methods mutually perpendicular orientations of temperature gradient and gravitational acceleration were used to separate different mechanisms of bubble motion, levitation due to buoyancy, and thermocapillary migration. In one case the bubbles were free to levitate in a vertical liquid layer heated on the side [Fig. 3(b)] and buoyancy-driven convection was reduced by making use of a liquid with very low thermal expansion coefficient. In another case the bubbles executed horizontal migration in a plane thin horizontal liquid layer subjected to a longitudinal temperature gradient [Fig. 3(c)], and convection was drastically reduced due to the small layer thickness.

II. THERMOCAPILLARY MIGRATION OF DROPS AND BUBBLES IN WEIGHTLESSNESS CONDITIONS (THEORY)

Let a large enough liquid contain a drop of another immiscible liquid. If the temperature gradient applied (at infinity), $\nabla T = \mathbf{A}$, is constant, the Marangoni effect at the drop surface gives rise to thermocapillary convection in both media, inside and outside the drop. Under steady-state conditions, if $\mathbf{A}$ is low enough, the velocity of the drop migration, u, is constant.

Let us analyze the problem in the simplest possible case [9]. All parameters of the liquids (densities, ρ_i; kinematic and dynamic viscosities, ν_i and η_i; heat diffusivity and heat conductivity, κ_i and χ_i; $i = 1, 2$, for the first and second medium, respectively) are assumed to be constant, except for the surface tension, $\sigma(T) = \sigma(T_0) - \sigma_1(T - T_0)$, which, as in earlier studied cases, for standard liquids is assumed to decrease linearly with temperature. Furthermore, the problem is axially symmetric, the motion is slow and steady, the migration velocity of the drop is constant, the shape of the drop does not significantly depart from that of a sphere (the corresponding Weber number is very low), the liquids are mutually insoluble, and the liquid at infinity is at rest.

The polar axis z is directed along the vector **A** and brought into coincidence with the expected trajectory of the drop inertia center. Assume that at the initial time the center of the drop is at point $z = 0$ and the reference temperature is taken to be the undisturbed temperature at the origin of the coordinate system. Then, the distribution of velocities, $\mathbf{v}_i$, pressures, p_i, and temperatures, T_i, in both media obeys the system of equations ($i = 1, 2$)

$$\operatorname{div} \mathbf{v}_i = 0 \tag{2}$$

$$\frac{\partial \mathbf{v}_i}{\partial t} + (\mathbf{v}_i \cdot \nabla)\mathbf{v}_i = -\frac{\nabla p_i}{\rho_i} + \nu_i \Delta \mathbf{v}_i \tag{3}$$

$$\frac{\partial T_i}{\partial t} + \mathbf{v}_i \cdot \nabla T_i = \chi_i \Delta T_i \tag{4}$$

At infinity $T_1 = Az, p_i = 0$, and $\mathbf{v}_i = 0$.

To these equations we must add the appropriate boundary conditions at the drop surface. Since the interface is considered to be impenetrable the normal components of velocities vanish there. The tangential components of velocities, as well as temperatures, normal components of heat fluxes, and normal and tangential components of stresses must be continuous.

Let us now assume that the drop has taken a (constant) velocity **u** along the gradient **A**. By passing to the drop moving frame, Eqs. (2)–(4) become stationary. The origin of the spherical coordinates (r, θ, φ) is then made coincident with the center of the drop mass and the direction of the z-axis is kept as before. The reference temperature is taken to be the undisturbed temperature at point $z = ut$, which at the time considered coincides with the center of the drop mass. Therefore primed ($\prime$) functions below depend only on the space coordinates:

$$z = z' + ut, \qquad \mathbf{v}_i(r, t) = \mathbf{v}_i'(\mathbf{r}') + \mathbf{u}, \qquad T_i(\mathbf{r}, t) = Aut + T_i'(\mathbf{r}') \tag{5}$$

To formulate the problem in dimensionless form we introduce the following scales: the radius of the drop, a, stands for the unit of length, $\sigma_1 aA/\eta_1$ for velocity, $\sigma_1 A$ for pressure, and aA for temperature. In the following we shall disregard primes. Then Eqs. (2) and (3) yield a system of equations for dimensionless quantities

$$\nabla \cdot \mathbf{v}_1 = 0, \qquad M(\mathbf{v}_1 \cdot \nabla)\mathbf{v}_1 = -\nabla p_1 + \Delta^2 \mathbf{v}_1,$$
$$PrM(u + \mathbf{v}_1 \cdot \nabla T_1) = \Delta T_1 \tag{6a}$$

$$\nabla \cdot \mathbf{v}_2 = 0, \qquad M(\mathbf{v}_2 \cdot \nabla)\mathbf{v}_2 = -\frac{1}{\rho}\nabla p_2 + \nu \Delta^2 \mathbf{v}_2,$$
$$PrM(u + \mathbf{v}_2 \cdot \nabla T_2) = \chi \Delta T_2 \tag{6b}$$

$$M = \frac{\sigma_1 A \rho_1 a^2}{\eta_1^2}, \qquad \rho = \frac{\rho_2}{\rho_1}, \qquad Pr = \frac{\nu_1}{\chi_1}, \qquad \nu = \frac{\nu_2}{\nu_1}, \qquad \chi = \frac{\chi_2}{\chi_1}$$

As a denotes the radius of the drop in this chapter we shall use the letter M rather than Ma to define the Marangoni number. Note that M appears as Reynolds number in the Navier–Stokes equations and PrM appears as the Peclet number in the heat equation.

Now, suppose that a homogeneous liquid flow is initiated at infinity in the direction of the drop. The flow develops, under the prescribed pressure and temperature, with the velocity

$$\mathrm{v}_1 = -u\cos\theta \mathbf{e}_r + u \sin\theta \mathbf{e}_\theta, \qquad T_1 = r\cos\theta, \qquad p_1 = 0 \tag{7}$$

where $\mathbf{e}_r$ and $\mathbf{e}_\theta$ are unit vectors in polar coordinates.

The drop surface is described by $r = R(\theta)$. Let ε be the angle between the radius vector and the outer normal to the drop surface, $\mathbf{e}_n$. Then the unit vectors of the normal, $\mathbf{e}_n$, and the tangent, $\mathbf{e}_\tau$, to the meridian section are $\mathbf{e}_n = \mathbf{e}_r \cos\varepsilon - \mathbf{e}_\theta \sin\varepsilon$, $\mathbf{e}_\tau = \mathbf{e}_r \sin\varepsilon + \mathbf{e}_\theta \cos\varepsilon$ and we now make partial use of primes ($'$) to denote derivatives with respect to θ, $\tan\varepsilon = R'/R$, $R' = \mathrm{d}R/\mathrm{d}\theta$.

With the given notation we write down the abovementioned conditions for velocities, temperatures, and heat fluxes at the boundary, $r = R(\theta)$, in spherical coordinates. Dividing each equation by $\cos\varepsilon$ gives

$$\mathrm{v}_{r1} - \frac{R'}{R}\mathrm{v}_{\theta 1} = 0, \qquad \mathrm{v}_{r2} - \frac{R'}{R}\mathrm{v}_{\theta 2} = 0, \qquad T_1 = T_2 \tag{8}$$

$$\frac{R'}{R}\mathrm{v}_{r1} + \mathrm{v}_{\theta 1} = \frac{R'}{R}\mathrm{v}_{r2} + \mathrm{v}_{\theta 2}, \qquad \frac{\partial T_1}{\partial r} - \frac{R'}{R^2}\frac{\partial T_1}{\partial \theta} = \kappa\left(\frac{\partial T_2}{\partial r} - \frac{R'}{R^2}\frac{\partial T_2}{\partial \theta}\right)$$

where v_{ri} and $\mathrm{v}_{\theta i}$ are the velocity components of the ith medium in the spherical coordinates and $\kappa = \kappa_2/\kappa_1$.

When writing equations for tangential and normal components of the stress tensor, the Marangoni effect is to be incorporated and hence, in addition to normal stress, it generates a tangential stress directed toward a decreasing temperature.

After dividing the equations by $\cos\varepsilon$ one obtains

$$p_1 - p_2 + \left(\frac{\sigma_0}{M} - T\right)\cdot 2H = -[L_1 - \eta L_2]\frac{R'}{R}$$

$$+ 2\left(\frac{\partial v_{r1}}{\partial r} - \eta\frac{\partial v_{r2}}{\partial r}\right) - \left(R'\frac{\partial T}{\partial r} + \frac{\partial T}{\partial \theta}\right)\frac{R'}{R}(R^2 + R'^2)^{-1/2} \tag{9a}$$

$$\left[p_1 - p_2 + \left(\frac{\sigma_0}{M} - T\right)\cdot 2H\right]\frac{R'}{R} = -[L_1 - \eta L_2]$$

$$+ \left[2\left(\frac{\partial v_{\theta 1}}{\partial \theta} + v_{r1}\right) - \eta 2\left(\frac{\partial v_{\theta 2}}{\partial \theta} + v_{r2}\right)\right]\frac{R'}{R}$$

$$+ \left(R'\frac{\partial T}{\partial r} + \frac{\partial T}{\partial \theta}\right)(R^2 + R'^2)^{-1/2} \tag{9b}$$

$$L_i \equiv \frac{1}{R}\frac{\partial v_{ri}}{\partial \theta} + \frac{\partial v_{\theta i}}{\partial r} - \frac{v_{\theta i}}{R}$$

$$H = \frac{1}{2}\left[\frac{2R^2 + 3R'^2 - RR''}{(R^2 + R'^2)^{3/2}} - \operatorname{ctg}\theta\frac{R'}{R(R^2 + R'^2)^{1/2}}\right]$$

where H is the mean surface curvature, and $\eta = \eta_2/\eta_1$ and $\sigma_0 = \sigma\rho_1 a/\eta_1^2$ are the dimensionless parameters of the system. Because the surface temperature is continuous and the tangential stresses only depend on tangential derivatives of the temperature, subscripts for temperature can be dropped. In addition to conditions (7)–(9) it is necessary to include the requirement that the function at the origin of the coordinates must be bounded.

The problem of Eqs. (6)–(9) involves only one control parameter, M. All other parameters define physical properties of the heterogeneous system. Therefore a solution to the problem (6)–(9) is sought in the form of series expansions in powers of M,

$$\mathbf{v}_i = \mathbf{v}_i^{(0)} + M\mathbf{v}_i^{(1)} + \cdots, \qquad p_i = p_i^{(0)} + Mp_i^{(1)} + \cdots,$$
$$T_i = T_i^{(0)} + MT_i^{(1)} + \cdots \tag{10}$$

$$u = U + Mu^{(1)} + \cdots, \qquad R(\theta) = 1 + M^2 f_2(\theta) + \cdots$$

Substituting Eqs. (10) into Eqs. (6)–(9) yields

$$\mathbf{v}_1^{(0)} = U\left[\left(-1+\frac{1}{r^3}\right)P_1\mathbf{e}_r + \left(-1-\frac{1}{2r^3}\right)r\nabla P_1\right] \tag{11a}$$

$$\mathbf{v}_2^{(0)} = \frac{3}{2}U[(1-r^2)P_1\mathbf{e}_r + (1-2r^2)r\nabla P_1] \tag{11b}$$

$$T_1^{(0)} = \left(r+\frac{1-\kappa}{2+\kappa}r^{-2}\right)P_1, \qquad T_2^{(0)} = \frac{3}{2+\kappa}rP_1 \tag{12}$$

$$p_1^{(0)} = 0, \qquad p_2^{(0)} = -15U\eta rP_1 + \frac{2\sigma_0}{M} \tag{13}$$

$$U = \frac{2}{(2+\kappa)(2+3\eta)}, \qquad f_2(\theta) = sP_2$$

$$s_2 = -\frac{3}{4\sigma_0}\left[\frac{5}{2}U^2\frac{(1-\rho)(1+\eta)}{4+\eta} + \frac{PrU}{63\chi}\frac{49\chi-7\chi\kappa+18\kappa}{(3+2\kappa)(2+\kappa)}\right]$$

where $P_1 = \cos\theta$ and $P_2 = \frac{1}{2}(3\cos^2\theta - 1)$ are Legendre polynomials.

As the velocity in the zeroth-order approximation is always positive, it can be concluded that the drop moves towards warmer liquid regions. The migration velocity in a given approximation is defined by only two dimensionless parameters, κ and η. With the growth of these parameters the value of U decreases. Indeed, an increase of the drop viscosity reduces thermocapillary convection and in the limiting case, at $\eta \to \infty$, the motion is suppressed. Increasing the heat diffusivity of the second medium the temperature gradient along the surface decreases. At $\kappa \to \infty$ the drop temperature is constant and there is no convection because no Marangoni effect survives. In the other limiting case (a gas bubble in a viscous medium at $\eta \to 0$ and $\kappa \to 0$) the dimensionless migration velocity is equal to 0.5. In dimensional form we have

$$U = \frac{a\sigma_1 A}{2\eta} \tag{14}$$

The shape of the drop in motion is affected by s_2 because it contains the Weber number. However, it can be said that the migrating drop tends to flatten when it has a lower density compared to the surrounding liquid, for example, a gas bubble ($\rho < 1$) and, on the otherhand, tends to elongate in the direction of motion when it has a higher density ($\rho > 1$). These effects might be imperceptible owing to the positive term proportional to the Prandtl number, *Pr*, in Eq. (11). It is only because of the increase in *Pr* defined by the parameters of the first medium that the drop tends to flatten. The amplitude of the deformation, s_2, is inversely pro-portional to the value of the surface tension (recall the earlier mentioned Weber number). Therefore for high enough values of the surface tension the deformation effects that appear only in the third approximation will be very weak. The influence of the other parameters on the drop shape is of a rather intricate nature.

III. BUBBLE MIGRATION UNDER CONDITIONS OF WEAK BUOYANCY-DRIVEN CONVECTION (GROUND-BASED EXPERIMENT)

Let us briefly recall experimental findings about bubble motion in a thin vertical liquid layer heated on the side [36,37]. Because the temperature gradient is perpendicular to the gravitational acceleration, the slow horizontal thermocapillary migration can be discriminated from the vertical fast buoyancy upward motion. A sketch of the experimental set-up is shown in Fig. 4.

A layer of distilled water was confined in a thin 7-mm gap between two vertical massive metal plates, 1, with transparent plates as other lateral walls. The

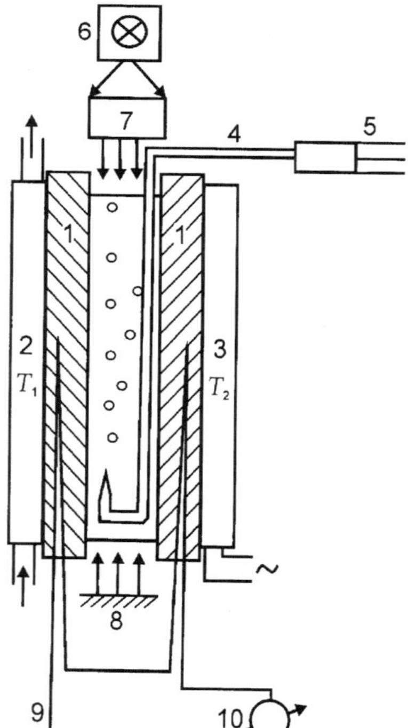

FIG. 4 Thermocapillary migration of bubbles and drops. Experimental set-up for thermocapillary migration of gas bubbles in a vertical water layer laterally heated. (From Ref. 37.)

temperature of one plate (cold) was maintained to within 0.1 K with the help of a heat exchanger, 2, connected to a thermostat. The other plate (hot) was heated by the electric heater, 3. Temperature was measured by thermocouples whose junctions were immersed in water. A glass tube, 3 mm in diameter, with curved lower end, was introduced into the container from above. The upper end of the tube was fitted to a syringe, 5. The syringe piston was set into motion by a synchronous motor and used to generate air bubbles of diameter 0.2–0.4 mm, which periodically detached from the tube end. The container was illuminated from above by a collimating system, 6–7, generating a parallel light beam. The light intensity was increased by reflection using a glass (mirror), 8, located beneath. The convective flow was visualized with the aid of aluminum powder added to the liquid. The bubble motion tracks were obtained by using a pulsed light source of 50–100 Hz frequency.

In the absence of temperature gradient, buoyancy led the bubbles introduced into the liquid to move vertically up along a straight line with a velocity of approximately 1 cm/s. With the heaters operating, the vertical motion of the bubbles was deflected towards the heated side because of thermocapillarity. Accordingly, the slow horizontal thermocapillary migration can be separated from the vertical fast buoyancy-driven upward motion to obtain the bubble migration velocity. Both components of the bubble velocity were determined from the bubble track images on the stroboscopic photographs shown in Fig. 5. The vertical component was then used to calculate the bubble radius using the second term in Eq. (1). The size of bubbles was experimentally determined from the large-scale photographs and disagreement between the bubble radii defined under similar conditions was found to be not more than 4–5%.

This fact should be viewed as indirect evidence for how clean the bubble surface was, which is very unusual as water surfaces are easily contaminated and hence water is, generally, considered to be unsuitable for studying thermocapillary effects. However, experiments exist providing successful observation of thermocapillary convection on the surface of a thin layer (film) of running water flowing very rapidly down an inclined surface [38]. In the experiment described above, the vertical velocity of the rising bubbles was just consistent with the drop velocity computed by Hadamard and Rybczynskii [4] rather than with the Stokes formula for a solid particle [3,4,39], as it should be in the case of a contaminated surface. This possibly implies that in the case considered the fast bubble immersion was accompanied by a surfactant drift toward the rear pole of the bubble and surface renovation. The measurements of thermocapillary migration velocity as a function of the bubble radius showed that at moderate temperature gradients the measured and theoretically predicted values are in good agreement (Fig. 6).

Although surface tension gradient (Marangoni) driven motion can be discriminated against the buoyancy (Rayleigh) driven migration, the latter has various undesirable effects such as trapping of the bubble by ascending or

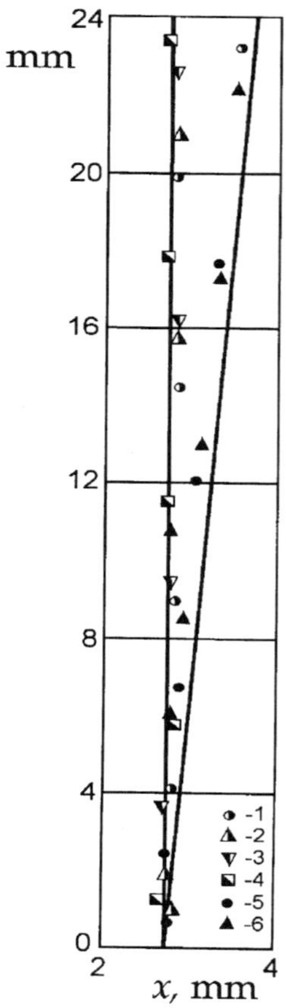

FIG. 5 Thermocapillary migration of bubbles and drops. Tracks of air bubbles moving upwards in a vertical liquid layer in the absence (points 1–4) and presence (points 5–6) of a lateral temperature gradient. (From Ref. 37.)

descending liquid layers and distortion of the temperature field, hence creation of a transverse force acting on the bubble in the direction of the higher flow velocity region as shown in Fig. 5. Therefore, the influence of buoyancy was eliminated by operating with liquid water held at a mean temperature of 4°C, which corresponds to zero thermal expansion coefficient and consequent inversion of its sign below this temperature. To estimate the influence of this side effect, thermogravitational convection in a vertical water layer in a horizontal temperature gradient was investigated [38,39], both for the case when the mean water temperature was held close to 4°C, and for normal temperature intervals. The velocity profiles (Fig. 7) were investigated with the help of tracers. The analysis allowed the determination

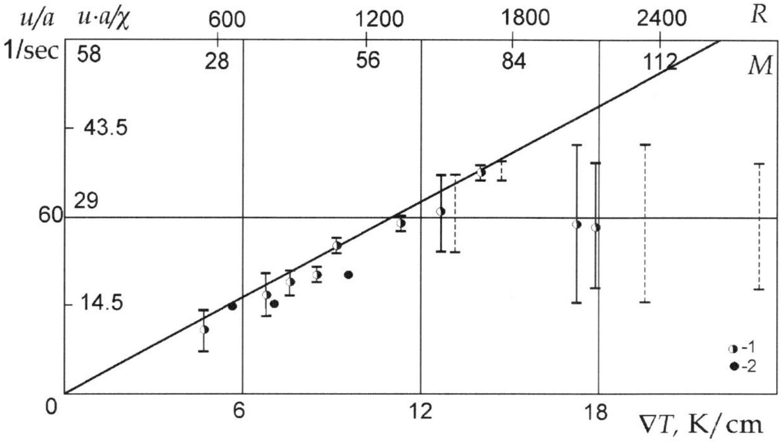

FIG. 6 Thermocapillary migration of bubbles and drops. Velocity of thermocapillary migration of an air bubble (divided by the bubble radius) as a function of the applied horizontal temperature gradient. (From Refs. 36 and 37.)

of the flows and temperature range, in which buoyancy-driven convection was a few orders of magnitude weaker (velocity profiles 1, 2) than under normal conditions (20°C) at the same temperature gradients (curve 4) and therefore exerted negligible effect on bubble migration. The results indicate that at Rayleigh numbers lower than 1500, buoyancy-driven convection will exert negligible effect on the data. Owing to this fact, good agreement was found between measured and calculated migration velocities (straight line in Fig. 6) in the region of small temperature gradients. However, at large temperature gradients the distortion of the temperature field caused by the flow must be taken into account (Fig. 8). Indeed, beginning from the value 15 K/cm, a marked nonlinearity is observed (curve 5 and following), i.e., there appears a core in which the temperature gradient is lower than in the boundary layers.

Thus, from the investigation of buoyancy-driven convection in water with mean temperature near 4°C it can be concluded that at moderate temperature gradients the data on the thermocapillary migration velocity as a function of the bubble radius and the predicted theoretical results are in good agreement.

IV. BUBBLE MIGRATION IN THIN HORIZONTAL LIQUID LAYERS (EXPERIMENT)

The thermocapillary migration of air bubbles in a thin horizontal liquid layer heated from the side has also been experimentally investigated in the set-up shown

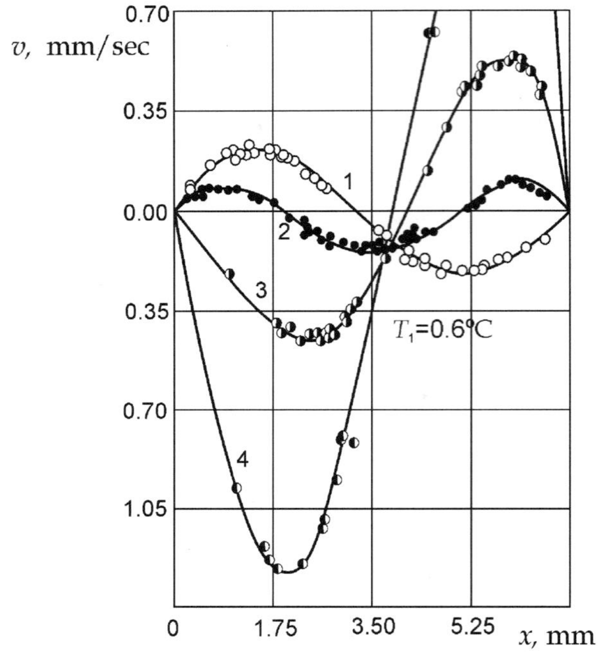

FIG. 7 Thermocapillary migration of bubbles and drops. Velocity profiles in a vertical liquid layer heated on the side to temperatures close to 4°C (curves 1–3) and to 20°C (curve 4) at the same temperature gradient. (From Ref. 37.)

in Fig. 9 [38,39]. The liquid layer, 1, with horizontal dimensions 35 mm × 35 mm and thickness $d = 2\,\text{mm}$, was sandwiched between two glass plates, 2 (Hele–Shaw cell). The lower glass was located on a metal plate, 3, in which a uniform horizontal temperature gradient was generated by two heat exchangers, 4, connected to thermostats. The temperature variations were measured by thermocouples whose junctions were immersed in the liquid.

Owing to the small layer thickness, even at considerable temperature gradients, buoyancy-driven convection was found to be negligible for all examined liquids. The air bubble was placed inside the layer and was naturally pressed to the upper glass plate as a result of buoyancy. In the absence of temperature gradient, the bubble was at rest, but with the horizontal temperature gradient it began moving towards the heater. The velocity of the bubble was determined from photographs of the bubble taken from above with a camera, 5, under stroboscopic illumination with a frequency in the range of 0.2–0.02 Hz. Ethanol, acetone, and n-heptane were used in the experiments as working liquids.

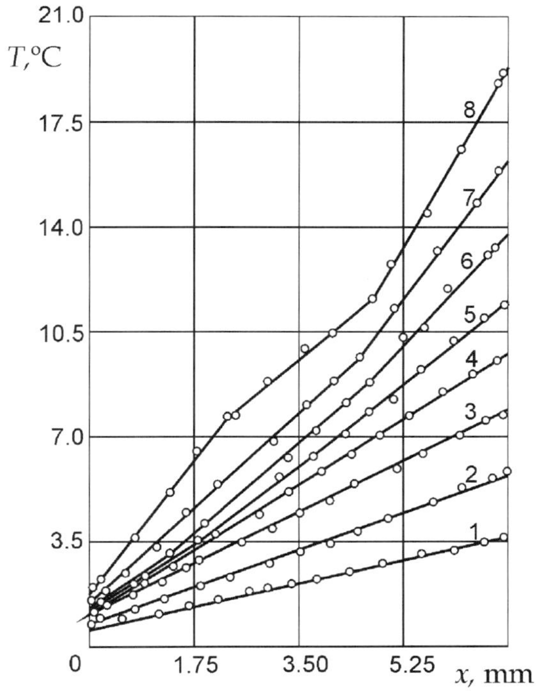

FIG. 8 Thermocapillary migration of bubbles and drops. Temperature distribution across the width of a vertical liquid layer heated on the side. (From Ref. 37.)

This experimental technique allowed the measurement of the velocities of the bubble thermocapillary migration over a wide range of temperature gradients and bubble sizes. The experimental data for spherical bubbles whose diameters are less than the layer thickness are shown in Fig. 10. Here the velocity of thermocapillary migration, which is related to the bubble radius divided by the dynamic viscosity of the liquid [$v = U\eta/a$, see Eq. (13)], is plotted on the ordinate. This made it possible to accommodate on the same curve the data obtained for bubbles of different sizes and liquids drastically differing in their viscosity. The migration velocity is proportional to the temperature gradient and inversely proportional to the dynamic (shear) viscosity.

Figure 11 [40] illustrates the dependence of the thermocapillary migration velocity on the bubble radius at constant temperature gradient. The migration velocity strongly depends on the ratio of the bubble diameter to the layer thickness. When the bubble diameter is much less than the layer thickness ($2a \leq d$), a spherical bubble touches only the upper wall. Then the velocity grows with an increase of the radius. When the bubble diameter increases and eventually exceeds

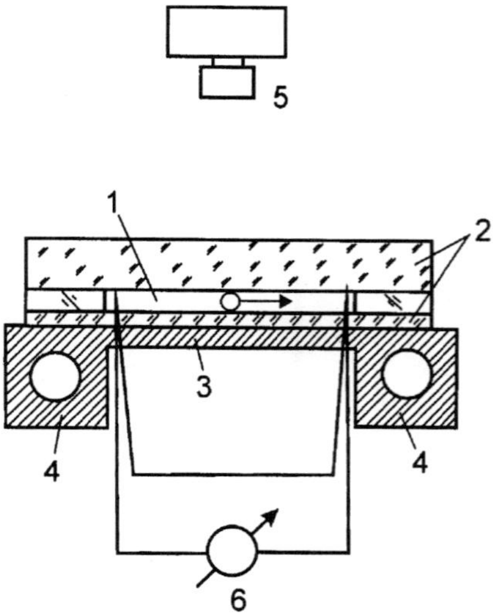

FIG. 9 Thermocapillary migration of bubbles and drops. Experimental apparatus for the investigation of thermocapillary migration of gas bubbles in a horizontal liquid layer heater on the side. (From Refs. 39 and 40.)

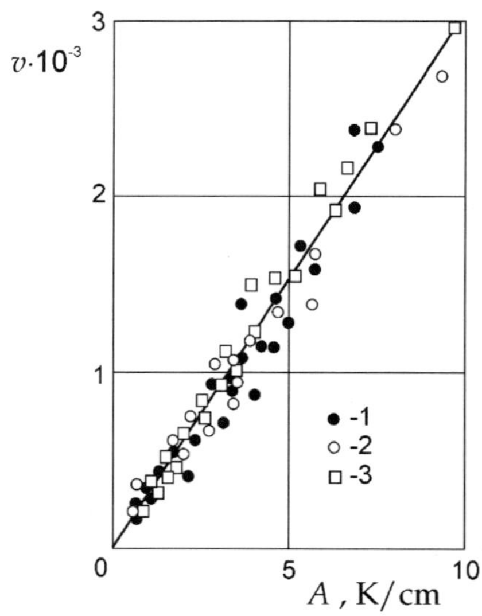

FIG. 10 Thermocapillary migration of bubbles and drops. Velocity of thermocapillary migration vs. temperature gradient for spherical bubbles in a thin horizontal layer. The symbols ●, ○, and □ denote experimental results with acetone, heptane, and ethanol, respectively. (From Refs. 39 and 40.)

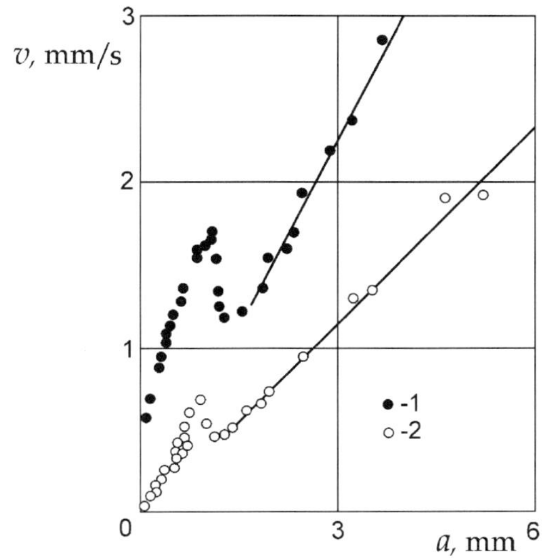

FIG. 11 Thermocapillary migration of bubbles and drops. Velocity of thermocapillary migration vs. radius in a thin horizontal layer. Experimental results with heptane (●) and ethanol (○). (From Ref. 40.)

the layer thickness ($2a \geq d$), it becomes squeezed between the plates and acquired the shape of a round cylinder confined between the walls. This is accompanied by a sharp decrease in the migration velocity but a further increase of the radius ($2a \gg d$) the velocity again begins to grow linearly. At this stage the bubble turns into a thin circular disc. Accordingly, the increase of the relative bubble dimension leads to changes in the bubble shape and in the character of the flow in the gap between the bubble and the cavity wall and also affects the bubble interaction with the walls.

Also, the measured value of velocity was an order of magnitude lower than that predicted by Eq. (13) for the velocity of free bubble migration in a large, and therefore unbounded space. Thus the interaction of the gas bubble with the walls and its friction on the walls seems to be a significant factor in the problem.

In order to construct a theoretical model of the bubble motion, it is necessary to know whether there is "dry" friction or whether the bubble is separated from the glass by a thin liquid layer. This is accomplished by using a special interferometric set-up that permits us to see that in all cases the bubble is indeed separated from the glass plate by a thin liquid film, whose thickness is a key parameter for the theoretical analysis. In this auxiliary experiment a parallel monochromatic light beam at an angle close to normal illuminates the glass plate at the top of the cavity. The light is reflected both from the upper and lower faces of the glass plate. When the cavity is not filled with liquid the reference interferometric pattern can be

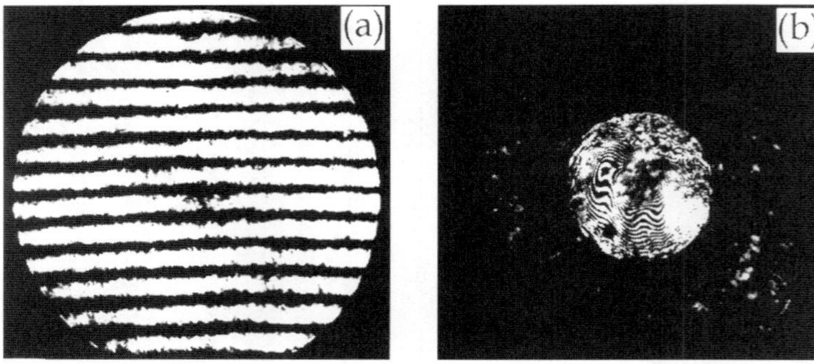

FIG. 12 Thermocapillary migration of bubbles and drops. Interferogram of bands equally spaced observed on the glass surface of the model with (a) an empty cavity, and (b) when the liquid contains an air bubble. (From Ref. 39.)

observed in the reflected light with a number of equally spaced bands [Fig. 12(a)]. When the cavity is filled with toluene, whose refraction index is equal to that of the glass, the beam is not reflected from the lower face and the interference bands disappear. The insertion of a bubble in the form of a round disk, 1 cm in radius, leads to the reappearance of the interference pattern. However, it differs from the former because this time the interference is generated by the beams reflected from the surface of the upper glass and from the film surface (liquid–air interface) [Fig. 12(b)]. If it were not for the liquid interlayer the original interference pattern would be restored. This fact shows that there exists "dry" friction of the bubble against the glass.

V. BUBBLE MIGRATION IN A THIN HORIZONTAL LIQUID LAYER (THEORY)

Let us consider the problem of thermocapillary migration of an air bubble of radius a in a plane horizontal layer of viscous liquid bounded by solid walls [40]. The upper surface of the layer is thermally insulated, while the lower conducts heat; the z-axis is directed downwards along the force of gravity, $\mathbf{g} = g\mathbf{k}$. As earlier, it is assumed that far from the bubble there is a constant temperature gradient, $\nabla T = \mathbf{A}$. Constraints on parameters here are the same as in the second section of this chapter.

Let us assume that at the initial time, $t = 0$, the motion is already steady, and hence the bubble migrates in the direction of the temperature gradient, $\mathbf{A}$, with velocity, u. We direct the x-axis (unit vector $\mathbf{i}$) of the Cartesian coordinate system along the vector $\mathbf{A}$ and make it coincide with the trajectory of the center of the bubble. We shall solve the problem in a frame of reference attached to the drifting bubble. We shall take the temperature from the unperturbed temperature of the point at which the center of the bubble is placed at a given time. We make quantities dimensionless using the scales introduced earlier. Then, if the values of the Marangoni number, $M = \sigma_1 A\rho a^2/\eta^2$, are small enough, we obtain the following system of equations for the velocity field, $\mathbf{v}$, the pressure, p, and the temperature, T, in the liquid:

$$\operatorname{div}\mathbf{v} = 0 \tag{15}$$

$$\nabla p = \Delta\mathbf{v} \tag{16}$$

$$\Delta T = 0 \tag{17}$$

In this approximation, only a single dimensionless adjustable parameter, the thickness, d, of the layer enters the problem through the boundary conditions. The dimensionless drift velocity, u, is a function of this parameter. We consider separately three possible ranges of variation of d: $d \sim 1, d \to 0$, and $d \to \infty$.

(a) The thickness of the layer is comparable with the bubble diameter i.e., d is of order unity. It can be assumed that the bubble is compressed by both surfaces and forms a cylindrical circular cavity as a disk of unit radius. In such a formulation Eq. (15) can be solved in a cylindrical coordinate system (r, φ, z) with center in the middle of the bubble, $r^2 = x^2 + y^2$. The lower surface $z = -d/2$ is heat conducting, while the upper surface $z = d/2$, and the surface of the bubble $r = 1$ are thermally insulated. We formulate the boundary conditions for the temperature, velocity, and tangential stresses:

$$r \to \infty, \qquad T = x, \qquad \mathbf{v} = -u\mathbf{i}$$

$$r = 1\colon \mathrm{v}_r = 0, \qquad \frac{\partial T}{\partial r} = 0, \qquad \left(\frac{\partial \mathrm{v}_\varphi}{\partial r} - \mathrm{v}_\varphi\right) = \frac{\partial T}{\partial \varphi}$$

$$z = -\frac{d}{2}\colon T = x, \qquad \mathbf{v} = -u\mathbf{i}, \qquad z = \frac{d}{2}\colon \frac{\partial T}{\partial z} = 0, \qquad \mathbf{v} - u\mathbf{i} \tag{18}$$

Approximating in the problem of Eqs. (15)–(18), the dependence on z by means of $\cos\kappa_1 z$, $\kappa_1 = \pi ld$ for the velocity and $\cos\kappa_2 z$, $\kappa_2 = 3\pi/2d$ for the temperature, we obtain for the functions averaged over z

$$\langle \mathrm{v}_r \rangle = \cos\varphi \left\{ -u + \frac{C_1}{r^2} + C_3 \frac{K_1(\kappa_1 r)}{\kappa_1^2 r} \right\}, \qquad \mathrm{v}_z = 0$$

$$\langle \mathrm{v}_\varphi \rangle = \sin\varphi \left\{ u + \frac{C_1}{r^2} + C_3 \left[K_0(\kappa_1 r) + \frac{K_1(\kappa_1 r)}{\kappa_1^2 r} \right] \frac{1}{\kappa_1} \right\}$$

$$\langle T \rangle = \cos\varphi \{ r + C_2 K_1(\kappa_2 r) \}, \qquad p = 0 \tag{19}$$

$$C_1 = u + \frac{4u - 1 + K_1(\kappa_2)/[K_1(\kappa_2)] + \kappa_2 K_0(\kappa_2)}{\kappa_1^2 + 2\kappa_1 K_0(\kappa_1)/K_1(\kappa_1)}$$

$$C_3 = \frac{1 - 4u + K_1(\kappa_2)/[K_1(\kappa_2)] + \kappa_2 K_0(\kappa_2)}{2K_0(\kappa_1) + \kappa_1 K_1(\kappa_1)}$$

$$C_2 = -\frac{1}{K_1(\kappa_2) + \kappa_2 K_0(\kappa_2)}, \qquad \langle \ldots \rangle = \frac{1}{d} \int_{-d/2}^{d/2} \cdots \mathrm{d}z$$

where $K_\nu(z)$ is a Bessel function of imaginary argument. We take into account the condition for the normal stresses, σ_{rr}, on the surface of the bubble integrally, equating the horizontal force

$$F_x = \oint \sigma_{rr}|_{r=1} \cos\varphi \, \mathrm{d}\varphi$$

to the "force of friction," f, of the bubble on the wall. To estimate f, we assume that as a disk the bubble slides between the two planes. If we assume that between it and the surfaces there is a liquid layer of thickness δ_1, then f can be approximated by $f = ua/\delta_1$, since the velocity changes by u in the layer δ_1 (hence δ_1 is an adjustable parameter). Equating F_x, calculated using Eq. (19), to the force f, we obtain

$$u = \frac{2 + \kappa_2 K_0(\kappa_2)/K_1(\kappa_2)}{1 + \kappa_2 K_0(\kappa_2)/K_1(\kappa_2)} \times \left[\left(4 + 2\kappa_1 \frac{K_0(\kappa_1)}{K_1(\kappa_1)} + \frac{\kappa_1^2}{2} \right) + \frac{a}{\delta_1 \pi \kappa_1} \left(\frac{K_0(\kappa_1)}{K_1(\kappa_1)} + \frac{\kappa_1}{2} \right) \right]^{-1} \tag{20}$$

(b) The bubble is compressed by the two nearby planes and takes the form of a thin disk of unit radius, i.e., $d \to 0$. It can be assumed that the liquid does not flow around the bubble but penetrates through the gaps between the bubble and the walls. We shall assume that the thicknesses of these gaps, which differ slightly, because of buoyancy are equal to δ_2 (δ_2 is then an adjustable parameter). It is sufficient to consider the motion in one gap. Suppose the motion in the layer δ_2 is two-dimensional. We choose a Cartesian coordinate system (z, x), making the plane of the wall coincide with the surface $z = 0$. The lower surface of the bubble is $z = \delta_2/2$. The surfaces $z = 0$ and $z = \delta_2$ are thermally insulated, and the velocity of the wall is $-u\mathbf{i}$. There are no tangential stresses on the surface $z = \delta_2$. Under these assumptions, the boundary conditions are

$$z = 0: \frac{\partial T}{\partial z} = 0, \qquad \mathbf{v} = -u\mathbf{i}, \qquad z = \delta_2: \frac{\partial T}{\partial z} = 0,$$
$$\frac{\partial \mathrm{v}_x}{\partial y} = \frac{\partial T}{\partial x}, \qquad \mathbf{v} = 0 \tag{21}$$

The solution of the problem of Eq. (15), Eq. (21), is

$$\mathrm{v_x} = -u + z; \qquad T = x; \qquad u = \delta_2 \tag{22}$$

and as one would expect, the dimensionless velocity u does not depend on the thickness, d, of the layer.

(c) The thickness of the layer is much greater than the bubble diameter, i.e., $d \to \infty$. It can be assumed that due to buoyancy the bubble is pressed to the surface $z = 0$, and the other plate is at "practical" infinity. We also assume that the capillary length is greater than the bubble diameter, so that the deformation of the bubble shape due to gravity and the viscous stresses can be ignored.

To further simplify the problem, we assume that the bubble rolls along a thin liquid layer of thickness δ_3 that remains between the glass and the bubble. We shall assume that the liquid flows past it but we take into account the interaction with the wall by means of integral conditions introducing an effective "force of friction" of the bubble on the wall. We shall distinguish the force of "dry" friction, f_1, from the force of "viscous" friction, f_2, which can be easily estimated. Thus, for the motion of a bubble along a rough surface, the curvature of the bubble changes slightly as it moves between humps of height b. This leads to the appearance of a force $f_1 = 2\pi\sigma_0 b/\sigma_1 A a^2$. The force of the viscous friction is $f_2 = u/\delta_3$. In these expressions, b and δ_3 are adjustable parameters.

Equating $f_1 + f_2$ to the force of the viscous stresses

$$F_x = \oint \sin^2 \vartheta \cos \varphi \sigma_{rr} \mathrm{d}\vartheta \mathrm{d}\varphi$$

we obtain the bubble drift velocity u

$$u = \frac{1}{2}\left(1 - \frac{\sigma_0 b}{\sigma_1 A a^2}\right)\left(1 + \frac{1}{4\pi\delta_3}\right)^{-1} \tag{23}$$

Figure 13 shows the theoretically predicted dimensionless drift velocity, u, vs. bubble radius Eq. (23) (curve 1, $b = 3.52 \times 10^{-5}$ mm, $\delta_3 = 6.40 \times 10^{-3}$), Eq. (18) (curve 2, $\delta_1 = 0.08$ mm), and Eq. (22) (curve 3, $\delta_2 = 2.25 \times 10^{-2}$). These three curves describe the three experimentally observed different motions of the

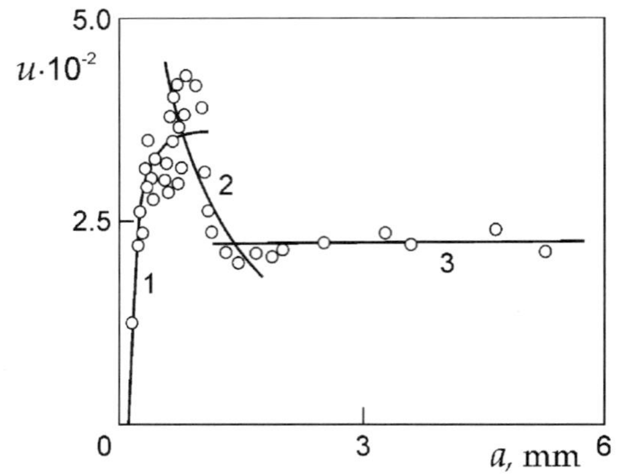

FIG. 13 Thermocapillary migration of bubbles and drops. Dimensionless thermocapillary migration velocity of bubbles with different diameters and geometry. The three curves describe the three experimentally observed different bubbles motions for $2a \leq d$, $2a \geq d$, and $2a \gg d$, respectively. (From Ref. 39.)

bubbles for, $2a \leq d, 2a \geq d$, and $2a \gg d$, respectively. We have plotted also the experimental points obtained with ethanol.

The plot in Fig. 13 shows good qualitative agreement between the theoretically predicted dimensionless drift velocities and the experimental data. To compare theoretical and experimental results an adjustable parameter is included into the theory, which is the thickness of the liquid film between the bubble and the cavity walls.

VI. INSTABILITY AND THERMOCAPILLARY MIGRATION OF ACTIVE DROPS (BUBBLES) IN A SURROUNDING HOMOGENEOUS MEDIUM

An *active* drop is, for example, a drop that reacts exo- or endothermically at the surface with reactant dissolved in the external medium or a drop that dissolves at its surface with heat effect or even a drop with internal heat sources. Under such circumstances, it is possible for a drop initially at rest to start moving as a consequence of an instability caused by the Marangoni effect. Indeed as Ryazantsev, Velarde and their collaborators [10–16] have shown under various circumstances an initial uniform temperature on the surface of an active drop can be disrupted due to a spontaneous flow fluctuation or to a spontaneous change in, say, the surface chemical reaction rate or the intensity in the internal heat generation. Once such a disturbance appears, the symmetry is broken, very much as in the cases studied earlier regarding liquid layers. It then suffices that the

Marangoni effect is strong enough to be able to sustain the initial motion and hence, thermocapillary migration results. In contrast with the YGB case, here there is no need for an imposed *external* temperature gradient. The theory predicts diversity of active drop behaviors depending on the values of the parameters. For instance, in the absence of buoyancy, an active drop can move autonomously with constant velocity in an arbitrary direction. In the presence of buoyancy an active drop tends to move along with the external force with a velocity that is quite different from that given in Eq. (1) or moves in a direction opposite to the buoyancy force. In nonzero buoyancy an active drop can also levitate if the thermocapillary force balances buoyancy.

For illustration let us consider that on the surface of the drop there is a chemical reaction releasing heat. The continuity, Navier–Stokes, heat (Fourier), and mass (Fick) diffusion equations describing a possible steady motion of such a reacting drop are

$$\operatorname{div} \mathbf{v}_i = 0 \tag{24}$$

$$-\frac{1}{\rho_i}\nabla p_i + \nu_i \Delta \mathbf{v}_i = 0 \tag{25}$$

$$\mathbf{v}_i \cdot \nabla T_i - \chi_1 \Delta T_i = 0 \tag{26a}$$

$$\mathbf{v}_i \cdot \nabla C_i - D_1 \Delta C = 0 \tag{26b}$$

where $(i = 1, 2)$, together with the boundary conditions far away from the drop and at its center

$$r \to \infty\colon \mathbf{v}_1 \to \mathbf{U}_\infty, \qquad T_1 \to T_\infty, \qquad C \to C_\infty, \tag{27}$$

$$r = 0\colon \mathbf{v} < \infty, \qquad T_2 < \infty \tag{28}$$

At the drop surface

$$\begin{gathered} r = a\colon \mathrm{v}_{r1} = \mathrm{v}_{r2} = 0, \qquad \mathrm{v}_{\theta 1} = \mathrm{v}_{\theta 2}, \\ \eta_1\left(\frac{\partial \mathrm{v}_{\theta 1}}{\partial r} - \frac{\mathrm{v}_{\theta 1}}{r}\right) - \eta_2\left(\frac{\partial \mathrm{v}_{\theta 2}}{\partial r} - \frac{\mathrm{v}_{\theta 2}}{r}\right) + \frac{1}{a}\frac{\partial \sigma}{\partial \theta} = 0, \\ T_1 = T_2, \qquad C = 0, \qquad \kappa_1 \frac{\partial T_1}{\partial r} - \kappa_2 \frac{\partial T_2}{\partial r} = -Q D_1 \frac{\partial C}{\partial r}, \\ \sigma = \sigma_0 - \sigma_1 (T - T_0), \qquad \sigma_1 > 0 \end{gathered} \tag{29}$$

In writing the above equations and conditions we have assumed the diffusion regime for the reaction and that Stefan flow can be neglected. For the coordinate system we use (r, θ) attached to the drop center. The point $r = a$, $\theta = -\pi$ denotes the front while $r = a$, $\theta = 0$ is the rear pole. In the equations is the concentration of a dissolving component, D_1 is the diffusion coefficient of the component dissolving in the outer fluid, and Q accounts for the heat release of the surface chemical reaction. The linearization of the problem is based on the following conditions for the Schmidt, Peclet, and Reynolds numbers:

$$Pe_{\mathrm{D}} = Sc = \frac{aU_\infty}{D_1} \ll 1, \qquad Pe_i = \frac{aU_\infty}{\chi_i} \ll 1, \qquad Re_i = \frac{aU_\infty}{\nu_i} \ll 1 \tag{30}$$

with χ_i denoting heat diffusivity ($i = 1, 2$).

The axisymmetry allows that we use a stream function to solve the problem and to determine in analytical form the flow, temperature and concentration fields, as well as the force acting on the moving drop as a result of the viscous and thermocapillary effects. We shall not provide details of the calculations involved. It has been found that the force has the form [4,12]

$$\mathbf{F} = -4\pi\eta_1 a A_F \mathbf{U}_\infty, \qquad A_F = \frac{3m + \frac{3}{2}\beta + 1}{1 + \beta + m} \tag{31}$$

where the following dimensionless groups and parameters have been used

$$m = -\frac{MPe(1 - Le)}{12(2 + \delta)}, \qquad M = \frac{-QC_\infty D_1 \sigma_1}{\eta_1 \kappa_1 U_\infty}, \qquad Le = \frac{\chi_1}{D_1},$$
$$\beta = \frac{\eta_2}{\eta_1}, \qquad \delta = \frac{\kappa_2}{\kappa_1}, \qquad \sigma_1 = -\frac{\partial \sigma}{\partial T} \tag{32}$$

M and m denote, respectively, the Marangoni number and a rescaled Marangoni number, both defined with the velocity at infinity. The quantity m is better suited for analyzing the stability results. Here for convenience we call Le the (inverse) of the earlier defined Lewis number, which is the way this parameter is defined when gases are involved. $\mathbf{U}_\infty$ denotes the velocity of the liquid at infinity in the coordinate system attached to the moving drop, hence the velocity of the drop is $-\mathbf{U}_\infty$. When $Q = 0$ we have a passive drop and the relationship of Eq. (31) reduces to the corresponding value in the Hadamard–Rybczynskii–Stokes theory [4].

The force $\mathbf{F}$ is the integral of the pressure, and the thermocapillary and the viscous stresses acting at the surface of the drop. If the drop is to be maintained in steady motion the thermocapillary stresses at the surface should be balanced with the viscous ones and hence the resulting force $\mathbf{F}$ should be zero or it should be in balance with an external force such as that due to buoyancy. Equations (31) and (32) show that drop activity disappears if $Le = 1$, when the distributions of temperature and concentration lose spherical symmetry due to the motion of a drop but retain similar form and hence the temperature at the surface of the drop remains uniform.

The quantity $A_F = A_F(m, \beta)$, and therefore $\mathbf{F}(m, \beta)$, is qualitatively the same for all reasonable values of β [12]. As the parameters Q, C_∞, and σ_1 appear in the expression for A_F in the form of the product $QC_\infty\sigma_1$, the dependence of A_F on the $QC_\infty\sigma_1$ is qualitatively similar to the dependence of $A_F(m)$. The function $A_F(m)$ corresponding to $\beta = 1$ is shown in Fig. 14. It appears that depending on the numerical values of the parameters the force $\mathbf{F}$ can be either directed opposite to the drop velocity ($A_F < 0$; drag) or directed along with drop velocity ($A_F > 0$), and in this case the drop experiences (self-propulsive) thrust. Equation (31) also permits the calculation of the velocity of the active drop motion for a given buoyancy level equal and opposite to $\mathbf{F}$. If $A_F < 0$ the active drop will move along with the mass force. If $A_F > 0$ the active drop will move against the mass force.

Two special cases appear in Eq. (31), one when $A_F(m_{\text{crit1}}) = 0$, at $m_{\text{crit1}} = \frac{1}{3} - \beta/2$, and the other when $A_F(m_{\text{crit2}})$ diverges to infinity at $m_{\text{crit2}} = -1 - \beta$. For $\beta = 1$ the critical values of m are $m_{\text{crit1}} = -0.83$ and $m_{\text{crit2}} = -2$, respectively.

The value $m = m_{\text{crit1}}$ corresponds to the case when the thermocapillary force fully compensates the viscous drag force and the moving drop does not experience resistance. To study the motion of a drop for a value of m that is close to m_{crit1}, a weakly nonlinear analysis was developed [13]. It was shown that in the absence of mass force this limit case corresponds to the possibility of self-sustained or autonomous motion of an active drop with finite velocity. On the other hand the

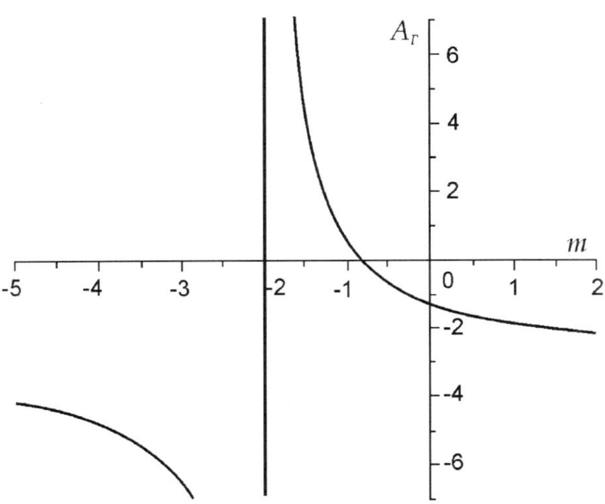

FIG. 14 Thermocapillary migration of bubbles and drops. The function A_F, defining sign and value of the force on an active drop [Eq. (30)], as a function of the (modified) Marangoni number, m [Eq. (32)], for $\beta = 1$. (From Ref. 12.)

singularity of A_F at $m = m_{crit2}$ corresponds to the possibility for levitation of an active drop in the presence of an external force.

Thrust appears if the parameter m takes a value between the above given two critical values $m_{crit2} < m < m_{crit1}$, while the force is resistance if $-\infty < m < m_{crit2}$, and $m_{crit1} < m < \infty$. The latter force is smaller than the drag force given by the Hadamard–Rybczynskii–Stokes theory [4] if $m_{crit1} < m < 0$ and it is greater when $-\infty < m < m_{crit2}$, $m > 0$. Furthermore this drag for an active drop is even greater than the Stokes force for a solid particle [4] when $A_F < -\frac{3}{2}$, hence we have enhanced (super) drag due to the Marangoni effect.

The linear approximation also permits us to obtain an estimate of velocity of an active drop under buoyancy, which is

$$\mathbf{U}_\infty = -\frac{(\rho_2 - \rho_1)a^2}{3A\eta_1}\mathbf{g} \tag{33}$$

where **g** denotes gravity or centrifugal acceleration.

To illustrate the above-recalled analytical results, let us consider the following parameter values: $\eta_1 = \eta_2 = 10^{-3}\,\text{kg/ms}$, $D_1 = 5 \times 10^{-10}\,\text{m}^2/\text{s}$, $\chi_1 = 10^{-7}\,\text{m}^2/\text{s}$, $\kappa_1 = 2 \times 10^{-1}\,\text{W/mK}$, $\kappa_2 = 10^{-1}\,\text{W/mK}$, $\sigma_1 = -0.01\,\text{mN/mK}$, $a = 10^{-3}\,\text{m}$, and $C = 10\,\text{kg/m}^3$.

Then the critical values $m_{crit1} = -0.83$ and $m_{crit2} = -2$ correspond to $Q_{crit1} = 42\,\text{J/g}$, and $Q_{crit2} = 128\,\text{J/g}$, respectively. Accordingly, for $0 < Q < 53\,\text{J/g}$, its activity reduces the experimental drag by the drop, and thus the velocity of the active drop becomes greater than the velocity corresponding to the Hadamard–Rybczynskii–Stokes value ($Q = 0$) [4]. For a given buoyancy level, $(\rho_2 - \rho_1)g$, increasing the heat release from zero up to $33\,\text{J/g}$ gives a twofold reduction of the resistance force, which is significant indeed.

REFERENCES

1. Young, N.O., Goldstein, J.S., Block, M.J. The motion of bubbles in a vertical temperature gradient. J. Fluid Mech. **1959**, *6*, 350–356.
2. Birkhoff, G. *Hydrodynamics. A Study in Logic, Fact and Similitude*; Princeton University Press: Princeton, N.J., 1960.
3. Levich, V.(B.)G. *Physicochemical Hydrodynamics*; Prentice Hall: Englewood Cliffs, N.J., 1962; 85–94.
4. Subramanian, R.S., Balasubramaniam, R. *The Motion of Bubbles and Drops in Reduced Gravity*; Cambridge University Press: Cambridge, 2001 (and references therein).
5. Hardy, S.C. The motion of bubbles in a vertical temperature gradient. J. Colloid Interface Sci. **1979**, *69*, 157–162.

6. Braun, B., Ikier, C., Klein, K., Woermann, D. Marangoni convection-induced motion of droplets during liquid/liquid phase separation. J. Colloid Interface Sci. **1993**, *159*, 515–516.
7. Harper, J.P., Moore, D.W., Pearson, J.R.A. The effect of the variation of surface tension with temperature on the motion of bubbles and drops. J. Fluid Mech. **1967**, *27*, 361–366.
8. Bratukhin, Yu.K. Thermocapillary drift of a viscous fluid droplet. Izv. Akad. Nauk SSSR, Mech. Zhidk. Gaza **1975**, *5*, 156–161 (in Russian).
9. Subramanian, R.S. The motion of bubbles and drops in reduced gravity. In *Transport Processes in Drops, Bubbles and Particles*; Hemisphere: New York, 1992; 1–42.
10. Ryazantsev, Y.S. On the thermocapillary motion of a reacting drop in a chemically active medium. Fluid Dyn. **1985**, *3*, 180–183.
11. Rednikov, A.Ye., Ryazantsev, Y.S. On thermocapillary instability of cooled or heated droplet. Microg. Quart. **1991**, *1*, 199–203.
12. Rednikov, A.Ye., Ryazantsev, Y.S., Velarde, M.G. Drop motion with surfactant in a homogeneous surrounding. Phys. Fluids **1994**, *6*, 451–468.
13. Rednikov, A.Ye., Ryazantsev, Y.S., Velarde, M.G. On the development of translational subcritical Marangoni instability for a drop with uniform internal heat generation. J. Colloid Interface Sci. **1994**, *164*, 168–180.
14. Rednikov, A.Ye., Ryazantsev, Y.S., Velarde, M.G. Drop motion with a surfactant transfer in an inhomogeneous medium. Int. J. Heat Mass Transfer **1994**, *37* (Suppl. 1), 361–374.
15. Rednikov, A.Ye., Ryazantsev, Y.S., Velarde, M.G. Active drops and drop motions due to nonequilibrium phenomena. J. Non-Equilibrium Thermodyn. **1994**, *19*, 95–113 (and references therein).
16. Rednikov, A.Ye., Kurdyumov, V.N., Ryazantsev, Y.S., Velarde, M.G. The role of time-varying gravity on the motion of a drop induced by Marangoni instability. Phys. Fluids **1995**, *7*, 2670–2678.
17. Subramanian, R.S., Balasubramaniam, R., Wozniak, G. Fluid mechanics of bubbles and drops. In *Physics of Fluids in Microgravity*; Monti, R., Ed.; Taylor & Francis: London, 2001; 149–177.
18. Wozniak, G., Siekman, J., Srulijes, J. Thermocapillary bubble and drop dynamics under reduced gravity. Survey and Prospects Z. Flugwiss. Weltraumforsch. **1988**, *12*, 137.
19. Sadhal, S.S., Ayyaswamy, P.S., Chung, J.N. *Transport Phenomena with Drops and Bubbles*. Springer: Berlin, 1997.
20. McGrew, J.L., Rehm, T.L., Griskey, R.G. The effect of temperature induced surface tension gradients on bubble mechanics. Appl. Sci. Res. **1974**, *29*, 195–210.
21. Merritt, R.M., Subramanian, R.S. The migration of isolated gas bubbles in a vertical temperature gradient. J. Colloid Interface Sci. **1988**, *125*, 333–339.
22. Lacy, L.L., Witherow, W.K., Facemire, B.R., Nishioka, G.M. *Optical Studies of a Model Binary Miscibility Gap System*; NASA TM–82494: Washington DC, 1982.
23. Langbein, D., Heide, W.Z. Entmischung von Flüssigkeiten aufgrund von Grenzflächenkonvektion. Z. Flugwiss. Weltraumforsch. **1984**, *8*, 192–199.

24. Langbein, D., Heide, W.Z. The separation of liquids due to Marangoni convection. Adv. Space Res. **1984**, *4*(5), 27–36.
25. Delitzsch, V., Eckelmann, H., Wuest, W. The influence of thermocapillarity on the migration of droplets in a liquid possessing a uniform temperature gradient. Proceedings 5th European Symposium Material Sciences under Microgravity, Schloss Elmau, 1984, pp. 245–249.
26. Hahnel, M., Delitzsch, V., Eckelmann, H. The motion of droplets in a vertical temperature gradient. Phys. Fluids A **1989**, *1*, 1460–1466.
27. Barton, K.D., Subramanian, R.S. The migration of liquid drops in a vertical temperature gradient. J. Colloid Interface Sci. **1989**, *133*, 211–222.
28. Smith, H.D., Mattox, D.M., Wilcox, W.R., Subramanian, R.S., Meyyappan, M. Experimental observation of the thermocapillary driven motion of bubbles in a molten glass under low gravity conditions. In *Materials Processing in the Reduced Gravity Environment of Space*; Rindone, G.E., Ed.; Elsevier: New York, 1982; 279–288.
29. Thompson, R.L., Dewitt, K.J., Labus, T.L. Marangoni bubble motion phenomenon in zero gravity. Chem. Eng. Commun. **1980**, *5*, 299–314.
30. Wozniak, G., Siekmann, J. Experimental investigation of thermocapillary droplet motion in reduced gravity—two sounding rocket experiments. Appl. Microg. Tech. **1989**, *II*(3), 173–175.
31. Wozniak, G., Siekmann, J. Experimental investigation of thermocapillary droplet motion in reduced gravity using sounding rockets and simulation liquids. Proceedings 30th Israel Annual Conf. Aviation and Astronautics, Haifa, Israel, 1989, pp. 125–130.
32. Nahle, R., Neuhaus, D., Siekmann, J., Wozniak, G. Srulijes, J. Separation of fluid phases and bubble dynamics in a temperature gradient—a Spacelab D1 experiment. Z. Flugwiss. Weltraumforsch. **1987**, *11*, 211–213.
33. Szymczyk, J.A., Wozniak, G., Siekmann, J. On Marangoni bubble motion at higher Reynolds and Marangoni numbers under microgravity. Appl. Mcrogravity Tech. **1987**, *I*(1), 27–29.
34. Balasubramanian, R., Lacy, C.E., Wozniak, G. Subramanian, S.R. Thermocapillary migration of bubbles and drops at moderate values of the Marangoni number in reduced gravity. Phys. Fluids **1996**, *8*, 872–880.
35. Viviani, A., Golia, C. Microgravity experiments on thermocapillary bubble migration in presense of non linear surface tension. IAF paper 98-J.3.09, Proceedings 49th Int. Astronational Congress, September 28–October 2, 1998, Melbourne, Australia, 1998, pp. 1–7.
36. Bratukhin, Yu.K., Evdokimova, O.A., Phenichnikov, A.F. Migration of gas bubbles in nonuniformly heated liquid. Izv. Akad. Nauk SSSR, Mech. Zhidk. Gaza **1979**, *5*, 55–57 (in Russian).
37. Bratukhin, Yu.K., Briskman, V.A., Zuev, A.L., Pshenichnikov, A.F., Rivkind, V.Ya. Experimental investigation of thermocapillary migration of gas bubbles in liquid. In *Hydromechanics and Heat/Mass Transfer in Microgravity*. Nauka: Moscow, 1982; 98–109 (in Russian).

38. Evdokimova, O.A., Petukhova, V.G. Heat transfer in a thin wire located on the surface of flowing water. In *Hydrodynamics 6*. Perm University: Perm, 1975; 107–112 (in Russian).
39. Briskman, V.A., Zuev, A.L. Terrestrial modeling of thermocapillary migration of bubbles in microgravity. In *Technological Experiments in Microgravity*, V.A. Briskman, G.Z. Gershuni and T. Lyubimova (editors), Ural Branch, Russian Academy of Sciences: Sverdlovsk 1983; 96–100 (in Russian).
40. Bratukhin, Yu.K., Zuev, A.L. Thermocapillary drift of air bubbles in a horizontal Hele–Shaw cell. Fluid Dyn. **1984**, *19*, 393–397.

11
Spreading and Layer Breaking Driven by the Marangoni Effect

In this chapter we shall consider phenomena in which nonuniform surface tension causes relatively strong deformations of liquid surfaces and dramatic displacements of liquid volumes. We shall give results about the deformation and eventual rupture of a thin liquid layer or a film under the influence of the Marangoni effect. We shall consider the effect of temperature gradients (thermocapillary convection) or concentration gradients with dissolved impurities or surfactants (generically called solutocapillary convection). For thermocapillary surface deformation we shall discuss how transverse high-frequency vibrations allow the delaying of deformation and avoid the rupture of liquid layers. We shall also study the phenomena occurring when a point-like heat source is located under the surface of a relatively large fluid volume.

I. DEFORMATION AND RUPTURE OF THIN LIQUID LAYERS CAUSED BY THE MARANGONI EFFECT

As noted earlier, in normal laboratory conditions, that is, under normal gravity on Earth, the experimental investigation of the Marangoni effect is made difficult by the existence of buoyancy-driven flows caused by liquid density inhomogeneity. Indeed the generation of the temperature gradient leading to thermocapillary instability inevitably produces buoyancy-driven convection, whose intensity is, as a rule, several orders of magnitude stronger than the former. To overcome this situation we already know that using thin liquid layers, the thinner the better, helps.

A. Deformation and Rupture of Liquid Layers Caused by Thermocapillary Convection

A sketch of an experimental set-up used to study deformation and rupture in thin liquid layers is presented in Fig. 1. A thin liquid layer, 1, was placed in a circular (diameter, 9 cm) or rectangular (7.0 cm × 7.4 cm) cuvette, with a metallic support, 3, of corresponding shape. The plate sharp edges prevented the liquid from overflowing. The depth of the liquid layer varied from 0.3 to 2 mm and hence buoyancy-driven convection was expected to be drastically reduced. Experiments were performed with several organic fluids of different viscosity and low saturated vapor pressures to prevent evaporation. The maximum temperature difference between heater and cooler, measured with thermocouples, did not exceed 30 K (within 0.1 K).

The cell was heated at the center and cooled around the perimeter by means of heat exchangers, 2. As a result, a stationary axisymmetric surface tension gradient (Marangoni) driven flow appeared over the surface in the direction of increasing surface tension, i.e., toward the relatively cooler edges of the cuvette, with the corresponding reverse flow underneath to the center. The surface relief was determined with reflected light by the shadowgraph method. A parallel light beam formed by the short-focus objectives, 8 and 5, was reflected from the liquid surface and, with the aid of the diaphragm, 7, and the semitransparent mirror, 8, was collected by the same objective, 5, and then by the camera, 10, focused on the

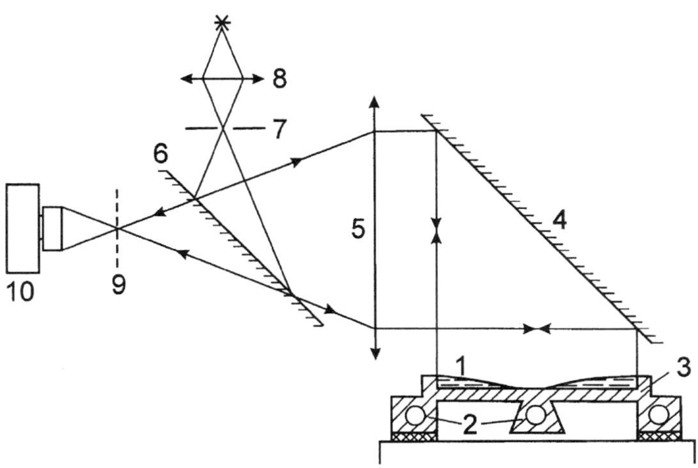

FIG. 1 Experimental investigation of thermocapillary-induced deformation. Set-up for ground-based study in a thin horizontal liquid layer with open upper surface. 1, liquid layer; 2, heat exchangers; 3, metal cuvette; 4, mirror; 5, objective; 6, semitransparent mirror; 7, diaphragm; 8, micro-objective; 9, grating diaphragm; 10, photocamera. (From Refs. 2 and 8.)

surface. By means of a concentric grating diaphragm, 9, with a 0.5-mm step, mounted in the focal plane of the camera and objective 5, photographs of the isolines of the inclination angle of the normal to the open surface were obtained.

Let us first recall the results found in experiments with the circular cuvette [1,2]. Typical shadow pictures of the deformed surface of an n-decane layer for $\Delta T = 14\,\text{K}$ and initial thicknesses, d_0, of the layer equal to 0.80, 0.60, and 0.45 mm, respectively, are given in Fig. 2. The change in radial distribution of the local thickness of the liquid layer due to the flow was determined by performing two successive graphic integrations in the case when the known initial liquid volume was kept constant. The shape of the liquid layer surface is shown in Fig. 3.

It was found that the thermocapillary flow significantly deforms the open surface. The layer depth over the cold region increased when liquid moved from the hotter to the colder region, as expected. Figure 3(a) shows the measured dependence of the local layer thickness on the radius measured from the heater center for a layer of n-decane for different initial liquid depths and fixed temperature gradient. The maximum height difference between the layer center and its edges was found to be inversely proportional to the initial layer thickness. The smallness of the layer initial thickness seems to be the crucial factor in triggering surface deformation. The reverse flow near the bottom has no time to pump the liquid back, hence leading to an increase of the horizontal component of the pressure gradient and a lift of the liquid surface over the cold region.

The surface deformation also increased with the applied temperature difference, as seen in Fig. 3(b). The curves correspond to fixed layer thickness and

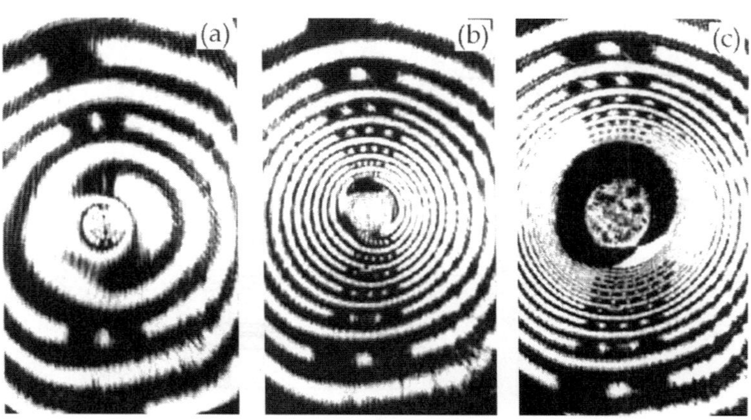

FIG. 2 Experimental investigation of thermocapillary-induced deformation. Isolines of the surface slope (circular liquid layer heated at the center) for n-decane. $\Delta T = 14\,\text{K}$ and d_0: (a) 0.80, (b) 0.60, (c) 0.45 mm. In the center of photograph (c) the dry spot is visible. (From Refs. 1 and 8.)

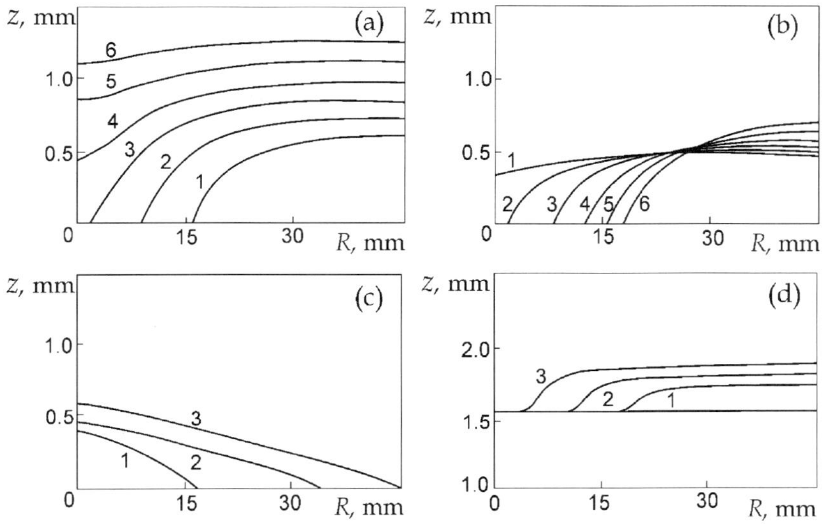

FIG. 3 Experimental investigation of thermocapillary-induced deformation. Shape of the liquid layer surface: (a) heated at its center, decane, $\Delta T = 14\,\text{K}$, $d_0 = 0.47, 0.63, 0.80, 0.94, 1.10$, and $1.26\,\text{mm}$; (b) heated at its center, decane, $d_0 = 0.47\,\text{mm}$, $\Delta T = 3.5, 6.1, 10.2, 16.1, 22.5$, and $30.0\,\text{K}$; (c) heated at its periphery, decane, $\Delta T = 14\,\text{K}$, $d_0 = 0.07, 0.45$, and $0.75\,\text{mm}$; (d) heated at its center, heptane over water, $\Delta T = 2.5\,\text{K}$, $d_0 = 0.16, 0.24$, and $0.30\,\text{mm}$. (From Refs. 2 and 8.)

varying temperature gradient. The height difference between the hottest and coldest regions appeared to be linearly dependent on the temperature drop between these surface areas. As was also expected, the change in the direction of the temperature gradient reversed the direction of the flow leading to the formation of an axisymmetric hill at the layer center over the cold region [Fig. 3(c)]. Thus the Marangoni effect was truly the mechanism underlying convection. As shown in Figs 3(a) and (b) it appears that for very thin layers (less than 1 mm) and large enough temperature gradients the surface depression reached the cuvette bottom, leading to the appearance of a dry spot over the heater. The radius of the dry spot increased monotonically with the temperature difference (Fig. 4). The liquid spread apart, allowing the cuvette bottom to be seen. The dynamic wetting angle seemed to be nonzero, even for wetting liquids. Similar results were obtained for the rectangular container and two-layer systems, in which a thin liquid layer is placed on the surface of another immiscible liquid [Fig. 3(d)] [2,3].

As already said, the maximum difference of height between the hottest and coldest regions appears to be proportional to the temperature difference and inversely proportional to initial thickness of the layer (Fig. 5); $\Delta\zeta$ is the height

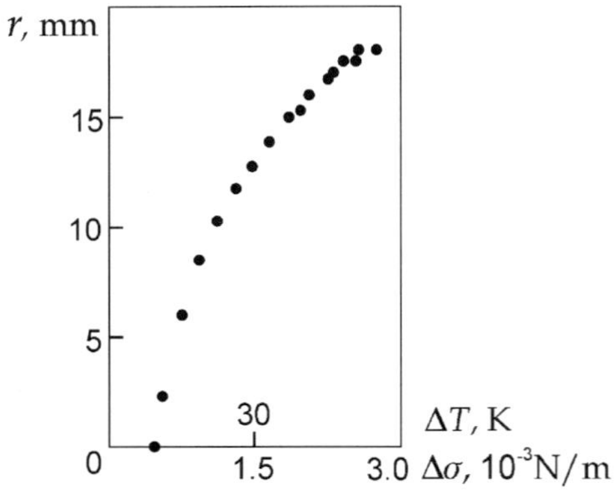

FIG. 4 Experimental investigation of thermocapillary-induced deformation. Dry spot radius vs. surface tension difference. Decane, $d_0 = 0.5\,\text{mm}$. (From Ref. 8.)

difference measured in units of the capillary length $(\sigma_0/\rho g_0)^{1/2}$, and $\Delta\theta = 3\Delta T\sigma_1/\sigma_0$ is the dimensionless temperature difference.

The measured critical values of the dimensionless temperature difference, $\Delta\theta_*$, which, for a given initial thickness, *d*, led to layer breakdown, are given in Fig. 6. The points along curve 1 correspond to experiments with the rectangular cuvette, while the points along curve 2 come from the circular one. The solid and dashed lines give theoretical estimates [4]. Clearly, the temperature difference is

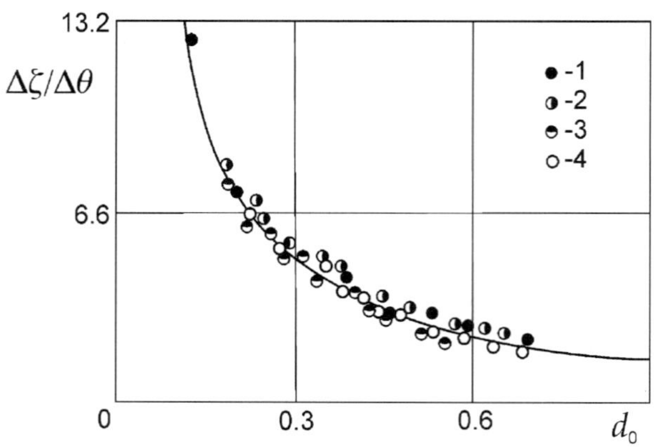

FIG. 5 Experimental investigation of thermocapillary-induced deformation. Dimensionless maximum height difference between the hottest and coldest layer edges normalized by their temperature difference. Ethanol, ΔT: ●, 5 K; ◑, 12.5 K; ◓, 14.2 K, and ○, 18.4 K. (From Refs. 1 and 2.)

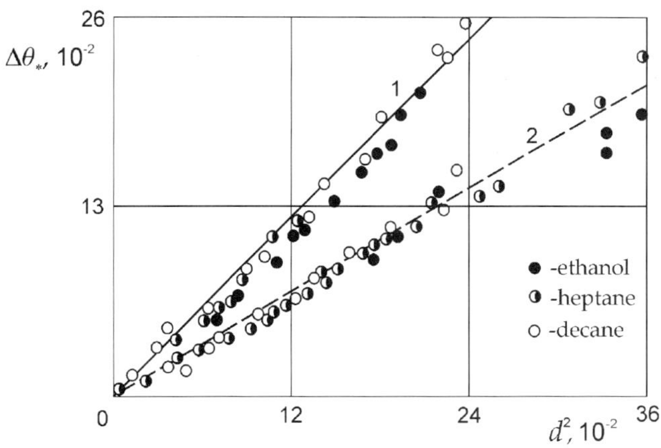

FIG. 6 Experimental investigation of thermocapillary-induced deformation. Critical values of the temperature difference between liquid layer edges leading to layer rupture. Points along curve 1 correspond to experiments with a rectangular cuvette, and those along curve 2 to experiments with a circular one. The solid and dashed lines depict theoretical estimates [1]. (From Ref. 2.)

directly proportional to the square of the layer thickness. Breaking in the cylindrical layer begins at lower values of the temperature difference than in rectangular layers. For liquid layers of 0.5–1 mm thicknesses, for instance, the critical temperature differences are in the range 30–40 K.

Similar effects were observed for thin liquid films with two free surfaces [1]. For the experiment, a liquid film was fixed surrounding a 3 mm × 30 mm horizontal wire frame (Fig. 7). The width of the film was smaller than the capillary length in the problem. A film of decane was stable and remained stable for several hours (the saturated vapor pressure of decane is very low, thus preventing evaporation). The film thickness was equal to the diameter of the wire and varied from 0.1 to 1.0 mm. The uniformity of the thickness was interferometrically controlled (Fig. 8). For plane films the nonuniformity of the thickness in the experiments never exceeded 0.01 mm [Fig. 8(b)]. When the ends of the film were differentially heated, thermocapillary motion appeared, thus deforming the film. The interference picture disappears and before the layer breaks in the thinner portions it returns as the thickness becomes of the order of the laser light wavelength. At a critical temperature difference the film broke [Fig. 8(c)]. The analytical treatment of the problem is similar to that of the liquid layer placed on a solid support [4]. The experimental data are in good agreement with theory when only the Marangoni effect is considered, neglecting the role of gravity, for small enough layer thicknesses (less than 0.3 mm), where the Rayleigh–Taylor

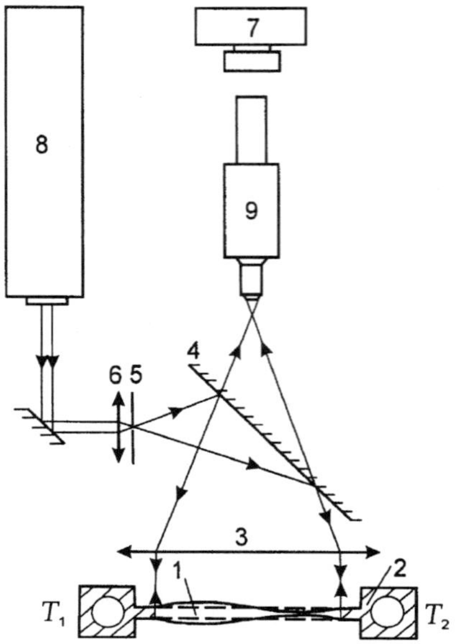

FIG. 7 Set-up for the ground-based investigation of the thermocapillary-induced deformation of a liquid film. 1, liquid film; 2, heat exchangers; 3, objective; 4, semi-transparent mirror; 5, diaphragm; 6, micro-objective; 7, photocamera; 8, laser; and 9, microscope.

instability is not expected. Figure 9 shows critical values of the temperature difference between the ends of the film of initially given thickness, which lead to film rupture [1]. For thick layers the experimental points deviate from the theoretical line, as expected.

B. Deformation and Rupture of Liquid Layers Caused by Solutocapillary Convection

Let us now consider the Marangoni effect due to a gradient of surfactant concentration on the free surface (solutocapillary convection). Most of the results about surface deformation are expected to be similar to those found with heat transfer, but return flows and other phenomena may change as the surface tension is more sensitive to changes in concentration than changes in temperature. Thus one expects that with solutocapillarity one could observe the same deformation of the layer free surface, and also the appearance of dry spots in much thicker layers than found with thermocapillary convection.

As an example, consider a needlelike monocrystal growth in isothermal conditions in a thin (about 100 μm thickness) horizontal film of $Ba(NO_3)_2$ 9.3%

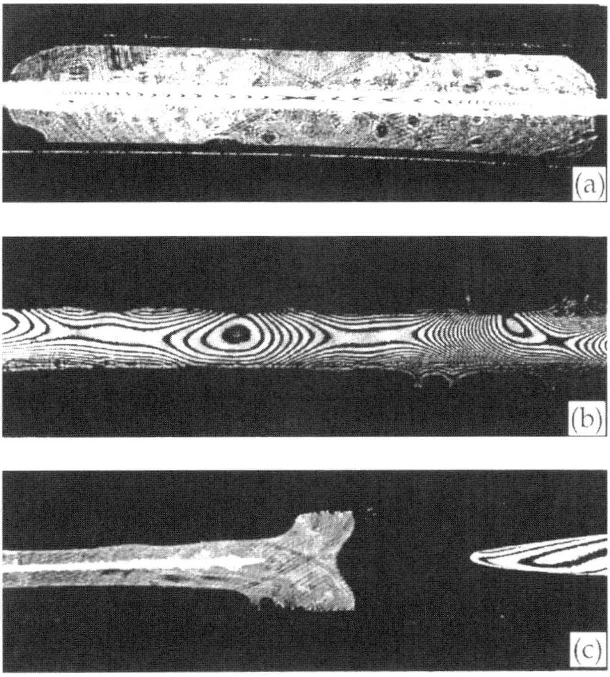

FIG. 8 Experimental investigation of thermocapillary-induced deformation. Interference bands of uniform thickness on the surface of the liquid film: (a) nonparallel surfaces, (b) plane film, and (c) film at the moment of rupture. (From Ref. 1.)

aqueous solution located on a glass substratum as has been described in Ref. 5. Owing to losses through evaporation of water from the open surface, the solution soon became supersaturated and the crystal starts to grow in the form of a long thin needle. Surface tension gradients induce oscillatory motions of the film surface, in response to which the liquid periodically recedes from the crystal, remaining on the dry place, and then returns, recontacting it. These ebbs naturally stop the growth of the crystal and, as a result, its configuration becomes wavy. Solutocapillary convective motion accounts for such behavior of the solution. Because the surface tension of the $Ba(NO_3)_2$ solution in water is proportional to its concentration, then from the onset of crystal growth there is local depletion of the solution in salt, thus lowering the surface tension. Flow appears directed along the surface away from the crystal, leading to surface depression and displacement of the liquid phase boundary. As soon as the liquid separates from the crystal, salt depletion finishes and diffusion brings back concentration homogeneity. The flow also vanishes and the liquid returns back towards the crystal.

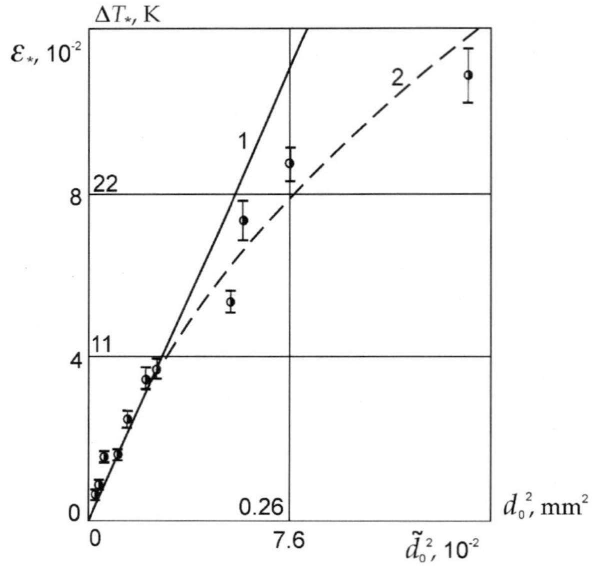

FIG. 9 Experimental investigation of thermocapillary-induced deformation. Critical values of the temperature difference between the ends of a film of an initially given thickness leading to film rupture in dimensional ΔT_* (d_0^2 in mm) and dimensionless ε_* ($\tilde{d}_0^2$ in units of the capillary length). (From Refs. 1 and 2.)

The experimental study of solutocapillary convection is more difficult than for thermocapillarity. The main problem is the necessity of controlling the local values of the surface tension of the liquid layer over the surface where the surfactant propagates, dissolves, and evaporates in an unsteady manner. For this reason, most experimental studies of solutocapillary convection usually offer qualitative rather than quantitative results.

A particular illustration of the phenomena under consideration is the spreading of an insoluble surfactant over the surface of a viscous liquid thin layer [6]. A surfactant drop of 3 μl volume was deposited with the aid of a syringe in the center of a thin layer of distilled water of 0.32 ± 0.01 mm thickness placed in a borosilicate glass Petri dish of 20 cm diameter. As a surfactant, a water solution of sodium dodecyl sulfate (SDS) of 20 g/l concentration was used. On touching the water surface, the drop spread very quickly. This process was registered by a video camera (25 Hz frequency). To make the flow visible, a small amount of talc powder was used as a marker.

The spreading of the surfactant caused a local decrease of the surface tension of the liquid at the center of the cuvette, which produced strong convection. As the liquid moved toward the cuvette edges, there was depression in its center and, eventually, part of the bottom appeared naked. The radius of the dry zone was found to be smaller than that of the spread surfactant spot and the boundary angle of the liquid contacting with the dry spot was nonzero. Both processes, the spreading of the surfactant and the surface deformation of the liquid layer, were studied theoretically and experimentally [6,7].

The solutocapillary deformation of a thin horizontal liquid layer has also been studied when a drop of a soluble liquid with surface tension lower than that of the substrate fluid was placed over the layer surface [8]. This drop initially tended to spread to maximum size, i.e., in the limit, a monomolecular film rapidly disappearing by dissolution or evaporation. In contrast to the thermocapillary case, the overall process was clearly nonstationary, thus allowing measurement of only the maximum deformation value and, in particular, the radius of the resulting dry spot. In this experiment, the variable quantities were the initial layer thickness, the volume of the surfactant drop, and the surface tension difference between the liquids, when using pairs of different soluble liquids.

In some experiments, distilled water was used as a substrate fluid and some organic liquids dissolved in the water as surfactants (ethanol or propan-2-ol/ isopropyl alcohol). Also, the couple of isopropyl alcohol and a saturated hydrocarbon, i.e., hexane, heptane, decane, etc., were used. The surface tensions of the liquids used are given in Fig. 10 (lines 1–7 correspond, respectively, to hexane, heptane, decane, isopropyl alcohol, undecane, tridecane, and hexadecane). It is seen, for example, that hexane and heptane are surfactants for isopropyl alcohol; on the other hand, isopropyl alcohol is itself a surfactant for undecane,

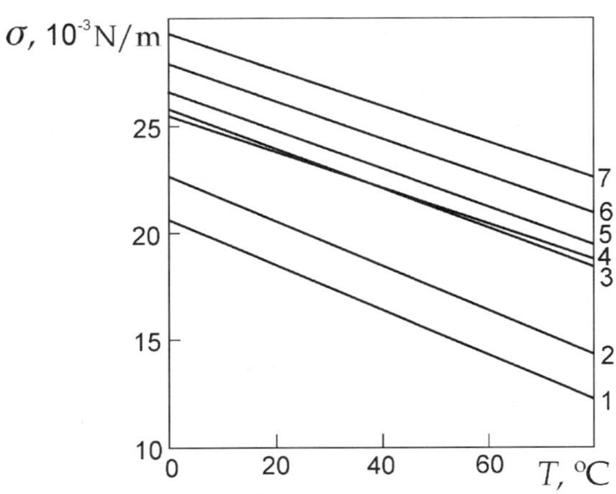

FIG. 10 Experimental investigation of solutocapillary-induced deformation. Variation of liquid surface tension with temperature. Lines 1–7 correspond, respectively, to hexane, heptane, decane, isopropyl alcohol, undecane, tridecane, and hexadecane. (From Ref. 8.)

tridecane, and hexadecane. It should be noted that although in some cases a drop of a heavier liquid was placed on the surface of a lighter one, rather than sinking the drop rapidly spread over the surface.

The experimental procedure was as follows. A liquid of volume $3-10\,cm^3$ was poured into a circular container of 80 mm diameter, forming a thin horizontal layer of 0.6–2.0 mm depth. The drop of a soluble surfactant of volume $5-100\,\mu l$ was deposited with the aid of a syringe in the center of the layer. The process was monitored using a 25 Hz video camera.

Because the surface tension of the surfactant drop was much lower than that of the substrate fluid, the drop became a spot of minimum thickness and maximum size. The film immediately started dissolving in the substrate fluid [photograph in Fig. 11(a)], thus reducing the local value of the surface tension under the spot. The surface tension gradient generated convection, which, in turn, led to the surface deformation with a decrease in the local layer thickness directly under the surfactant film. This deformation was just analogous to that observed in the case of thermocapillary convection when the center part of the liquid layer was heated. For rather thin substrate fluid layers, a dry spot appeared [Fig. 11(b)], whose diameter first increased rapidly during the spreading of the drop and then, due to the dissolution and evaporation of the surfactant, the dry spot disappeared [Fig. 11(c)].

The radius of the dry spot depended primarily on the amount of surfactant used. Figure 12 illustrates the observed dependence of the extent of the deformation with the volume of the surfactant drop. The area of the surfactant film spread was proportional to the drop volume. This fact suggested that the drop volume remain unchanged in further experiments.

The evolution of the dry zone radius is represented graphically in Fig. 13. It is seen that the deformation process terminates very quickly and, as in the

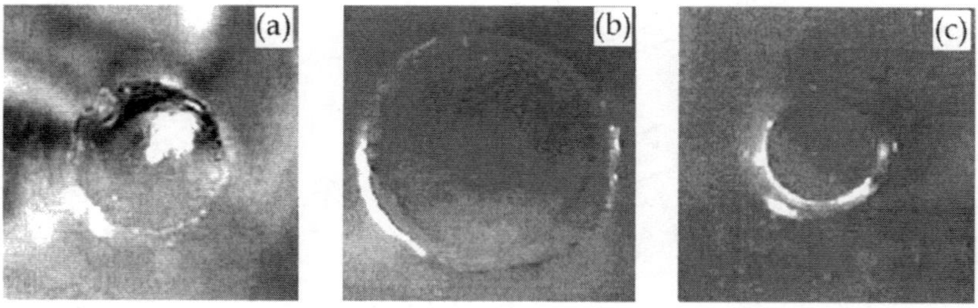

FIG. 11 Experimental investigation of solutocapillary-induced deformation. Liquid layer rupture due to the action of a surfactant: hexane drop over isopropyl alcohol. Time evolution of the dry spot: (a) film is starting to dissolve in the substrate fluid; (b) a dry spot appears; (c) because of the dissolution and evaporation of the surfactant the dry spot disappears. (From Ref. 8.)

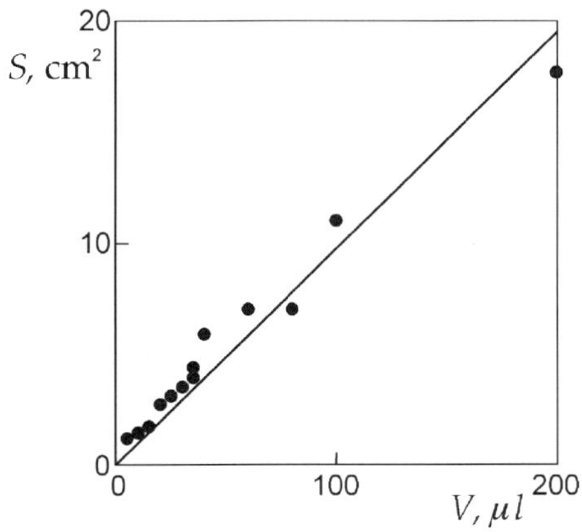

FIG. 12 Experimental investigation of solutocapillary-induced deformation. Area of dry spot vs. drop volume. Hexane drop over isopropyl alcohol, $d_0 = 0.8\,\text{mm}$. (From Ref. 8.)

thermocapillary case, crucially depends on the initial layer thickness. The thinner the liquid layer, the more pronounced is the surface depression.

The dependence of the maximum radius of the dry spot on the initial liquid layer thickness for a hexane drop of 20 μl fixed volume on the surface of an isopropyl alcohol liquid substrate is given in Fig. 14 (line 1). The experimental points can be fitted well by a polynomial of second degree. The extrapolation of the data allows us to estimate the value of the critical initial layer thickness at which the depression of the surface under the spreading drop grows until it reaches the layer bottom (vanishing dry zone radius). This also yields the maximum size of the dry spot for vanishing layer thickness. For a hexane drop over an isopropyl alcohol surface, these values are $h_0 \sim 1.2\,\text{mm}$ and $r \sim 30\,\text{mm}$, respectively.

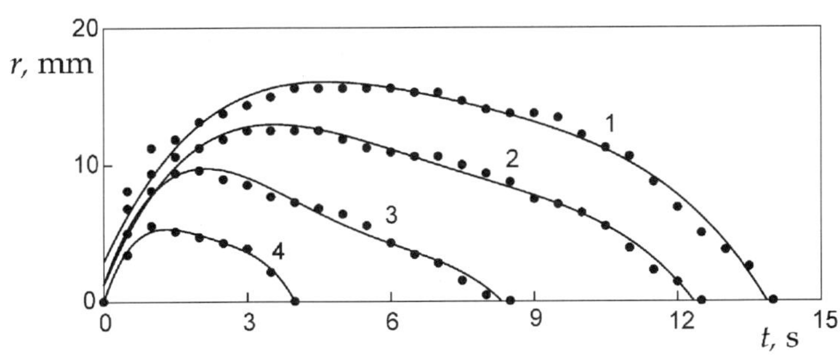

FIG. 13 Experimental investigation of solutocapillary-induced deformation. Evolution of the dry spot radius with time. Lines 1–4 correspond to experiments with d_0 equal to 0.6, 0.7, 0.8, and 1.0 mm, respectively. (From Ref. 8.)

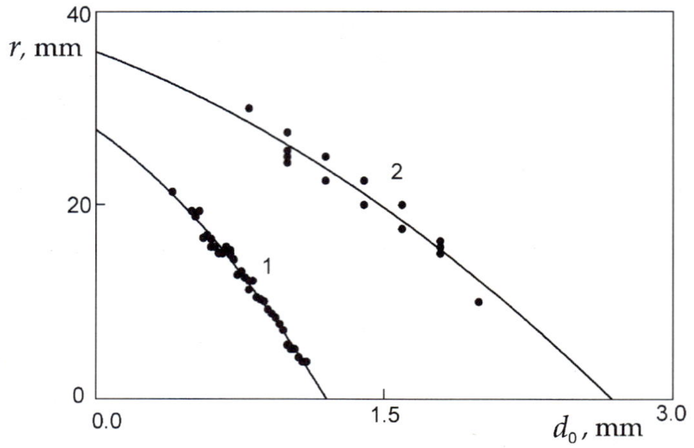

FIG. 14 Experimental investigation of solutocapillary-induced deformation. Dry spot radius vs. initial layer thickness: line 1, hexane over isopropyl alcohol; line 2, isopropyl alcohol over water. (From Ref. 8.)

Similar results for isopropyl alcohol drops on a water surface were obtained (Fig. 14, line 2). In this case, approximately the same deformation can be observed in much thicker fluid layers (up to 2.5 mm), because the difference in the surface tensions of water and isopropyl alcohol is larger than between the alcohol and hexane. Thus, surface deformation due to concentration gradients can be observed in liquid layers several times thicker than in the case of thermocapillary convection.

An attempt was also made (Fig. 15) to estimate the influence of the difference in surface tensions of liquid–surfactant pairs on the maximum radius of the dry zone. This experiment was analogous to the thermocapillary case (Fig. 4) in which the radius increased monotonically with the surface tension change (proportional to the temperature difference). Unfortunately, no regularity was found. Presumably, this was due to the fact that each experimental point corresponds to a couple of liquids differing not only in the value of surface tension but also in other physico-chemical properties.

II. SUPPRESSION OF THERMOCAPILLARY SURFACE DEFORMATION BY MEANS OF HIGH FREQUENCY VIBRATIONS

Let us now focus attention on the results found about the influence of transverse vibration on the deformation of free surfaces and on the rupture of liquid layers caused by thermocapillary convection [9].

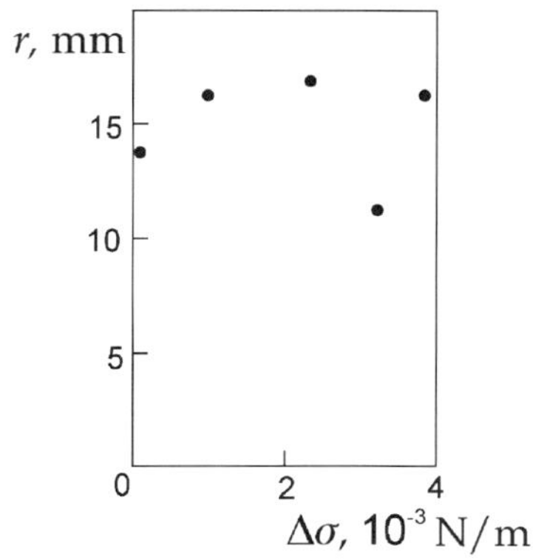

FIG. 15 Experimental investigation of solutocapillary-induced deformation. Dry spot radius vs. surface tension difference. Isopropyl alcohol over different liquids (hexane, pentane, decane, etc.), $d_0 = 0.8\,\mathrm{mm}$. At variance with Fig. 4, no clear results were obtained, probably because of the disparity of values in the physical properties of the liquid pairs used. (From Ref. 8.)

A. Experimental Set-Up and Procedure

The shape of a thin liquid layer subjected to a horizontal thermal gradient and high frequency vibration was observed with the help of the Schlieren apparatus sketched in Fig. 1 [2]. The liquid with an open upper surface was placed in a cylindrical or rectangular metal cuvette. The cuvette was placed on a vibrating platform, subjected to vertical oscillations with frequency in the range $10-1000\,\mathrm{Hz}$ and amplitude in the range $0-8\,\mathrm{mm}$. The amplitude was measured within $\pm 0.01\,\mathrm{mm}$ with an optical cathetometer, and the frequency within $\pm 0.1\,\mathrm{Hz}$ with a digital frequency meter.

The diameter of the circular liquid layer was 90 mm, and the rectangular layer was $70\,\mathrm{mm} \times 74\,\mathrm{mm}$ in size. The liquid volume introduced was measured within $\pm 1\,\mathrm{mm}^3$, which made it possible to vary the thickness of the layer within $\pm 0.02\,\mathrm{mm}$. The layer was heated at its center and cooled around the perimeter by means of heat exchangers, as a result of which the temperature in the liquid decreased from the center to the edges of the cuvette in accordance with the logarithmic law for the axisymmetric case. It was possible also to reverse the direction of the temperature gradient by heating at the periphery and cooling at the center.

Experiments were performed with organic liquids of different viscosities: ethanol, heptane, decane, hexadecane, and silicone oil. The liquid layer thickness, d, was allowed to vary from 0.3 to 1.5 mm. The maximum temperature difference between heater and cooler, measured with thermocouples to within $\pm 0.1\,\mathrm{K}$, did not exceed 30 K.

The surface relief was recorded by reflected light using a short-focus shadowgraph instrument. Photographs of the isolines of the angle of inclination of the surface in the interval 10^{-1}–10^{-3} rad were obtained. The radial distribution of the local thickness of the layer was determined by two successive graphic integrations subject to the condition of conservation of the known initial liquid volume.

The mirror periscope 4 (Fig. 1) was so oriented that the direction of vibration coincided with the optical axis of the system. Because the amplitude of the vibrations was small relative to the 250 mm focal length, 5, of the objective, the instrument enabled measurement of the surface relief of the liquid in the vertically vibrating cell, provided the relief was steady.

B. Effect of Transverse Vibration on the Surface Layer Deformation Caused by Thermocapillary Convection

It was found that vibration orthogonal to the liquid layer led to the suppression of thermocapillary deformation, thus smoothing the surface. It also led to a decrease in the radius, r, of the dry zone, which was measured as a function of the vibration amplitude, b, for various values of the frequency, ω. With the vibrator platform switched on at a fixed frequency, the layer surface turned out not to be stationary enough (beyond spontaneous equilibrium surface tremor) to use the very sensitive Schlieren method, even with stroboscopic light. Therefore only the influence of vibrations on the thermocapillary breakdown of the layer was studied and only the radius of the "dry zone" was chosen as the quantity of interest. The design of the vibrator platform allowed the observers to carry out all measurements by swiftly changing the vibration amplitude at fixed frequency and given temperature difference between heater and cooler.

For illustration of the results let us recall the thermocapillary deformation of a 0.5 mm thick decane layer with a temperature difference of 24 K. In the absence of vibrations the expected radius of the dry zone in the center was equal to 16 mm, but under the action of vibrations it was seen to become smaller.

In Fig. 16, the ratio r/r_0 is plotted as a function of the ratio of the vibrational and gravitational accelerations, $\Omega = b\omega^2/g_0$, for the case $r_0 = 16$ mm when $\Delta T = 24$ K. All the experimental points obtained at frequencies of 10, 20, and 40 Hz (points ●, ×, and ○, respectively) and different vibration amplitudes lie on the same curve. As the vibrational acceleration increases from zero to 3g, the ratio r/r_0 decreases from unity to zero. Accordingly, at large enough (critical) vibration amplitudes, the continuity of the liquid layer was restored.

To answer the question of which one of two vibrational parameters, vibration rate, $b\omega$, or acceleration, $b\omega^2$, is responsible for the leveling effect, the critical values of the amplitude, b, and frequencies $f = \omega/2\pi$ of the vibrator needed to restore the integrity of an initially broken layer were measured. It can be seen from

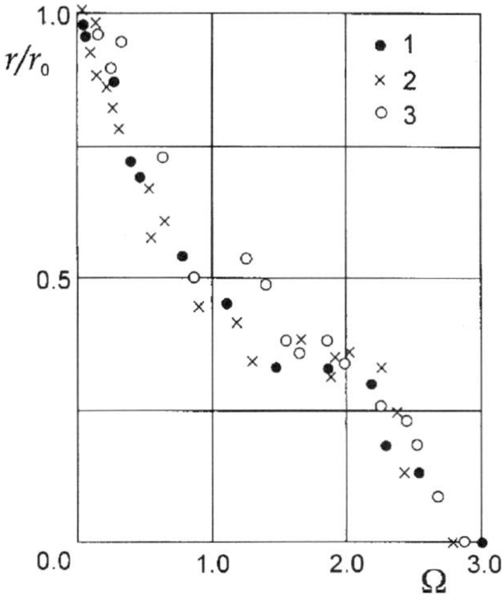

FIG. 16 Suppression of thermocapillary-induced surface deformation in a liquid layer by means of high-frequency vibrations. Dry zone radius vs. dimensionless frequency, Ω, for different values of *f*: ●, 10 Hz; ×, 20 Hz, and ○, 40 Hz. (From Ref. 2.)

the left graph in Fig. 17 that at high frequencies the reduction of the thermocapillary convection becomes appreciable, already at very low amplitudes. On the right graph this critical curve is shown in logarithmic coordinates. In log–log coordinates all the experimental points lie close to a single straight line corresponding to a constant value of the vibrational acceleration $(b\omega^2)_* = 794 \pm 4$. This experimental result, obtained for various values of the frequency and amplitude of vibrations, gives evidence that the restoration of the layer unity is practically controlled by the vibrational acceleration.

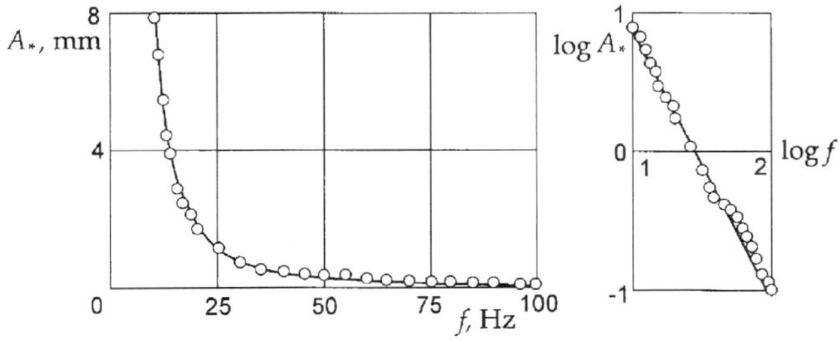

FIG. 17 Suppression of thermocapillary-induced surface deformation in a liquid layer by means of high-frequency vibrations. Critical values of amplitudes and frequencies needed to restore the integrity of the layer. (From Ref. 2.)

In order to show how thermocapillary convection is reduced by vibrations a modified Marangoni number is useful. This quantity was determined not through the real temperature difference, but using the temperature difference that produces a similar deformation in the absence of vibrations. Figure 18 shows how this "effective" Marangoni number, characterizing the magnitude of the thermocapillary deformation, significantly decreases with the growth of vibrational acceleration from zero to the value $3g_0$.

Thus, within the error bounds of the experiment, the restoration of the continuity of the layer and all other manifestations of the effect of vibration on the thermocapillary deformation of the surface seem to depend only on the combination of vibration parameters defining the vibrational acceleration, $b\omega^2$, and not on the vibrational rate, $b\omega$, as in the earlier studied problem of high-frequency vibrational convection.

Two remarks are now pertinent. As will be shown later, vertical vibrations may lead to parametric resonance, i.e., the occurrence of parametric waves, similar to Faraday ripples, on the surface of the nonisothermal liquid. With increasing frequency the threshold values of a vibration velocity, $(b\omega)_*$, for Faraday ripples also increase, while the vibration rate necessary for suppressing deformations of the surface at a constant value of the vibrational acceleration falls. Hence, in the high-frequency interval there is a possibility of suppressing deformation without exciting parametric waves. This effect was observed, for example, using the liquid PMS-200 at frequencies higher than 200 Hz. However, even when Faraday ripples were present at low-frequency vibrations with low-viscosity liquids, this side effect did not have a significant influence on the phenomenon investigated.

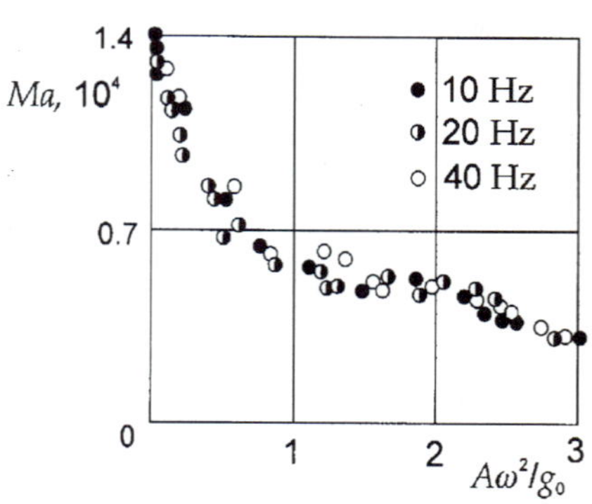

FIG. 18 Suppression of thermocapillary-induced surface deformation of liquid layer by means of high-frequency vibrations. Decrease of the effective Marangoni number under the action of vibrations. (From Ref. 2.)

III. THERMOCAPILLARY CONVECTION INDUCED BY A POINT-LIKE HEAT SOURCE

Let us now examine the flow triggered in a large fluid volume by a point-like heat source located below the open surface with a size that is small compared to the layer depth [9]. We shall see that thermocapillary effects can drastically alter an existing flow structure and intensify the heat transfer in a deep layer with already existing natural convection. We shall also see that the perturbations associated with oscillatory surface instability and hence leading to the formation of surface waves of various types are found to be the most dangerous.

A. Experimental Set-Up

The experiments were performed with a set-up shown schematically in Fig. 19. It consisted of a (10 cm × 10 cm) convection chamber or cuvette filled with liquid to a depth of 11 cm. A transparent glass frame, 2, specifying chamber dimensions, is attached to a textolite base, 1. A cylindrical heat source, 3, very much like a point heat source was located inside the chamber. In the experiments, heaters of the same height (2 mm) and outer diameters (5 and 7 mm) were used. This corresponded to 5% and 7% of the characteristic linear dimension of the chamber, respectively. The heat source was attached to a glass tube, 4, which was also used for leading wires from the power source and thermocouples. The opposite end of the tube was brought out of the chamber through a hole in the base and connected to a micrometer slide used to displace the heating element vertically. By construction the slide allowed the depth of the heat source to be described relative to the open liquid surface within 0.1 mm accuracy.

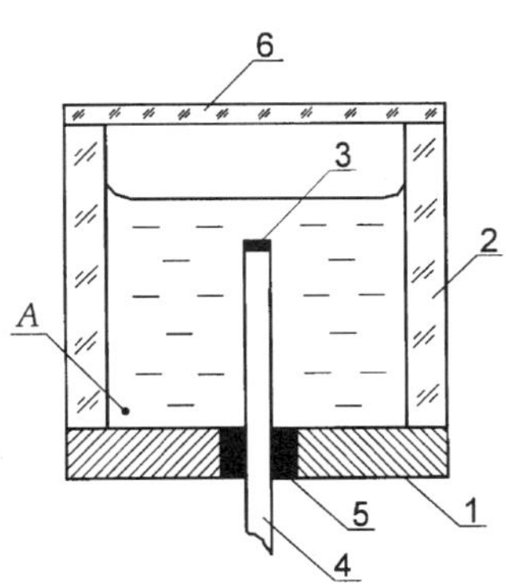

FIG. 19 Thermocapillary convection induced by a pointlike heat source. Sketch of the experimental set-up: 1, textolite base; 2, glass frame; 3, heat source; 4, glass pipe; 5, fluoroplastic gasket, and A, location of the thermocouple junction. (From Ref. 9.)

The liquid used in these experiments was decane. The temperature difference between the heat source and some point A (Fig. 19) in the remote region of the fluid was measured with a copper–constantan thermocouple with one of the junctions located on the body of the heat source. This made it possible to monitor the heat transfer. During the experimental runs the temperature distribution throughout the liquid and the variation in the shape of the liquid surface were visualized with a holographic focused image interferometer. A small quantity of light-scattering particles (aluminum powder) was added to the liquid to ensure simultaneous observation of the structure and velocity of the flows. During the experiments the change in the thickness of the liquid layer over the heat source due to a liquid flow out of the gasket, 5, did not exceed 0.1%.

The convection cavity was closed at the top by a transparent glass plate. The presence of the plate prevented random air flows capable of causing nonuniformity of the surface tension as a result of nonuniform evaporation over the fluid surface. The experiments were carried out for an open and closed liquid surface (free and solid surface, respectively). In the latter case the liquid surface was covered by a thin glass plate to eliminate capillary effects completely.

B. Heat Transfer Modes

The experiments showed that as the heat source approaches the open surface the temperature over the heat source decreases significantly compared to the temperature at deeper liquid layers. This indicates that the heat transfer from the heater increases. For a closed surface the heat release decreases. In Fig. 20 the surface temperature of the heating element is plotted as a function of its depth below surface for a given heat source power. It is readily seen that at a considerable distance from the liquid surface the heat release is identical for both the open (○)

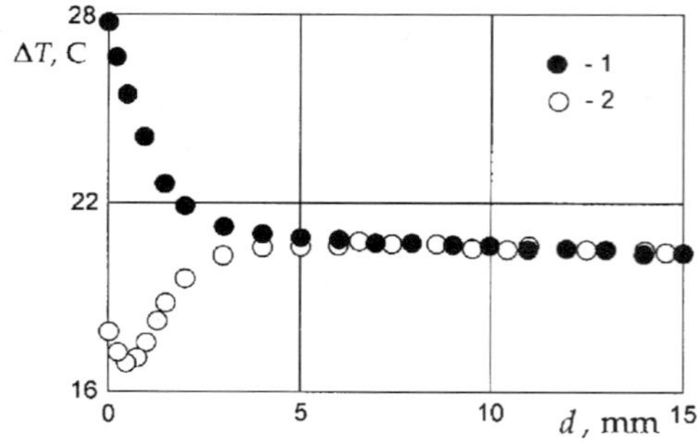

FIG. 20 Thermocapillary convection induced by a pointlike heat source. Heater element surface temperature as a function of its depth below the surface level (heater power 0.77 W; source diameter 5 mm): points ● and ○ correspond to experiments with closed and open fluid surfaces, respectively. (From Ref. 9.)

and closed (●) liquid surfaces. In both cases flow was observed in the form of a thermal jet with a rather low characteristic velocity. However, when the distance to the surface becomes comparable with the diameter of the heating element, the nature of the heat transfer to the fluid is significantly altered. As the heat source approaches the open fluid surface, its heat transfer increases appreciably, mainly because of the onset of thermocapillary flow along the free surface. Furthermore, as the heat source approaches the free surface, the convective flow region is displaced to the near-surface layer and a toroidal convective cell forms. The characteristic value of the velocity, with a maximum on the fluid surface, increases appreciably, reaching several centimeters per second.

The data obtained were used to represent graphically the dependence of the intensity of the thermocapillary motion with the dimensionless depth of the heater immersion for various values of the heater power. The depth below level was nondimensionalized using as unit length the diameter of the heating element. The intensity of the thermocapillary flows was defined in terms of the dimensionless relation $K = (T_o - T_c)/T_c$, where T_o and T_c are the temperatures of the heating element for open and closed fluid surfaces, respectively. This quantity is here the analogue to the Nusselt number in thermoconvective flows and is introduced as the ratio of the heat release due to thermocapillary effects to the sum of the heat releases due to molecular transfer and convective motion minus unity. In Fig. 21 this dependence is plotted for heating elements of various diameters. Thermocapillary effects occur at a distance from the surface smaller or approximately equal to the source diameter. In this case, the heat transfer due to

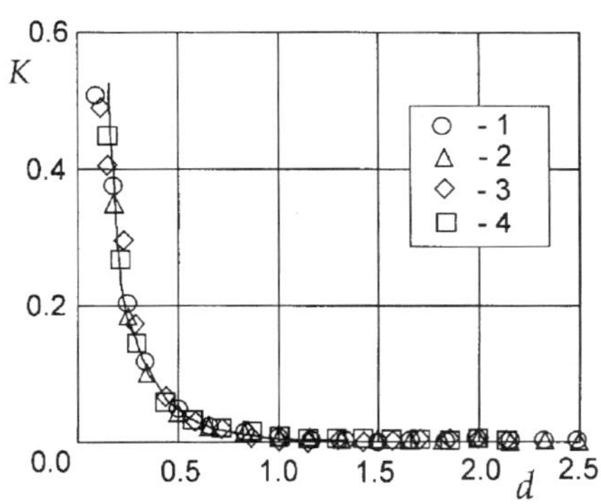

FIG. 21 Thermocapillary convection induced by a pointlike heat source. Dimensionless intensity of the thermocapillary motion as a function of the dimensionless depth of immersion of heat sources, 5 mm in diameter with 0.77 W and 1.32 W power rate (points ○ and △) and 7 mm in diameter with 0.6 W and 1.25 W power rate (points ◇ and □). (From Ref. 9.)

thermocapillary convection is 1.5 times greater than the heat transfer due to buoyancy-driven convection.

C. Flow Structure and Surface Shape

Observations of the liquid surface over the heat source were made in parallel with the temperature measurements. There was evidence indicating that the fluid surface was deformed and took various shapes in accordance with the nature and intensity of the flows developing in the bulk. Both steady and oscillatory modes of instability of the fluid surface were detected.

At a relatively large immersion depth of the heat source when the thermocapillary flow intensity was still very low, buoyancy-driven flows were predominant. In the case considered, the characteristic feature of this flow was the presence of a thermal jet, which propagated from the heat source to the fluid surface, causing surface bending in the direction of the gas phase. This curvature was seen to depend on the distance from the heat source to the surface and the heater power, but it did not change with time.

Because the heat source gradually approached the liquid surface, the contribution of the thermocapillary convection to the overall flow pattern started to increase. An axisymmetric radial flow initiated in this case led to a decrease in the curvature of the fluid surface. As the immersion depth was decreased further, such that the intensity of thermocapillary flow became comparable with that of the buoyancy-driven flow, the surface curvature reversed sign, and the surface became curved in the direction of the liquid phase, i.e., funnel-shaped. Up to a certain value of the immersion depth this surface shape was also independent of time.

When the immersion depth reached a certain value the surface started to vibrate. Then the curvature of the liquid surface above the heater alternated periodically with time around an equilibrium value. This led to the formation of surface waves propagating away from the heat source. The configuration of the induced waves depended on the distance to the surface and the heater power. The first wave mode generated by the surface oscillations consisted of concentric wave crests [Fig. 22(a)].

As the gap between the heat source and the liquid surface decreased, the wavenumber increased, and at a certain distance to the surface the wave shape became unstable. Then a new configuration appeared, namely, a right-hand or left-hand rotating spiral wave [Fig. 22(b)]. A decrease in the heater depth led to an increase in both the radial and azimuthal wavenumbers. This was manifested in an increase in the number of arms of the spiral wave. The evolution of a spiral wave as the heat source approached the free surface is shown in Figs 23(a)–(c).

When after the next rise the shell of the heat source came into contact with the fluid surface the oscillations of the fluid surface ceased and a steady-state wave structure with a raylike configuration appeared at the surface [Fig. 22(c)]. At the

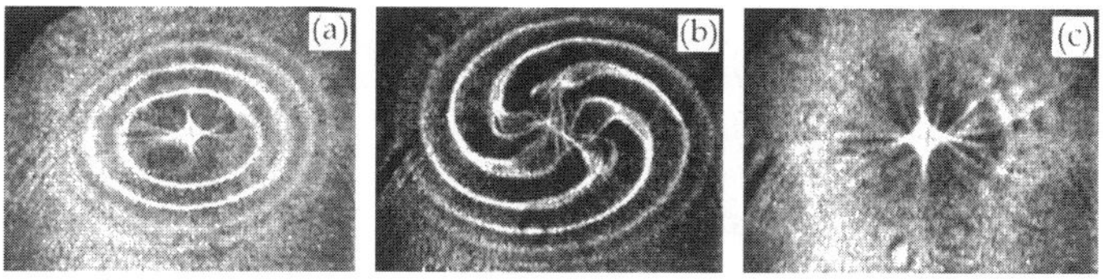

FIG. 22 Thermocapillary convection induced by a pointlike heat source. Schlieren photographs of the shape of the liquid free surface: (a) circular wave; (b) spiral wave; and (c) standing wave with star-shaped configuration. (From Ref. 9.)

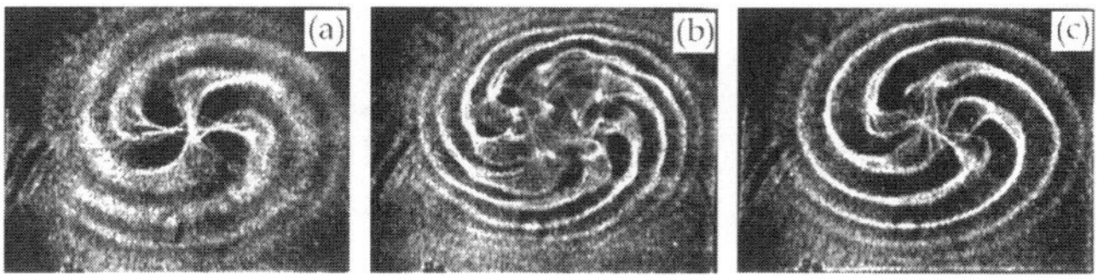

FIG. 23 Thermocapillary convection induced by a pointlike heat source. Schlieren photographs of the shape of the liquid free surface illustrating the evolution of the spiral wave configuration as the heat source approaches the free surface. (From Ref. 9.)

instant of contact the heat release from the heat source decreased. This was made clear by observing the behavior of the temperature at the heater shell, which varied with the depth of the heat source in the case of an open surface in the region of small immersion depths (see Fig. 20). The heat release decreased in proportion to the increase in the contact area between the heater shell and the gas phase, whose thermal conductivity was lower than that of the liquid phase.

Observation of the flow structure and measurements made simultaneously led to the following conclusions. The structure of the radial flow, which was independent of the azimuthal direction and typical of thermocapillary convection in the chosen geometry, remained unaltered whatever the values of the immersion depth and heat source power, as well as for any surface configuration and surface wave shape. In this case, the quantities that varied with the immersion depth and the power of the heat source were the size of the convective cell, the penetration depth, and the flow velocity characteristics.

REFERENCES

1. Zuev, A.L., Pshenichnikov, A.F. Deformation and rupture of fluid films under thermocapillary convection. Appl. Mech. Thech. Phys. **1987**, *3*, 90–95 (in Russian).

2. Briskman, V.A., Zuev, A.L. Influence of different factors on the thermocapillary deformation of a thin liquid layer. In *Hydromechanics and Heat and Mass Transfer in Microgravity*; Gordon and Breach: N.Y., 1992; 139–144.
3. Briskman, V.A., Zuev, A.L., Lyubimova, T.P., Nepomnyashchy, A.A. Thermocapillary flows and deformations of the surface in systems of fluid layers with longitudinal temperature gradient in microgravity. In *Microgravity Fluid Mechanics*. Springer-Verlag: Berlin-Heidelberg, 1992, 145–152.
4. Pschenichnikov, A.F., Tokmenina, A.G. The deformation of a free liquid surface by thermocapillary motion. Izv. Akad. Nauk SSSR, Mech. Zhidk. Gaza **1983**, *3*, 150–153 (in Russian).
5. Wang, Mu, Ming, N.-B. In situ observation of surface-tension-induced oscillation of aqueous-solution films in needlelike crystal growth. Phys. Rev. A **1991**, *44*, 7898–7901.
6. Starov, V., de Ryck, A., Velarde, M.G. On the spreading of an insoluble surfactant over a thin viscous liquid layer. J. Colloid Interface Sci. **1997**, *190*, 104–113.
7. Nepomnyashchy, A.A., Velarde, M.G., Colinet, P. *Interfacial Phenomena and Convection*; Chapman & Hall/CRC: Boca Raton, 2002.
8. Zuev, A.L., Briskman, V.A., Legros, J.-C., Velarde, M.G., Viviani, A. Experimental studies of Marangoni convection in thin liquid layers caused by soluble surfactant. Proceedings 1st International Symposium on Microgravity Research and Applications in Physical Sciences and Biotechnology, Sorrento, Italy, 2000; ESA SP-454, 2001, pp. 207–214.
9. Bratukhin, Yu.K., Makarov, S.O., Mizyov, A.I. Oscillating thermocapillary convection regimes driven by a point heat source. Fluid Dyn. **2000**, *35*, 232–241.

12
Parametric Wave Excitation in Nonisothermal Liquid Layers

In earlier chapters we studied the influence of two factors—vibration and the Marangoni effect—acting on free surfaces or interfaces. We have seen that, taken separately, each has significant effects on the equilibrium state of a liquid layer with an open surface. In this chapter we proceed to consider their interaction as either cooperation or competition. We shall consider parametric vibrational action with finite frequency on a nonisothermal liquid layer susceptible to thermocapillary instability and parametric excitation of surface tension gradient (Marangoni) driven convection when the heat flux through the free surface of a deep liquid layer is modulated in time.

I. INTERACTION BETWEEN PARAMETRIC WAVE EXCITATION AND OVERSTABILITY DUE TO THE MARANGONI EFFECT

As described earlier, under suitable conditions vertical oscillations of a container partially filled with liquid can excite and sustain capillary-gravity surface waves with frequency half the driving frequency (Faraday ripples). As already discussed, a key feature of this phenomenon is that parametric wave excitation occurs past a threshold, i.e., the amplitude of vibrations must exceed some critical value, in accordance with the dissipation level resulting from the viscosity of the liquid.

Another instability of the open surface of a liquid (also studied earlier in this book) is that caused by the variation of surface tension with temperature or surfactant concentration. In particular, at high enough values of a thermal gradient placed orthogonal and hence transverse to the liquid deformable surface, the motionless equilibrium can be made unstable to oscillatory disturbances identical in wave nature to the Faraday ripples. We disregard here longitudinal dilational waves that can also be excited by the Marangoni effect.

Parametric wave resonance and the abovementioned thermocapillary-induced overstability share common features. Both are characterized by instability of the quiescent state, and surface deformability is needed for both. Yet these mechanisms display essential differences. In the first case (Faraday ripples) the instability is caused by variable external forces (inertial forces in the vibrating vessel) and excitation occurs as a result of parametric resonance with the externally imposed frequency. In the second case the waves are initiated by a time-independent, constant thermocapillary force and their frequency is defined by intrinsic liquid properties. Let us discuss the interaction between these two agents, namely parametric and thermocapillary mechanisms simultaneously generating and sustaining surface waves [1]. We exclude from our consideration the instability caused by buoyancy forces, which is a reasonable approximation if we have moderate heating, a thin enough liquid layer, and a small enough coefficient of thermal expansion.

A. Equations and Boundary Conditions

Let a vessel containing liquid be subjected to harmonic vertical oscillations of amplitude a and frequency 2ω. In the frame of reference attached to the container we have an effective, modulated gravitational acceleration:

$$\mathbf{g}_{\text{eff}} = \mathbf{g}(1 - b\cos 2\omega\tau) \tag{1}$$

where

$$b = \frac{4\alpha\omega^2}{g} \tag{2}$$

and where, again, $\mathbf{g}$ is the gravity vector on Earth, pointing vertically downwards. The liquid in the container is subjected to a constant vertical temperature gradient. As noted earlier, the analysis of the continuity Navier–Stokes, and heat equations for an incompressible liquid, with corresponding boundary conditions, shows that when the temperature is a linear function of the vertical coordinate the liquid is in a quiescent state. We proceed now with the examination of the stability of this motionless state when $\mathbf{g}_{\text{eff}}$ is variable.

To further simplify the problem, without significant loss of generality, as in cases studied earlier in this book, we assume that the horizontal dimensions of the container and its depth are large compared to the characteristic dimension of the expected hydrodynamic excitations, e.g., wavelength. With this assumption we may neglect the boundary effects at the walls of the container.

Let us introduce a Cartesian coordinate system with z-axis directed vertically upward. The plane $z = 0$ is aligned with the free, level liquid surface (in the frame of reference attached to the oscillating container). At the steady state the liquid occupies the half-space $z < 0$ under the modulated gravitational field [see Eq. (1)] and a constant vertical temperature gradient denoted here by A.

For two-dimensional oscillations, depending only on x and z, the linearized disturbance equations for the vertical, w, and the horizontal, u, components of liquid velocity, the temperature, T, and pressure, p, are [1]:

$$\frac{\partial w}{\partial z} + \frac{\partial u}{\partial x} = 0 \tag{3}$$

$$\frac{\partial u}{\partial t} = -\frac{1}{\rho}\frac{\partial p}{\partial x} + \nu \Delta u \tag{4}$$

$$\frac{\partial w}{\partial t} = -\frac{1}{\rho}\frac{\partial p}{\partial z} + \nu \Delta w \tag{5}$$

$$\frac{\partial T}{\partial t} = Aw + \chi \Delta T \tag{6}$$

$$\Delta \equiv \left(\frac{\partial^2}{\partial x^2} + \frac{\partial^2}{\partial z^2}\right)$$

The equilibrium temperature gradient, denoted here by A, is taken positive when heating from the liquid side (heating from below) and negative when heating from the air side (heating from above). Note that here A rather than β denotes the temperature gradient. β will instead be used for the thermal expansion coefficient, as α will account for vibration amplitude. Disturbances are initiated close to the liquid surface and, because of viscosity, decay away from it, hence

$$u,\ w,\ \rho,\ T \to 0 \tag{7}$$

at $z \to -\infty$.

The deformed liquid surface is defined by the equation

$$z = \zeta(x, t) \tag{8}$$

where the function $\zeta(x, t)$ is yet to be determined. The balance of normal and tangential stresses at the free surface yields

$$p - \rho g(1 - b\cos 2\omega t)\zeta + \sigma\frac{\partial^2 \zeta}{\partial x^2} = 2\eta\frac{\partial w}{\partial z} \tag{9a}$$

$$\frac{d\sigma}{dT}\left(\frac{\partial T}{\partial x} - A\frac{\partial \zeta}{\partial x}\right) = \eta\left(\frac{\partial u}{\partial z} + \frac{\partial w}{\partial x}\right) \tag{9b}$$

Furthermore we have the kinematic condition

$$\frac{\partial \zeta}{\partial t} = w \tag{10}$$

The boundary conditions of Eqs. (7)–(10) are given in linearized form at $z = 0$. The condition (9a) brings the modulation of the gravitational force and the effect of surface tension. Let the air heat conductivity be negligible relative to the heat conductivity of liquid. This suggests that the free surface of the liquid may be considered thermally "insulated"; hence the fixed flux condition at $z = 0$

$$\frac{\partial T}{\partial x} = 0 \tag{11}$$

In view of the comments made above concerning the length scales of the problem, to write Eqs. (3)–(6) and the boundary conditions (7)–(11) in dimensionless form we choose the capillary length $l = (\sigma/\rho g)^{1/2}$ as a unit of length, the combination $\tau = (\sigma/\rho g^3)^{1/4}$ as a time unit, and l/τ, $\rho l^2/\tau^2$, and Al, respectively, as units of velocity, pressure, and temperature. We assume further that along the horizontal axis, x, quantities are periodic with the wavenumber k.

Then, using a standard Fourier mode decomposition, $\exp(ikx)$, the problem of Eqs. (3)–(11) reduces to

$$\frac{\partial w}{\partial z} + iku = 0 \tag{12}$$

$$\frac{\partial u}{\partial t} = -ikp + \tilde{\nu}\Delta u \tag{13}$$

$$\frac{\partial w}{\partial t} = -\frac{\partial p}{\partial z} + \tilde{\nu}\Delta w \tag{14}$$

$$\frac{\partial T}{\partial t} = w + \tilde{\chi}\Delta T \tag{15}$$

$$\Delta \equiv \left(\frac{\partial^2}{\partial z^2} - k^2\right)$$

with boundary conditions at $z \to -\infty$:

$$u,\ w,\ \rho,\ T \to 0 \tag{16}$$

and at $z = 0$:

$$\frac{\partial \zeta}{\partial t} = w \tag{17}$$

$$p - (1 + k^2 - b\cos 2\tilde{\omega}t)\zeta = 2\tilde{\nu}\,\frac{\partial w}{\partial z} \tag{18}$$

$$ik\,Ma(T - \zeta) = \tilde{\nu}\left(\frac{\partial u}{\partial z} + ikw\right) \tag{19}$$

$$\frac{\partial T}{\partial z} = 0 \tag{20}$$

The following dimensionless parameters have been introduced: $Ma = (Al/\sigma)(\mathrm{d}\sigma/\mathrm{d}T)$ is here the Marangoni number, $\tilde{\nu} = \nu\tau/l^2$ and $\tilde{\chi} = \chi\tau/l^2$ are dimensionless viscosity and heat diffusivity, respectively, b accounts for the dimensionless vibration amplitude, defined by Eq. (2), and $\tilde{\omega} = \omega\tau$ is the dimensionless vibration frequency. Note that $b = 4\tilde{a}\tilde{\omega}^2$, where $\tilde{a} = a(\rho g/\sigma)^{1/2}$ is the vibration amplitude, measured in capillary units. As usual, for standard liquids $(\mathrm{d}\sigma/\mathrm{d}T < 0)$, and hence in this chapter the Marangoni number, as defined above, is positive when heating from the air side and negative otherwise. From now on we suppress the tilde over the quantities ν, χ, and ω. Our choice of parameters is made to emphasize the fact that the excitation threshold for Faraday ripples and for thermocapillary waves are both directly related to the level of dissipation.

B. Instability Threshold

To obtain the stability boundaries we now use a series expansion in a dimensionless small parameter ε. For a low viscous liquid, with finite Prandtl number, $Pr = \nu/\chi$, the quantity ε accounts for the smallness of dissipative effects, e.g., $\varepsilon = (\nu k^2/\omega)^{1/2}$, which is the ratio of viscous vortex layer thickness to the expected wavelength of the disturbances.

Solution of the problem of Eqs. (12)–(20) can be sought as a series expansion

$$Q = Q_0 + \varepsilon Q_1 + \varepsilon^2 Q_2 + \cdots \tag{21}$$

where Q is any of the quantities u, w, p, and ζ. To solve the problem we can distinguish between relatively fast motions with the frequency of capillary-gravity waves and a slowly varying amplitude. This naturally suggests the application of a multiscale method and hence the time derivative is also represented as a series expansion [2]:

$$\frac{\partial}{\partial t} = \frac{\partial}{\partial t_0} + \varepsilon \frac{\partial}{\partial t_1} + \varepsilon^2 \frac{\partial}{\partial t_2} + \cdots \tag{22}$$

where t_0 is the "fast time" with characteristic scale defined by the wave frequency and $t_1, t_2, \ldots$ are "slow times" with scales characteristic of the amplitude evolution.

As already mentioned, at low values of the viscosity a viscous vortex skin layer is generated close to the surface. It is characterized by large gradients in the variables and therefore the derivative in z demands a stretched scale

$$\frac{\partial}{\partial z} = \varepsilon^{-1} \frac{\partial}{\partial z_{-1}} + \frac{\partial}{\partial z_0} + \varepsilon \frac{\partial}{\partial z_1} + \varepsilon^2 \frac{\partial}{\partial z_2} + \cdots \tag{23}$$

Such expansion is common for equations with small parameters involving high-order derivatives. We also assume

$$b = b_0 + \varepsilon b_1 + \varepsilon^2 b_2 + \cdots \tag{24}$$

As ε is also an ordering parameter, the substitution of expansions (21)–(24) into Eqs. (12)–(20) yields a hierarchy of equations for Q_0, Q_1, and Q_2, each demanding a solvability condition (the Fredholm alternative).

In the zeroth-order, the equations are integrated in general form and, as expected, the problem is reduced to the Mathieu equation [3] for the deviation of the liquid surface from the level position, ζ_0:

$$\frac{\mathrm{d}^2\zeta_0}{\mathrm{d}t_0^2} + (k + k^3)\left(\frac{\mathrm{d}\zeta_0}{\mathrm{d}t_0}\right) - (b_0 k \cos 2\omega t)\zeta_0 = 0 \tag{25}$$

where the differentiation is with respect to the "fast" time, t_0. For $b_0 \neq 0$ the solution (25) then diverges to infinity when

$$\omega^2 = k + k^3 \tag{26}$$

Because in the spectrum, k, there always exist values satisfying Eq. (26) we must assume that

$$b_0 = 0 \tag{27}$$

to avoid unlimited growth. Then, from Eq. (25), it follows that

$$\zeta_0 = f \exp i\Omega t_0 + h \exp(-i\Omega t_0) \tag{28}$$

where the frequency, Ω, is related to the wavenumber, k, by the dispersion relation for capillary-gravity waves of Eq. (26). We use Ω to distinguish the response of the system (the solution of the problem) from the driving frequency ω.

The amplitudes f and h, are independent of the "fast" time t_0, and only depend on the "slow" scales $t_1, t_2, \ldots$ Function (28) allows us to define time and coordinate dependence of all zeroth-order quantities, u_0, w_0, p_0, and T_0. The first-order ε calculation does not determine the instability threshold and it only yields $b_1 = 0$.

For the second order, ε^2, we get an evolution equation for the surface deformation, ζ:

$$\frac{d_2^2\zeta}{dt_2^2} + \Omega^2\zeta_2 = \left[-2i\Omega\frac{df}{dt_2} - 4i\Omega\nu k^2 f - \frac{k^3 Maf}{\lambda}\right]e^{i\Omega t}$$
$$+ \left[2i\Omega\frac{dh}{dt_2} + 4i\Omega\nu k^2 h - \frac{k^3 Mah}{\lambda^*}\right]e^{-i\Omega t}$$
$$+ b_2 k(fe^{i\Omega t} + he^{-i\Omega t})\cos 2\omega t \qquad (29)$$

where the superscript asterisk (*) denotes complex conjugate, and $\lambda = \sqrt{i\Omega/\chi}$.

The right-hand side of Eq. (29) involves resonance terms. If they do not vanish, the solution will have secular terms of the form $t_0\, e^{i\Omega t_0}$, which diverge with time and are unacceptable. Therefore the resonant terms in the right-hand side of Eq. (29) must be set equal to zero. As a result we obtain evolution equations for the quantities defining the amplitude, f and h, of the surface wave. It is readily seen from Eq. (29) that such evolution equations essentially differ in form depending on whether the condition of Eq. (26) is satisfied or not.

For disturbances not satisfying Eq. (26) the last term in Eq. (29) is nonresonant and the condition of absence of secular terms for ζ_2 takes the form

$$\frac{df}{dt_2} + 2\nu k^2 f - \frac{ik^3 Maf}{2\lambda\Omega} = 0 \qquad (30a)$$

$$\frac{dh}{dt_2} + 2\nu k^2 h + \frac{ik^3 Mah}{2\lambda^*\Omega} = 0 \qquad (30b)$$

Because f and h obey separate equations, assuming that both f and h behave exponentially with time, $\exp \Lambda t$, we obtain from Eq. (30)

$$\Lambda_f = \frac{k^3 Ma}{2\Omega}\sqrt{\chi/2\Omega} - 2\nu k^2 + \frac{ik^3 Ma}{2\Omega}\sqrt{\chi/2\Omega}$$

and

$$\Lambda_h = \Lambda_f^*.$$

Then for

$$Ma > Ma_* = \frac{4\nu\Omega}{k}\sqrt{\frac{2\Omega}{\chi}} \qquad (31)$$

the equilibrium is unstable. Ma_* denotes the value for the instability threshold. The prediction of Eq. (31) is independent of vibration parameters and agrees, except for notation, with the results given in Chapter 2, obtained for the vibrationless problem. The minimum of the neutral curve of Eq. (31) is at

$$k_c = 1/\sqrt{5} \tag{32}$$

Thus when the Marangoni number is higher than $Ma_c = Ma_*(k_c)$, with

$$Ma_c = \frac{2^{1/4} \cdot 3^{3/4}}{5^{5/8}} \frac{8\nu}{\chi^{1/2}} \tag{33}$$

the liquid surface is unstable, hence the appearance of surface waves with wave number k_c and frequency $\sqrt{k_c + k_c^3}$. The fact that the growth rates Λ_f and Λ_h involve an imaginary term means that the dispersion relation of the excited waves differs from Eq. (26). Their frequency is higher than that defined by Eq. (26) by a factor $(k^3 Ma/2\Omega)\sqrt{\chi/2\Omega}$. Note that these results have been obtained for low enough Marangoni numbers and therefore the wave frequency correction is small.

Let us now turn to a discussion of the role of vibrations when $Ma < Ma_c$ as we expect that the equilibrium may then become unstable due to the periodic external mechanical drive. Indeed, in the spectrum of wavenumbers there always exist values satisfying Eq. (26). For a disturbance $\Omega = \omega$, the last term in the right-hand side of Eq. (29) is of resonant character and the absence of secular solutions yields

$$\frac{\mathrm{d}f}{\mathrm{d}t_2} + 2\nu k^2 f - \frac{ik^3 Ma}{2\lambda\omega} f + \frac{ib_2 k}{4\omega} h = 0 \tag{34a}$$

$$\frac{\mathrm{d}h}{\mathrm{d}t_2} + 2\nu k^2 h + \frac{ik^3 Ma}{2\lambda^*\omega} h - \frac{ib_2 k}{4\omega} f = 0 \tag{34b}$$

In contrast to Eq. (30), here the amplitudes f and h, are no longer independent and have therefore a common growth rate:

$$\Lambda = \frac{k^3 Ma}{2\omega}\sqrt{\frac{\chi}{2\omega}} - 2\nu k^2 \pm \sqrt{\left(\frac{b_2 k}{4\omega}\right)^2 - \left(\frac{k^3 Ma}{2\omega}\sqrt{\frac{\chi}{2\omega}}\right)^2} \tag{35}$$

From Eq. (35) it follows that when

$$Ma > Ma_1 = \frac{4\nu\omega}{k}\sqrt{\frac{2\omega}{\chi}} \tag{36}$$

the real part of the growth rate is positive for all vibration amplitudes, b, and thus the equilibrium is unstable. Then, whatever the vibration frequency, the value of the Marangoni number of Eq. (36) is greater than Ma_c [Eq. (33)]. These two values coincide only at $k = k_c$ [Eq. (32)] where $\omega^2 = 6/(5\sqrt{5})$.

Vibrations may indeed render the equilibrium unstable at $Ma < Ma_c$. Assuming for Eq. (35) that $\Lambda = 0$, let us define the critical vibration amplitude at which parametric waves are initiated with frequency ω_1, i.e., with half the imposed vibration frequency:

$$b_*^2 = 16\left[\frac{Ma^2\chi k^4}{8\omega} + \frac{\omega^2}{k^2}\left(2\nu k^2 - \frac{k^3 Ma}{2\omega}\sqrt{\frac{\chi}{2\omega}}\right)^2\right] \tag{37}$$

The relation (37) is meaningful as the stability boundary of the equilibrium only for Ma lower than the Ma values derived from Eq. (36). At $b > b_*$. Faraday ripples appear on the liquid surface.

C. Cooperative Combined Effect of Thermocapillarity and Vibration and Instability Region

Rewriting Eqs. (30) and separating real and imaginary parts of λ, we obtain

$$\frac{df}{dt_2} + \left(2\nu k^2 - \frac{k^3 Ma}{2\Omega}\sqrt{\frac{\chi}{2\Omega}}\right)f - \frac{ik^3 Ma}{2\Omega}\sqrt{\frac{\chi}{2\Omega}}f = 0 \tag{38a}$$

$$\frac{dh}{dt_2} + \left(2\nu k^2 - \frac{k^3 Ma}{2\Omega}\sqrt{\frac{\chi}{2\Omega}}\right)h - \frac{ik^3 Ma}{2\Omega}\sqrt{\frac{\chi}{2\Omega}}h = 0 \tag{38b}$$

From Eq. (38) it is readily seen that the Marangoni effect plays a dual role. For $Ma > 0$ (which corresponds to heating from the liquid side), and sufficiently large Marangoni numbers, there is "negative" friction because thermocapillarity apparently reduces dissipation [in the second terms in Eq. (38)]. Furthermore, for neutral disturbances, when the viscosity and the thermocapillary effect involved in the friction term compensate for one another, both amplitudes f and h vary with the frequency $(k^3 Ma/2\Omega)\sqrt{\chi/2\Omega}$. Accordingly, neutral oscillations of this kind are oscillations with the frequency of Eq. (26) modulated in time with a small correction, as earlier noted. In this situation the amplitudes f and h are independent and hence the linear analysis fails to discriminate neutral disturbances as they may involve either stationary or running waves with frequency Ω and wavenumber k.

Let us further analyze the case of resonance. One of the amplitudes is easily eliminated from the system [Eq. (34)] and we readily obtain an evolution equation for, say, f. We have

$$\frac{d^2 f}{dt_2^2} + 2\left(2\nu k^2 - \frac{k^3 Ma}{2\omega}\sqrt{\frac{\chi}{2\omega}}\right)\frac{df}{dt_2} + \left[\left(2\nu k^2 - \frac{ik^3 Ma}{2\omega}\sqrt{\frac{\chi}{2\omega}}\right)^2 + \left(\frac{k^3 Ma}{2\omega}\sqrt{\frac{\chi}{2\omega}}\right)^2 - \left(\frac{b_2 k}{4\omega}\right)^2\right]f = 0 \tag{39}$$

which is the equation for a harmonic oscillator with friction. From Eq. (39) it is readily seen that here both mechanisms of instability are present. The thermocapillary effect, as in the nonresonant case, leads, at sufficiently high Marangoni numbers, to an apparent negative friction, i.e., the bracketed expression of the damping term in Eq. (39) becomes zero or negative. This mechanism of instability is unaffected by vibrations since the above equation does not include their amplitude. When the Marangoni number exceeds the value Ma_1 [Eq. (35)], instability occurs because of the apparent negative friction. Note that whatever the vibration frequency, Ma_1 is higher than the critical value, Ma_c, from Eq. (33), except when $\omega = (k_c + k_c^3)^{1/2}$, when they are equal. Hence vibrations are unable to stop the instability growth for $Ma > Ma_c$. However, as Eq. (39) suggests, they may be responsible for instability at $Ma < Ma_c$. In this case the mechanism of instability is essentially different from that found in the previous case, i.e., a sufficiently large amplitude of pumping, not affecting the friction, makes the oscillator frequency in Eq. (39) imaginary and at $b_2 > b_*$ [Eq. (37)] the flat liquid surface is unstable even when $Ma < Ma_c$.

For a more detailed analysis of the neutral curve it is convenient to put it in the form

$$\alpha = \left(\frac{k}{\omega}\right)\left[\frac{m^2k^2}{\omega^3} + \left(1 - \frac{mk}{\omega^{3/2}}\right)\right]^{1/2} \tag{40}$$

where use has been made of a newly modified vibration amplitude, α, and a redefined Marangoni number, m:

$$\alpha = \frac{b}{2\nu}, \qquad m = \frac{Ma}{4\nu}\sqrt{\frac{\chi}{2}} \tag{41}$$

With such parameters the neutral curves become universal for all liquids.

In Fig. 1 the critical vibration amplitude, α, is plotted against the new Marangoni number, m, at fixed wavenumber, k, i.e., at given corresponding vibration frequency. The vertical line depicts the threshold of the surface tension gradient (Marangoni) driven motions. The nonresonant region of thermocapillary-induced oscillatory motions lies to the right of the vertical line

$$m = m_c = \frac{6^{3/4}}{5^{5/8}} \approx 1.4 \tag{42}$$

The resonant neutral curves, $\alpha(m)$, in this region are depicted by broken lines. The function $\alpha(m)$ is nonmonotonic. At $m = 0$ the critical value is $\alpha_F = k/\omega$. With the appropriate notation this value coincides with the result known for Faraday ripples. With increasing value of m the threshold of parametric waves first decreases and then after reaching a minimum it grows. This can be

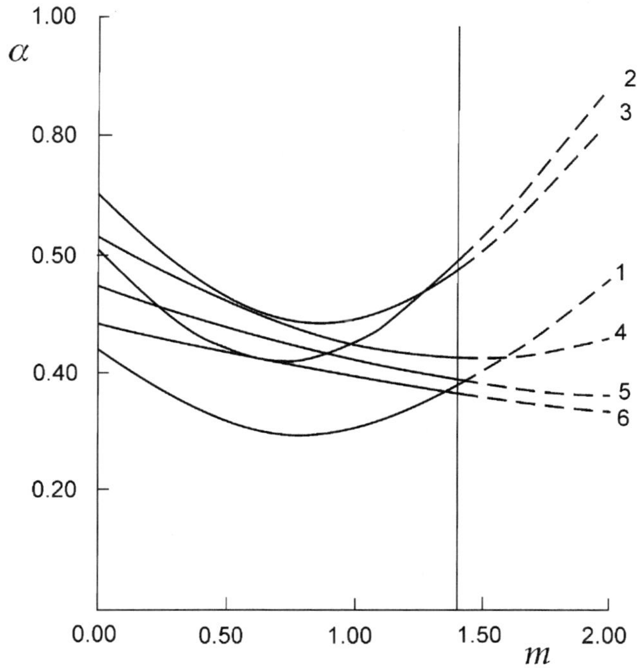

FIG. 1 Interaction between parametric wave excitation and overstability due to the Marangoni effect. Variation of the critical vibration amplitude with the (redefined) Marangoni number, m, for various resonant wavenumbers (and corresponding vibration frequencies): 1, $k = 0.2$; 2, $k = 1/\sqrt{5}$; 3, $k = 1$; 4, $k = 2$; 5, $k = 3$; and 6, $k = 4$. (From Ref. 1.)

readily seen in Fig. 1 for curves 1, 2, and 3, corresponding to $k = 0.2$, $1/\sqrt{5}$, and 1, respectively. Note that the minimum threshold amplitude is equal to $\alpha_F/\sqrt{2}$. Thus the lowering in the threshold for Faraday ripples caused by the thermocapillary effect is large enough to be detected by experiment.

When k is further increased (curves 4, 5, 6 for $k = 2, 3, 4$) the minimum is shifted to the region $m > m_c$ where instability is of nonresonant character. Hence, at vibration frequencies corresponding to such wavenumbers the function $\alpha(m)$ is nonmonotonic in the region of nonresonant instability. Formally, at $m = m_1 = \omega^{3/2}/k$ the critical vibration amplitude grows to the value corresponding to $m = 0$. However, as already mentioned, $m_1 > m_c$ for all frequencies except for frequencies corresponding to $k = 1/\sqrt{5}$, which is the critical wavenumber for thermocapillary instability (curve 2 in Fig. 1 intersects the straight lines $m = 0$ and $m > m_c$ at the same values of α).

To give proper explanation of the results obtained let us concentrate again on Eq. (39). Here thermocapillarity has a twofold effect. On the one hand, it apparently reduces friction, which results in the initial lowering of the critical vibration level. On the other hand, from Eq. (39) it follows that thermocapillarity increases, through the spring constant, the "elasticity" of the system, thus leading to growth of the critical amplitude of pumping. An interaction between these two mechanisms accounts for the nonmonotonic behaviour of the function $\alpha(m)$. Furthermore, the dependence of α on the vibration frequency is also nonmonotonic. For the smaller frequencies ($k = 0.2$, curve 1 in Fig. 1) the critical amplitude is the lowest. With k increasing (curve 2, $k = 1/\sqrt{5}$, and curve 3, $k = 1$) the critical amplitude grows and then at high vibration frequencies, and corresponding wavenumbers, it begins to decrease (curves 4, 5, 6). This effect can be clearly demonstrated by using alternative plots. Figures 2 and 3 give $\alpha(k)$ and the related $\alpha(\omega)$ function at fixed values of the Marangoni number [recall that in the resonant case ω and k satisfy Eq. (26)].

These figures readily demonstrate the effect of the nonmonotonic dependence of the critical vibration amplitude on the frequency. Curve 1 in Figs

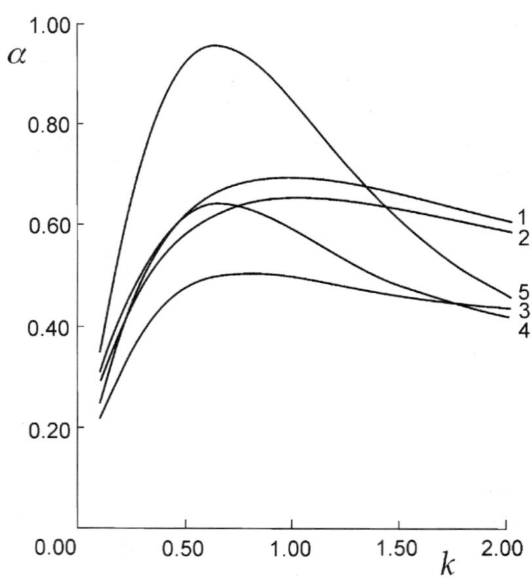

FIG. 2 Interaction between parametric wave excitation and overstability due to the Marangoni effect. Critical vibration amplitude vs. wavenumber at fixed values of the (redefined) Marangoni number: 1, $m = 0$; 2, $m = 0.2$; 3, $m = 1$; 4, $m = 1.4$; and 5, $m = 2$. (From Ref. 1.)

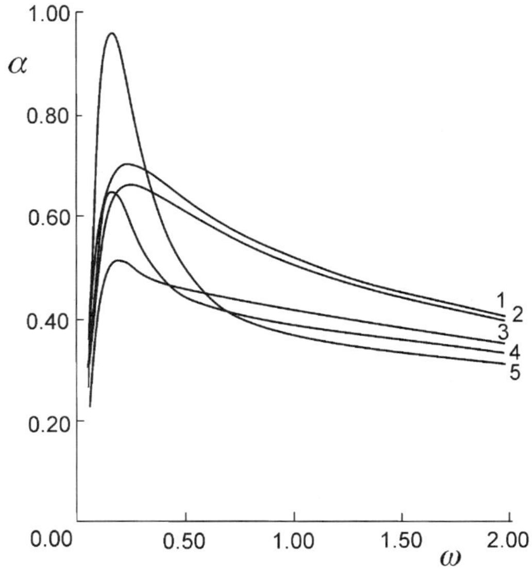

FIG. 3 Interaction between parametric wave excitation and overstability due to the Marangoni effect. Critical vibration amplitude vs. frequency at different values of the (redefined) Marangoni number: 1, $m = 0$; 2, $m = 0.2$; 3, $m = 1$; 4, $m = 1.4$; and 5, $m = 2$. (From Ref. 1.)

2 and 3 corresponds to isothermal Faraday ripples. Curves 2, 3, and 4 lie below curve 1 and hence, in the resonance region, the excitation threshold of parametric waves in a nonisothermal liquid is always lower than the standard isothermal threshold. Note that the curves 1 ($m = 0$) and 4 ($m = m_c = 1.4$) are tangent at $k = 1/\sqrt{5}$, as discussed earlier. Curve 5 ($m = 2$) in Figs 2 and 3 is above the isothermal curve 1 that formally corresponds to the stabilizing influence of the Marangoni effect on the excitation of parametric waves. Note, however, that at $m = 2$ the liquid surface is unstable to thermocapillary waves, whatever the level of vibration.

The neutral curves, $m(k)$, for different fixed values of the vibration amplitude, α, are presented in Fig. 4. Curve 1 refers to nonresonant disturbances. Following Eq. (24) the equation describing this curve is

$$m = \frac{\Omega^{3/2}}{k} \tag{43}$$

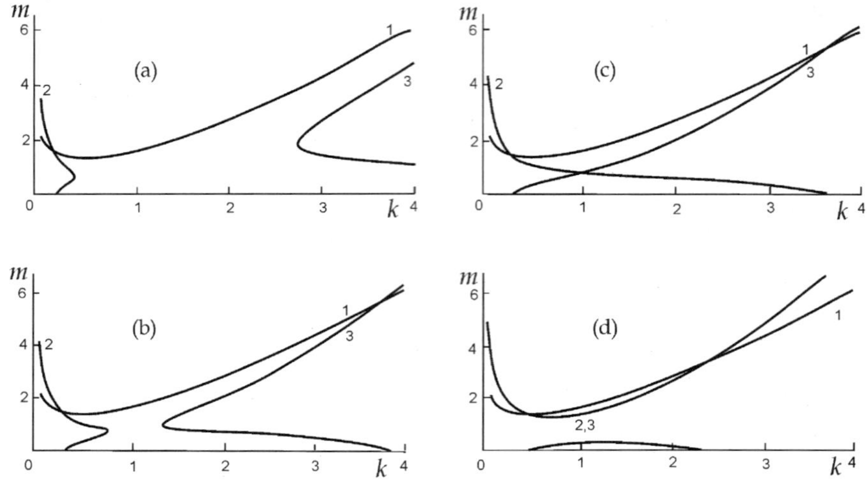

FIG. 4 Interaction between parametric wave excitation and overstability due to the Marangoni effect. Stability domain of the plane liquid surface. Marangoni number vs. wavenumber at fixed values of vibration amplitude: (a) $\alpha = 0.3$; (b) $\alpha = 0.4$; (c) $\alpha = 0.49$; and (d) $\alpha = 0.5$. (From Ref. 1.)

Above curve 1 the liquid surface is unstable to surface tension gradient (Marangoni) driven waves. This curve is common for all figures of the series 4 because it is independent of the vibration level.

The other curves are obtained from Eq. (31) using

$$m = \frac{\omega^{3/2}}{k}\left[\frac{1}{2} \pm \sqrt{\frac{1}{2}\left(\frac{\alpha\omega}{k}\right)^2 - \frac{1}{4}}\right] \tag{44}$$

For these curves the wavenumber of disturbances is a parameter related to the vibration frequency [Eq. (26)]. Figure 4(a) corresponds to $\alpha = 0.3$. Parametric resonance occurs on curve 2 (for low frequencies and the corresponding long waves) and curve 3 (for high frequencies and the corresponding short waves). The flat liquid surface is stable between curves 1, 2, 3 and the k-axis. The appearance of two regions of resonant instability is related to the nonmonotonic dependence of the critical vibration amplitude on the frequency (and wavenumber) as indicated in Figs 2 and 3. For a given vibration level (see for example Fig. 3), instability exists in the range of short and long waves while in the range of intermediate frequency values there is no unstable region. In Fig. 4(b) the vibration level is $\alpha = 0.4$. Both parametric instability regions (to the left of curve 2 and to the right of curve 3, respectively) approach each other, hence reducing the stability region relative to the case $\alpha = 0.3$. As α further increases, the short and long wave instability regions come closer (Fig. 4(c), $\alpha = 0.49$). Then at $\alpha = 0.5$ the instability regions merge [Fig. 4(d)] and the liquid surface becomes parametrically unstable for all vibration frequencies. As α continues to increase the stability region disappears. The vibration amplitude is so strong that parametric waves are excited whatever the value of the vibration frequency and the Marangoni number.

Another peculiarity of parametric waves is worth noting. The amplitudes f and h are related as on the neutral curve such that

$$h = ie^{i\varphi}f \tag{45a}$$

with

$$\varphi = \arg\left[2\nu k^2 - (1+i)\frac{k^2 Ma}{2\omega}\sqrt{\frac{\chi}{2\omega}}\right] \tag{45b}$$

From Eq. (45) it follows that for neutral disturbances, f and h have equal absolute value. This implies that neutral disturbances in this case must necessarily be of standing wave character, with the wavelength determined by Eq. (26).

Finally, Eq. (40) gives the growth of α at $m < 0$, but this result is of formal character because at $m < 0$ (liquid is cooled from above) another instability mechanism, not related to surface deformation, comes into play. In the absence of vibrations this instability has a long wave character; the most dangerous are perturbations with $k = 0$ and for the case considered (infinite depth) the critical Marangoni number vanishes. This no-threshold result is caused by neglecting gravity and, consequently, the hydrostatic contribution in the normal stress balance, thus leading to the capillary length, l, diverging to infinity. In the presence of vibrations this mode of instability may be analyzed using the method described above by requiring that the surface disturbances be of higher order than the velocity disturbances. It can be shown that vibrations change the neutral curve, but do not affect the critical values of the Marangoni number and the wavenumber. We have earlier described a similar effect of high-frequency suppression of Rayleigh–Taylor instability [3].

Thus it may be said that when heating from the liquid side, the equilibrium is unstable for arbitrary small values of the temperature gradient. A large scale flow is initiated, leaving the equilibrium surface flat. Equation (29) provides only a qualitative estimate of conditions responsible for the onset of Faraday ripples on the background of an already existing thermocapillary flow.

II. PARAMETRIC EXCITATION OF THERMOCAPILLARY INSTABILITY DUE TO A MODULATED HEAT FLUX

Let us now consider the problem of surface tension gradient (Marangoni) instability, parametrically excited with a time-periodic heat flux on the free surface.

A. Formulation of the Problem

Let us, for simplicity, consider a horizontal semi-infinite liquid volume, in other words, a liquid layer unbounded below, $0 < z < \infty$, with a free surface at $z = 0$ (z is taken positive downwards). In the absence of buoyancy (the thermal expansion coefficient is set to zero), at this surface we impose an orthogonal heat flux harmonically varying with time but otherwise uniform along the surface:

$$\kappa\left(\frac{\partial T}{\partial z}\right)_{z=0} = Q_0 \cos \omega t \tag{46}$$

where Q_0 and ω are the amplitude and the frequency of the heat flux oscillation, respectively, and κ denotes the heat conductivity of the liquid. The temperature inside the liquid, at large distances from the free surface, is chosen as the reference temperature:

$$(T)_{z\to\infty} = 0 \tag{47}$$

Under such heating conditions, mechanical equilibrium is possible with an unsteady temperature field, $T_0(z, t)$, satisfying the one-dimensional heat diffusion equation:

$$\frac{\partial T_0}{\partial t} = \chi \frac{\partial^2 T_0}{\partial z^2} \tag{48}$$

with χ denoting the heat diffusivity of the liquid.

The distribution of temperature and its gradient can be written in the following form:

$$T_0 = -\frac{Q_0 \delta}{2\kappa}[\cos(\omega t - z/\delta) + \sin(\omega t - z/\delta)]\exp(-z/\delta) \tag{49a}$$

$$\frac{\partial T_0}{\partial z} = \frac{Q_0}{\kappa}\cos(\omega t - z/\delta)\exp(-z/\delta), \qquad \delta = \sqrt{2\chi/\omega} \tag{49b}$$

The solution of Eq. (48) yields a temperature boundary layer or "skin-layer" with characteristic length scale δ.

Let us now consider the stability of the unsteady mechanical equilibrium of Eq. (49). Let ($\mathbf{v}$, θ, p', ζ) denote, respectively, small disturbances of velocity, temperature, pressure, and surface deformation. These disturbances obey the following evolution equations [4–7]:

$$\operatorname{div} \mathbf{v} = 0 \tag{50}$$

$$\frac{\partial \mathbf{v}}{\partial t} = -\frac{1}{\rho} \nabla p' + \nu \Delta \mathbf{v} \tag{51}$$

$$\frac{\partial \theta}{\partial t} + \mathrm{v}_z \frac{\partial T_0}{\partial z} = \chi \Delta \theta \tag{52}$$

The problem is isotropic in the plane of the undeformed liquid surface, hence disturbances depend on the x-coordinate only. The subscript z denotes the vertical velocity component. In what follows only the (x, z) two-dimensional problem is considered.

For the boundary conditions, first, we assume that the normal heat flux on the disturbed surface is fixed by the condition:

$$\kappa \left(\frac{\partial T}{\partial n} \right)_\zeta = Q_0 \cos \omega t \tag{53}$$

with

$$\left(\frac{\partial T}{\partial n} \right)_\zeta = \left(\frac{\partial T}{\partial x} \right)_\zeta n_x + \left(\frac{\partial T}{\partial z} \right)_\zeta n_z$$

where n_x and n_z are components of the unit vector normal to the surface. In the linear approximation, we have

$$n_x = \frac{\partial \zeta}{\partial x}, \qquad n_z = 1 \tag{54a}$$

$$(T)_\zeta = (T)_0 + \left(\frac{\partial T}{\partial z} \right)_0 \zeta = (T_0 + \theta)_0 + \left(\frac{\partial T_0}{\partial z} \right)_0 \zeta + \cdots \tag{54b}$$

$$\left(\frac{\partial T}{\partial x} \right)_\zeta = \left(\frac{\partial \theta}{\partial x} \right)_0 + \left(\frac{\partial T_0}{\partial z} \right)_0 \frac{\partial \zeta}{\partial x} + \cdots \tag{54c}$$

$$\left(\frac{\partial T}{\partial n} \right)_\zeta = \left(\frac{\partial T}{\partial z} \right)_\zeta = \left(\frac{\partial T_0}{\partial z} \right)_0 + \left(\frac{\partial \theta}{\partial z} \right)_0 + \left(\frac{\partial^2 T_0}{\partial z^2} \right)_0 \zeta + \cdots \tag{54d}$$

Equations (46) and (53) impose that the disturbances of the temperature gradient on the open surface vanish. In the first order in ζ we may write

$$\left(\frac{\partial \theta}{\partial z} \right)_0 = -\left(\frac{\partial^2 T_0}{\partial z^2} \right)_0 \zeta \tag{55}$$

Using Eq. (50) for the undisturbed equilibrium gradient, we obtain

$$z = 0: \frac{\partial \theta}{\partial z} = \frac{Q_0}{\kappa \delta}(\cos \omega t - \sin \omega t)\zeta \tag{56}$$

The vertical component of velocity obeys the kinematic condition

$$\mathrm{v}_z = \frac{\partial \zeta}{\partial t} \tag{57}$$

Once more, if the surface temperature variation is small, the surface tension σ is a linear function of the temperature

$$\sigma = \sigma_0 - \sigma_1 T, \qquad \sigma_1 > 0 \tag{58}$$

However, when considering the case of strong heating at the surface, we will consider that σ is the following function of the temperature:

$$\sigma = \sigma_0 \exp\left(-\frac{\sigma_1}{\sigma_0} T\right) \tag{59}$$

which for small $(\sigma_1 T/\sigma_0)$ yields Eq. (58).

In line with the preceding assumptions, the boundary conditions for tangential and transverse stresses may be linearized at the deformable surface. For the strong heating case:

$$\sigma = \sigma_0 \exp\left(-\frac{\sigma_1}{\sigma_0} T_0\right)\left[1 - \frac{\sigma_1}{\sigma_0}\left(\theta - \frac{\partial T_0}{\partial z}\zeta\right)\right] \tag{60}$$

and then the balance of viscous and thermocapillary forces on the free surface for tangential stresses yields

$$\eta\left(\frac{\partial^2 \mathrm{v}_z}{\partial z^2} - \frac{\partial^2 \mathrm{v}_z}{\partial x^2}\right) = -\sigma_1 \exp\left(-\frac{\sigma_1 T_0}{\sigma_0}\right)\frac{\partial^2}{\partial x^2}\left(\theta + \frac{\partial T_0}{\partial z}\zeta\right) \tag{61}$$

while the normal stress balance is

$$p + \rho g \zeta - \sigma_0 \exp\left(-\frac{\sigma_1 T_0}{\sigma_0}\right)\frac{\partial^2 \zeta}{\partial x^2} = 2\eta \frac{\partial \mathrm{v}_z}{\partial z} \tag{62}$$

The second and third terms in the left-hand side of Eq. (62) are the hydrostatic pressure and the Laplace overpressure, respectively, both a result of the deformation ζ. Note that buoyancy has been neglected, but not gravity.

For a semi-infinite layer, in accordance with the assumptions given earlier we consider that disturbances decay at a large distance from the open surface, such that

$$(\mathbf{v}, \theta)_{z \to \infty} \to 0 \tag{63}$$

To formulate the problem in dimensionless form we here use the following units: δ for length; $\delta^2/\chi = 2/\omega$ for time; χ/δ for the velocity; $\eta\chi/\delta^2$ for the pressure; and $\Theta = Q_0\delta/\kappa$ for the temperature. We obtain the following dimensionless amplitude problem [6]:

$$\text{div } \mathbf{v} = 0 \tag{64}$$

$$\frac{1}{Pr}\frac{\partial \mathbf{v}}{\partial t} = -\nabla p + \Delta \mathbf{v} \tag{65}$$

$$\frac{\partial \theta}{\partial t} + f_1 \mathrm{v}_z = \Delta\theta \tag{66}$$

together with

$$z = 0 : \mathrm{v}_z = \frac{\partial \zeta}{\partial t} \tag{67}$$

$$p + Ga\zeta - Ca \exp(Maf_2(t)/Ca)\frac{\partial^2 \zeta}{\partial x^2} = 2\frac{\partial \mathrm{v}_z}{\partial z} \tag{68}$$

$$\left(\frac{\partial^2 \mathrm{v}_z}{\partial z^2} - \frac{\partial^2 \mathrm{v}_z}{\partial x^2}\right) = -Ma \exp(Maf_2(t)/Ca)\frac{\partial^2}{\partial x^2}(\theta + \zeta \cos 2t) \tag{69}$$

$$\frac{\partial \theta}{\partial z} = (\cos 2t - \sin 2t)\zeta \tag{70}$$

$$z \to \infty : \theta = 0, \mathbf{v} = 0 \tag{71}$$

where $f_1 = \exp(-z)\cos(2t - z)$ and $f_2 = -T_0(0, t) = (\cos 2t + \sin 2t)/2$ are the dimensionless time-periodic equilibrium temperature gradient in the liquid and the surface temperature, respectively.

The dimensionless parameters of the problem are the familiar Marangoni number, $Ma = \sigma_1\Theta\delta/\eta\chi$, Prandtl number, $Pr = \nu/\chi$, capillary number $Ca = \sigma_0\delta/\eta\chi$, and Galileo number, $Ga = g\delta^3/\nu\chi$.

Let us analyze the periodic coefficient $\exp(Maf_2/Ca)$, where f_2 is a time-periodic function with amplitude of order unity, which appears in the boundary conditions for tangential viscous stresses. This coefficient describes the change of the surface tension connected with a temperature change uniformly along the surface. In the case of constant heating we may choose the temperature on the surface as the reference temperature, and then f_2 is equal to unity. For parametric excitation, if $Ma/Ca \gg 1$ this quantity f_2 is large over a certain interval of time, hence the possibility of thermocapillary instability even without surface deformation.

Eliminating the horizontal velocity component, v_x, and pressure disturbance, p, we can formulate the problem in terms of the vertical velocity component, v_z, the temperature disturbance, θ, and the deformation, ζ, by again using Fourier normal modes,

$$\{v_z, \theta, \zeta\} = \{w(z, t), \vartheta(z, t), \xi(t)\} \exp(ikx) \tag{72}$$

where k is the wavenumber. We obtain the boundary problem for the amplitudes w, ϑ, and ξ,

$$\frac{1}{Pr}\frac{\partial}{\partial t}\left(\frac{d^2 w}{dz^2} - k^2 w\right) = \frac{d^4 w}{dz^4} - 2k^2\left(\frac{d^2 w}{dz^2}\right) + k^4 w \tag{73}$$

$$\frac{\partial \vartheta}{\partial t} + f_1 w = \frac{d^2 \vartheta}{dz^2} - k^2 \vartheta \tag{74}$$

$$z = 0: \frac{d\vartheta}{dz} = (\cos 2t - \sin 2t)\xi, \qquad w = \frac{\partial \xi}{\partial t} \tag{75}$$

$$\frac{d^2 w}{dz^2} + k^2 w = k^2 Ma(\vartheta + \cos 2t\xi)\exp\left(\frac{Ma f_2(t)}{Ca}\right) \tag{76}$$

$$\frac{1}{Pr}\frac{\partial^2 w}{\partial t \partial z} - \left[\frac{d^3 w}{dz^3} - 3k^2\frac{dw}{dz}\right] - Gak^2\xi - Cak^4 \exp\left(\frac{Ma f_2(t)}{Ca}\right)\xi = 0 \tag{77}$$

$$z \to \infty: w = \frac{dw}{dz} = 0, \qquad \vartheta = 0 \tag{78}$$

The system of equations, boundary conditions (73)–(78), and the time periodicity for all the variables define the eigenvalue problem (Floquet theory or a quasi-steady approximation are to be used) and provide the threshold value of the Marangoni number, Ma, as a function of the other parameters, Pr, Ca, Ga, and the wavenumber k. Two types of periodic solution are possible. The first has a period twice the period of the heat flux (half the frequency of the drive) on the boundary ("half-integer" solutions). The other has the same period as the heat flux ("integer" solutions). The minimum of the function $Ma(k)$ at fixed values of other parameters determines the critical value of Ma for the parametric excitation of Marangoni instability.

B. Solution of the Problem

The periodic solutions of the eigenvalue problem can be Fourier analysed:

$$w(z, t) = \sum_m w_m(z)\exp(imt)$$

$$\vartheta(z, t) = \sum_m \vartheta_m(z)\exp(imt)$$

$$\xi(t) = \sum_m \xi_m \exp(imt) \tag{79}$$

where $m = \pm 1, \pm 3, \ldots$ stands for the "half-integer" mode and $m = 0, \pm 2, \pm 4, \ldots$ for the "integer" mode. Substituting Eq. (79) into Eq. (73), we obtain an infinite system of independent ordinary differential equations for the amplitudes, $w_m(z)$,

$$\frac{im}{Pr}\left(\frac{\mathrm{d}^2 w_m}{\mathrm{d}z^2} - k^2 w_m\right) = \frac{d^4 w_m}{\mathrm{d}z^4} - 2k^2\left(\frac{\mathrm{d}^2 w_m}{\mathrm{d}z^2}\right) + k^4 w_m \tag{80}$$

Owing to the decay of disturbances at large distances from the open surface [Eq. (78)], we can write the solution of this system as

$$w_0 = X_0 \exp(-kz) - \frac{Y_0}{2k} z \exp(-kz)$$

$$w_m = X_m \exp(-kz) - \frac{\mathrm{Pr}\, Y_m}{im} \exp(-r_m z), \qquad m \neq 0 \tag{81}$$

with $r_m = \sqrt{k^2 + im/Pr}$, and where X_m and Y_m are coefficients yet to be determined. The relationship between the amplitude of the surface deformation and that of the vertical velocity on the surface leads to the following relations:

$$\xi_0 = 0, \qquad m \neq 0 : \xi_m = w_m(0)/im \tag{82}$$

Substitution of the expressions (81) and (82) into Eq. (74) and boundary conditions for temperature disturbances [Eqs. (75) and (78)], allows us to formally solve the infinite system of differential equations for the amplitudes, $\vartheta_m(z)$. We have

$$\begin{aligned}\vartheta_m = &\frac{1}{2}\{\exp(-s_m z)[A_m X_{m-2} + B_m Y_{m-2} + C_m X_{m+2} + D_m Y_{m+2}]\\ &+ \exp[-z(1 + k + i)]E_m X_{m-2} + \exp[-z(1 + i + r_{m-2})]F_m Y_{m-2}\\ &+ \exp[-z(1 + k - i)]G_m X_{m+2}\\ &+ \exp[-z(1 - i + r_{m+2})]H_m Y_{m+2}\}, \qquad m \neq 2\end{aligned} \tag{83}$$

where

$$s_m = \sqrt{k^2 + im}, \qquad A_m = -\frac{1}{s_m}\left[\frac{1+k+i}{(1+k+i)^2 - s_m^2} + \frac{1-i}{m-2}\right] \tag{84a}$$

$$B_m = \frac{1}{s_m}\left[\frac{iPr(1+i+r_{m-2})}{(m-2)[(1+i+r_{m-2})^2 - s_m^2]} + \frac{Pr(1+i)}{(m-2)^2}\right] \tag{84b}$$

$$C_m = -\frac{1}{s_m}\left[\frac{1+k-i}{(1+k-i)^2 - s_m^2} - \frac{1+i}{m+2}\right] \tag{84c}$$

$$D_m = \frac{1}{s_m}\left[\frac{iPr(1-i+r_{m+2})}{(m+2)[(1-i+r_{m+2})^2 - s_m^2]} + \frac{Pr(1-i)}{(m+2)^2}\right] \tag{84d}$$

$$E_m = \left[(1+i+k)^2 - s_m^2\right]^{-1} \tag{84e}$$

$$F_m = -\frac{iPr}{(m-2)\left[(1+i+r_{m-2})^2 - s_m^2\right]} \tag{84f}$$

$$G_m = [(1-i+k)^2 - s_m^2] \tag{84g}$$

$$H_m = -\frac{iPr}{(m+2)[(1-i+r_{m+2})^2 - s_m^2]} \tag{84h}$$

If $m = \pm 2$, the amplitudes ϑ_m are

$$\begin{aligned}\vartheta_2 = \frac{1}{2}\{&\exp(-s_2 z)[B_2 Y_0 + C_2 X_4 + D_2 Y_4] \\ &+ \exp[-z(1+i+k)]\left[\frac{i(1+i+k)}{8k^3} - \frac{z}{4k^2(1+i)}\right]Y_0 \\ &+ \exp[-z(1-i+k)]G_2 X_4 + \exp[-z(1-i+r_4)]F_2 Y_4\}\end{aligned} \tag{85a}$$

$$\begin{aligned}\vartheta_{-2} = \frac{1}{2}\{&\exp(-s_{-2} z)[A_{-2} Y_{-4} + B_{-2} Y_{-4} + D_{-2} Y_0] \\ &+ \exp[-z(1-i+k)]\left[\frac{-i(1-i+k)}{8k^3} - \frac{z}{4k^2(1-i)}\right]Y_0 \\ &+ \exp[-z(1+i+k)]E_{-2} X_{-4} \\ &+ \exp[-z(1+i+r_{-4})]F_{-2} Y_{-4}\}\end{aligned} \tag{85b}$$

where the coefficients $C_2, D_2, G_2, H_2, A_{-2}, B_{-2}, E_{-2}$, and F_{-2} are defined in Eq. (84) and

$$B_2 = -\left(\frac{1-i}{8k^2} + \frac{(1+i+k)^2}{8k^3}\right), \qquad D_{-2} = (B_2)^*$$

We expand the periodic function $\exp(Ma f_2(t)/Ca)$ in the boundary conditions (76) and (77) into Fourier series such that

$$\exp(Ma f_2(t)/Ca) = \sum_n \gamma_n \exp(2int) \tag{86}$$

with the coefficients γ_n being expressed by means of modified Bessel functions $I_n(s)$ [8]:

$$\gamma_n = \left(\frac{-1+i}{\sqrt{2}}\right)^n I_n(s) \tag{87}$$

$$I_n(s) = \sum_{k=0}^{\infty} \left(\frac{s}{2}\right)^{2k+n} \frac{1}{k!(k+n)!}, \qquad s = -\frac{\sqrt{2}\,Ma}{2\,Ca}$$

Substituting Eqs. (82) and (86) into Eqs. (76) and (77) yields a condition relating $w_m(0)$ and $\vartheta_m(0)$:

$$\frac{\mathrm{d}^2 w_m}{\mathrm{d}z^2}(0) + k^2 w_m(0) = k^2 Ma \sum_{k=-\infty}^{\infty} \gamma_{m-2k}\left(\vartheta_k(0) + \frac{w_{k-2}}{2i(k-2)} + \frac{w_{k+2}}{2i(k+2)}\right) \tag{88}$$

and

$$\frac{1}{Pr}\frac{\partial^2 w}{\partial t \partial z} - \left[\frac{\mathrm{d}^3 w_m}{\mathrm{d}z^3} - 3k^2 \frac{\mathrm{d}w_m}{\mathrm{d}z}\right] - Ga\,k^2 \frac{w_m}{im} - Ca\,k^4 \sum_{k=-\infty}^{\infty} \gamma_{m-2k}\frac{w_k}{ik} = 0$$

Substituting Eqs. (81), (83), and (85) into Eq. (88) brings an infinite system of linear homogeneous algebraic equations for the coefficients X_m and Y_m. This system can be split into two independent subsystems for "half-integer" and "integer" disturbances, respectively. Each subsystem has nontrivial solutions if its corresponding determinant is set to zero. This condition determines the eigenvalue $Ma(Pr, Ca, Ga, k)$. Retaining N harmonics for each mode yields a $2N \times 2N$ matrix of the coefficients for the unknowns X_m and Y_m. $N = 10$ is good enough, although 6–7 harmonics may often be sufficient.

C. Regions of Parametric Instability

Numerical computations show that there are two types of parametrically excited oscillations: parametric oscillations bearing no relation to the deformation of the free surface and oscillations for which the surface deformation is an essential factor. The shape of the neutral curves for the longitudinal oscillations (first type) strongly depends on the Prandtl number. Shown in Fig. 5 are the neutral curves for $Pr = 1$. The solid and dashed lines denote the stability boundaries for half-integer and integer modes, respectively (the instability region is that enclosed by the curves). A half-integer mode with wavenumber $k_m = 0.63$ gives the lowest

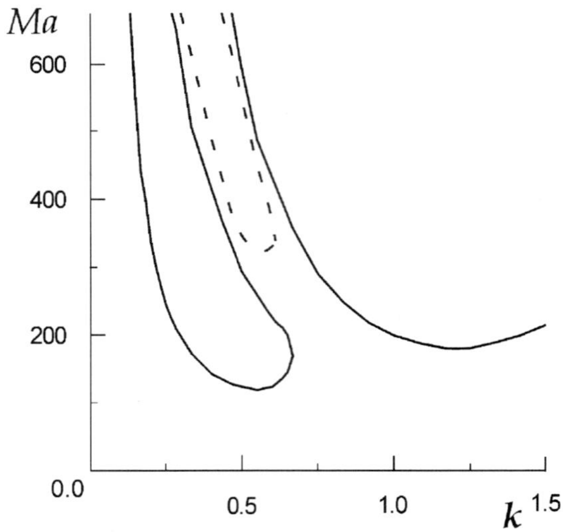

FIG. 5 Parametric excitation of the thermocapillary convective instability. Neutral curves for the longitudinal oscillations; $Pr = 1$. (From Ref. 7.)

Marangoni number $Ma_m = 110$. It is seen that the instability regions of integer and half-integer modes alternate. This situation is exceptional for longitudinal disturbances and can be observed only for liquids with Prandtl numbers close to unity. The region of low Prandtl numbers provides instability for the first half-integer mode excited in a wide range of the wavenumbers. For $Pr = 0.1$ the minimum critical Marangoni number of 176 is reached at $k_m = 0.97$. The region of large Prandtl numbers ($Pr > 5$) is characterized by integer mode instability, which is also excited over the whole range of wavenumbers. For $Pr = 7$ the corresponding values of the critical parameters are $Ma_m = 81$ and $k_m = 0.62$.

The parametric oscillations demanding surface deformation are transverse motions and their dispersion relation is indeed close to that of capillary-gravity waves: either

$$(Ga\,k + Ca\,k^3)Pr = 1 \tag{89}$$

for the half-integer mode, $\xi \sim \exp(i\omega t)$, or

$$(Ga\,k + Ca\,k^3)Pr = 4 \tag{90}$$

for the integer mode, $\xi \sim \exp(i2\omega t)$. The frequency of the drive is $2\omega = 2$.

Figure 6 shows neutral curves of parametric instability for a deformable surface at $Ca = 10^5$, $Ga = 0$, and $Pr = 0.1$. For the chosen values of parameters, very narrow regions of capillary parametric instability exist in the plane (k, Ma).

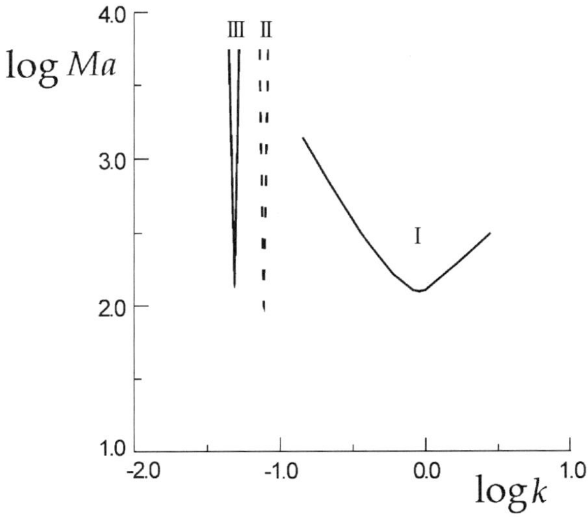

FIG. 6 Parametric excitation of the thermocapillary convective instability. Typical neutral curves for the transverse and longitudinal disturbances. I, longitudinal half-integer mode; II, transverse integer mode; and III, surface half-integer mode. (From Ref. 6.)

The "integer" disturbances grow in region II and the "half-integer" ones grow in region III. The critical Marangoni number in these regions strongly depends on the Galileo number. When the Galileo number increases (the surface deformation loses importance), the boundary of instability moves to infinity ($k \to 0$, $Ma \to \infty$).

The increase of the Prandtl number yields a narrowing of the parametric excitation regions and an increase of the corresponding critical Marangoni number (Fig. 7). These regions do not exist for an undeformable surface and hence this type of instability is the transverse instability.

The dependence of the critical Marangoni number on the Prandtl number for longitudinal and transverse modes of instability is shown in Fig. 8. The wavenumbers for surface modes are determined by the relation of Eqs. (89) or (90). At low Prandtl numbers ($Pr < Pr_A$) surface disturbances with frequency half of the drive are excited at lower Marangoni numbers than all other disturbances.

The most dangerous types of instability of the unsteady mechanical liquid equilibrium are different for different values of the Prandtl number, namely:

Transverse half-integer mode if $Pr < Pr_A$;
Transverse integer mode if $Pr_A < Pr < Pr_B$;
Longitudinal half-integer mode if $Pr_B < Pr < Pr_C$;
Longitudinal integer mode if $Pr > Pr_C$.

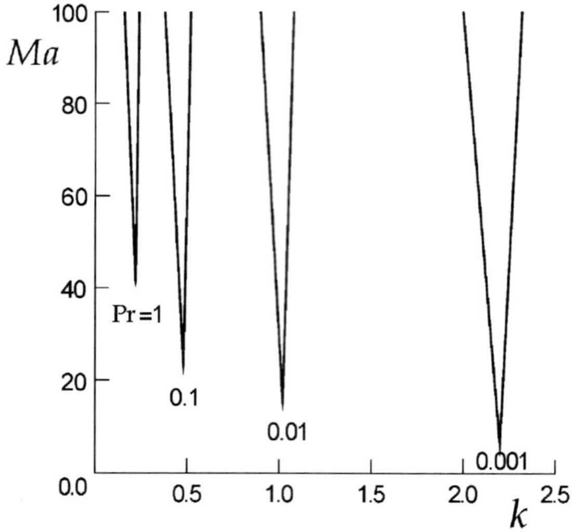

FIG. 7 Parametric excitation of the thermocapillary convective instability. Neutral curves of the surface parametric instability at different Prandtl numbers; $Ca = 10^2$. (From Ref. 6.)

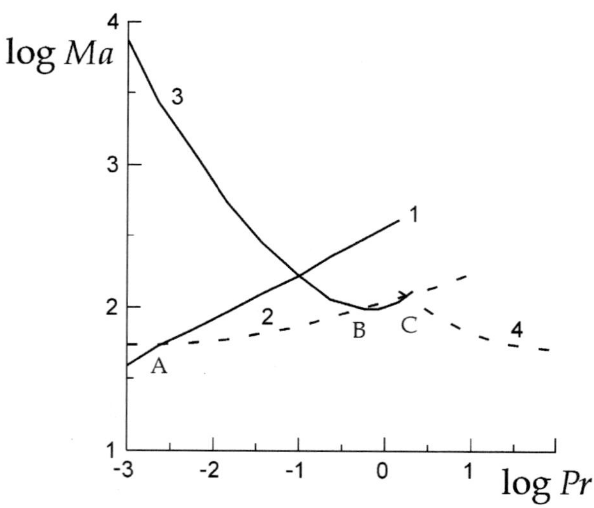

FIG. 8 Parametric excitation of the thermocapillary convective instability. Parameters of critical disturbances vs. Prandtl number; $Ca = 10^5$ and $Ga = 0$. Curves 1 and 2 are the instability boundaries for transverse half-integer and integer modes; 3 and 4 are the instability boundaries for longitudinal half-integer and integer modes, respectively. (From Ref. 6.)

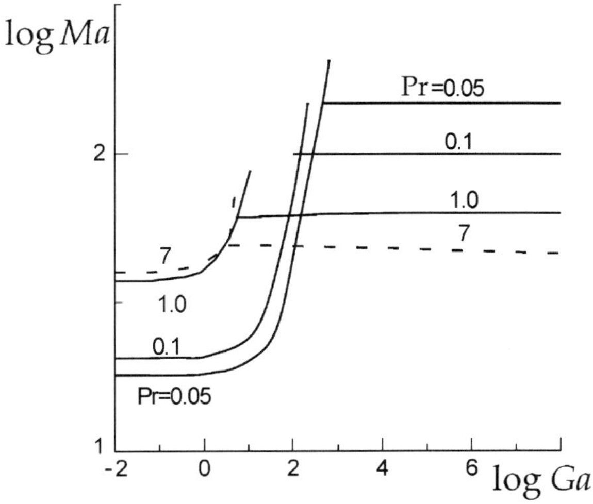

FIG. 9 Parametric excitation of the thermocapillary convective instability. Critical Marangoni number vs. Galileo number for several Prandtl numbers: $Pr = 0.05, 0.1, 1$, and 7; $Ca = 10^2$. (From Ref. 6.)

The competition of the longitudinal and transverse modes is shown in Fig. 9. At large values of the Galileo number, $Ga \to \infty$, the surface is flat for all practical purposes and the critical disturbances correspond to the longitudinal mode, whose corresponding critical Marangoni number is little if any affected by the decrease of Ga. If the Galileo number is lower than $Ga_*(Pr, Ca)$, the transverse instability mode is the most dangerous. The critical Marangoni number for the latter decreases as the Galileo number decreases.

REFERENCES

1. Birikh, R.V., Briskman, V.A., Cherepanov, A.A., Velarde, M.G. Faraday ripples, parametric resonance, and the Marangoni effect. J. Colloid Interface Sci. **2001**, *238*, 16–23.
2. Nayfeh, A.H. *Perturbation Methods*; Wiley: New York, 1973.
3. Drazin, P.G., Reid, W.H. *Hydrodynamic Stability*; Cambridge University Press: Cambridge, 1981.
4. Gershuni, G.Z., Nepomnyashchy, A.A., Velarde, M.G. On dynamic excitation of Marangoni instability. Phys. Fluids A **1992**, *4*, 2394–2398.
5. Gershuni, G.Z., Nepomnyashchy, A.A., Smorodin, B.L., Velarde, M.G. On parametric excitation of thermocapillary and thermogravitational convective instability. Microgravity Q. **1994**, *4*, 215–220.

6. Gershuni, G.Z., Nepomnyashchy, A.A., Smorodin, B.L., Velarde, M.G. On parametric excitation of Marangoni instability in a liquid layer with free deformable surface. Microgravity Q. **1996**, *6*, 203–209.
7. Gershuni, G.Z., Nepomnyashchy, A.A., Smorodin, B.L., Velarde, M.G. The dynamic excitation of Marangoni convection under the influence of varying heat flux on free surface. In *Vestnik Permskogo Universiteta*; 1995, Fizika, N 4; 17–28 (in Russian).
8. Yanke, E., Emde, F., Losch, F. *Tafeln Hoherer Functionen*; Teubner Verlag: Stuttgart, 1960.

13

Thermocapillary Instability of a Liquid Interface Under the Joint Action of High-Frequency Vibration and the Marangoni Effect

We considered earlier the resonant response of a nonisothermal liquid surface to a periodically variable force. In this chapter we shall study an averaged liquid response to high-frequency effects, i.e., to the phenomena resulting from the influence of high-frequency vibrations on the mechanical equilibrium of a nonisothermal liquid system with free surfaces. By high frequency we mean that the vibration period is considerably shorter than any characteristic time of the liquid layer. As shown earlier, vibrations increase the stability of an isothermal free liquid surface and help suppress Rayleigh–Taylor instability. The experiments described earlier showed that cross-vibrations suppressed thermocapillary flow caused by a longitudinal temperature gradient, and hence prevented rupture of the liquid layer.

I. SUPPRESSION OF MONOTONIC SURFACE TENSION GRADIENT (MARANGONI) DRIVEN INSTABILITY BY HIGH-FREQUENCY VIBRATIONS

Let us assume that in weightlessness conditions a liquid layer of depth d experiences harmonic oscillations with frequency Ω and displacement amplitude b in the direction of the vector $\mathbf{n}$. Then in the frame of reference related to the oscillating layer the vibrational acceleration

$$a = b\Omega^2 \cos \Omega t \tag{1}$$

plays the role of a modulated gravity field.

Although we have neglected gravity, there is buoyancy in the system due to inertial forces. Then, for the averaged components of velocity, **v**, temperature, T, and pressure, p, in dimensionless form we have:

$$\text{div}\ \mathbf{v} = 0, \qquad \text{div}\ \mathbf{w} = 0, \qquad \text{rot}(T\,\mathbf{n} - \mathbf{w}) \tag{2}$$

$$\frac{\partial \mathbf{v}}{\partial t} + \frac{1}{Pr}(\mathbf{v}\cdot\nabla)\mathbf{v} = -\nabla p + \Delta\mathbf{v} + Ra_{\text{v}}(\vec{\mathbf{w}}\cdot\nabla)(T\,\mathbf{n} - \mathbf{w}) \tag{3}$$

$$Pr\frac{\partial T}{\partial t} + (\mathbf{v}\cdot\nabla)T = \Delta T \tag{4}$$

$$Ra_{\text{v}} = \frac{(\alpha\Theta b\Omega d)^2}{\nu\chi}, \qquad Pr = \frac{\nu}{\chi}$$

where Ra_{v} is the vibrational analog of the thermal Rayleigh number, and Pr is the Prandtl number. To rescale length, time, velocity, pressure, and temperature we have adopted, respectively, the new units d, d^2/ν, χ/d, $\rho\nu\chi/d^2$, and Θ, where ρ, ν, χ, and α are density, kinematic viscosity, thermal diffusivity, and thermal expansion coefficient, and Θ is the characteristic temperature difference. The vector field $T\mathbf{n}$ is divided into two parts $T\mathbf{n} = \mathbf{w} + \nabla\phi$ where **w** is the amplitude of the pulsating component of the velocity field.

Because there is no isothermal stirring due to vibration, Eqs (2)–(4) may be used to treat the thermocapillary problem, with appropriate boundary conditions on the free surface. For the amplitude of pulsating velocity component it suffices to impose only one condition on the open boundary owing to the inviscid description of the pulsating motion. We have the following alternative: either (i) the boundary is undeformable, which demands a very high surface tension and, consequently, the normal component of the pulsation velocity is zero; or (ii) the liquid boundary is deformed by the pulsations with a pulsating pressure practically constant along the surface and, consequently, the tangential component of the pulsating velocity vanishes.

Because the averaged velocity component is characterized by its time scale, one may assume that the free surface is not deformed by the averaged flow. Then the normal velocity component on the surface must vanish. Furthermore, assuming that the surface tension depends on temperature in the usual way for standard liquids, then, due to the Marangoni effect, a tangential stress exists on the free surface in balance with the viscous stress only, but in terms of the inviscid approximation assumed for the pulsating flow the thermocapillary forces do not exert direct influence on the pulsation velocity. The tangential stress balance involves the Marangoni number, $Ma = \sigma_1\Theta d/\eta\chi$; $\eta = \rho\nu$. On the solid surfaces the averaged velocity and normal component of pulsating velocity vanish.

For a plane liquid layer confined between a hot solid bottom and having a cold free surface experiencing transverse oscillations there exists a steady solution of Eqs (1)–(4) in which the averaged and pulsating motion are absent and temperature is distributed according to Fourier's diffusion law.

The linear stability analysis of such a state demands consideration of the boundary conditions for disturbances. Let the coordinate system be such that the z-axis is aligned with the vector $\mathbf{n}$; $z = 0$ and $z = 1$ correspond, respectively, to the hot and cold layer boundaries.

Let us assume that on the free boundary the additional heat flux density from the surface is proportional to the temperature disturbance, whereas at the solid boundary two situations may be apply: (a) the temperature is held fixed or (b) heat flux is held fixed. Then the disturbances expressed as Fourier normal modes,

$$(\mathrm{v}_z, T, \mathrm{w}_z) = [\mathrm{v}(z), \theta(z), \mathrm{w}(z)] \exp[-\lambda t + i(k_x x + k_y y)]$$

obey the following system of differential equations and boundary conditions

$$\frac{\mathrm{d}^4\mathrm{v}}{\mathrm{d}z^4} - 2k^2\left(\frac{\mathrm{d}^2\mathrm{v}}{\mathrm{d}z^2}\right) + k^4\mathrm{v} + Ra_\mathrm{v}k^2\mathrm{w} = 0 \tag{5}$$

$$\frac{\mathrm{d}^2\theta}{\mathrm{d}z^2} - k^2\theta + \mathrm{v} = 0 \tag{6}$$

$$\frac{\mathrm{d}^2 w}{\mathrm{d}z^2} - k^2 w + k^2\theta = 0 \tag{7}$$

$$z = 0: \mathrm{v} = 0, \qquad \frac{\mathrm{d}\mathrm{v}}{\mathrm{d}z} = 0, \qquad w = 0, \qquad \text{(a) } \theta = 0, \qquad \text{(b) } \frac{\mathrm{d}\theta}{\mathrm{d}z} = 0 \tag{8}$$

$$z = 1: \mathrm{v} = 0, \qquad \frac{\mathrm{d}^2\mathrm{v}}{\mathrm{d}z^2} = -Ma\, k^2\theta, \qquad \frac{\mathrm{d}\theta}{\mathrm{d}z} = -Bi\theta \tag{9}$$

$$\text{(i) } w = 0, \qquad \text{(ii) } \frac{\mathrm{d}w}{\mathrm{d}z} = 0 \tag{10}$$

where Bi is the Biot number, $Bi = qd/\kappa$, and here q is a heat transfer coefficient and κ is the heat conductivity of liquid.

The four linear independent solutions to the system of Eqs (5)–(7) satisfying boundary conditions (8)–(10) can be obtained by means of the Runge–Kutta–Merson method. The parameter values providing nontrivial solution of the boundary-value problem are determined by the boundary conditions (9) and (10) [1].

The neutral stability curves are given in Fig. 1 for the case when the tangential component of pulsating velocity on free surface is equal to zero, a thermally insulated solid boundary, and $Bi = 0.1$ for several values of the vibrational Rayleigh number, Ra_v. Curve 1 corresponds to the case of instability in the absence of vibration ($Ra_\mathrm{v} = 0$), and curves 2–5 to the values

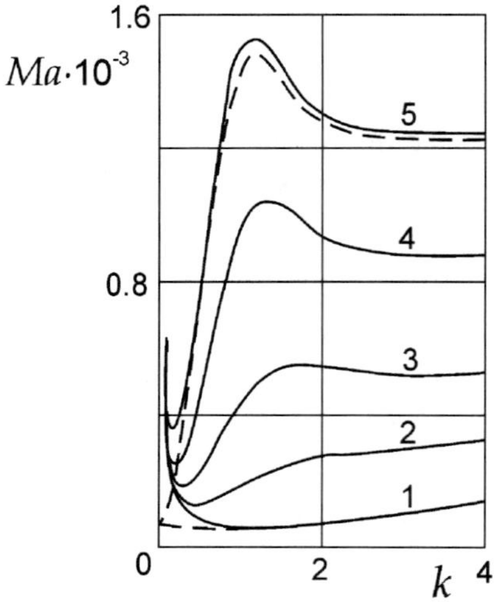

FIG. 1 Effect of high-frequency vibration on the surface tension gradient (Marangoni) driven instability of a free surface. Neutral curves at different values of vibrational Rayleigh number, Ra_v, for $Bi = 0.1$ and for a solid boundary where the heat flux is held fixed. Curves 1–5 correspond, respectively, to $Ra_v = 0$, 3×10^3, 10^4, 3×10^4, and 6×10^4. The broken lines correspond to $Bi = 0$. (From Ref. 1.)

$Ra_v = 3 \times 10^3$, 10^4, 3×10^4, and 6×10^4. The broken lines correspond to $Bi = 0$. Instability is predicted for Marangoni number values above the curves of given values of Ra_v.

For an open surface where the heat flux is held fixed the long-wavelength disturbances with $k = 0$ are the most dangerous. The critical Marangoni number at $k = 0$ is independent of the vibration. Its value in the vicinity of $k = 0$ is $Ma = 48[1 + k^2(616 + Ra_v)/840 + \cdots]$ [1].

If the heat flux from the free surface is nonzero the line $k = 0$ asymptotes the neutral curves and the most dangerous disturbances have finite wavelength. As seen from the figure the vibrations exert a strong stabilizing effect on the equilibrium against the disturbances with the wavenumber $k \sim 1$.

The effect of high-frequency vibration on the instability in a layer bounded by an isothermal solid boundary is illustrated in Fig. 2 (curves 1–6 correspond to $Ra_v = 0$, 10^4, 3×10^4, 10^5, 2.4×10^5, and 4.8×10^5). The vibration has negligible effect on the long-wavelength disturbances, and maximum stabilizing effect again occurs at $k \sim 1$.

Figure 3 displays the stability in the (Ma, Ra_v) plane for various heat transfer conditions. The values of the corresponding wavenumbers are also shown (ordinate on the right side of the figure). The solid lines represent the stability

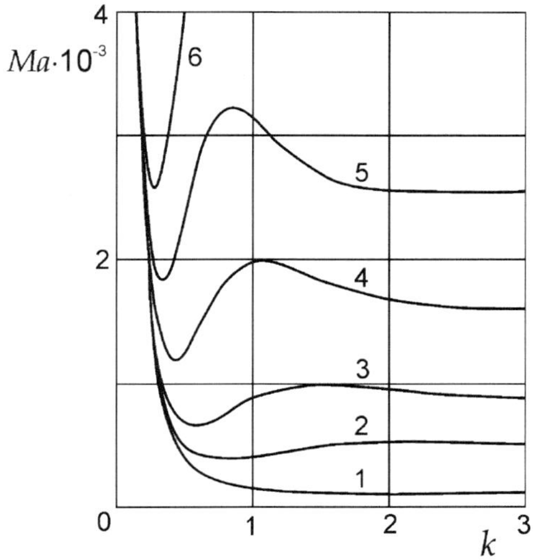

FIG. 2 Effect of high-frequency vibration on the surface tension gradient (Marangoni) driven instability of a free surface. Neutral curves at different values of the vibrational Rayleigh number (curves 1–6 correspond to $Ra_v = 0$, 10^4, 3×10^4, 10^5, 2.4×10^5, and 4.8×10^5) for a solid boundary where the temperature is held fixed. (From Ref. 1.)

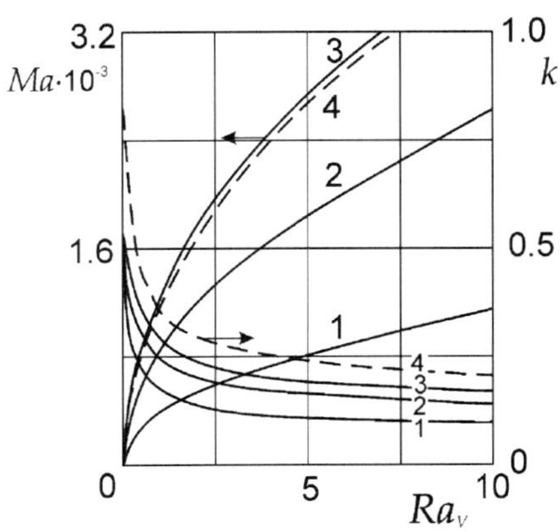

FIG. 3 Effect of high-frequency vibration on the surface tension gradient (Marangoni) driven instability of a free surface. Critical values of the Marangoni number and the wavenumber as a function of vibrational Rayleigh number. Solid lines, poorly conducting solid bottom boundary: 1, $Bi = 0.1$; 2, $Bi = 0.5$; and 3, $Bi = 1$. The broken lines, 4, represent isothermal solid bottom boundary ($Bi = 0$). (From Ref. 1.)

limits for a layer with a relatively poor heat-conducting solid boundary (curves 1, 2, and 3 correspond to $Bi = 0.1$, 0.5, and 1), and the broken lines refer to a layer with an isothermal solid boundary and $Bi = 0$. As usual, instability is predicted above the corresponding curve. With increase in the Biot number, i.e., with increase in the heat-conducting properties of the free surface, the influence of the Marangoni effect is reduced and the boundary of the region of instability is displaced towards higher values of the Marangoni number.

As expected, it appears that an increase of the Marangoni number has a destabilizing effect on the equilibrium due to the growth of the thermocapillary force. However, the growth of Ra_v stabilizes the steady state. Note, however, that Ra_v involves both the vibration parameters and the temperature gradient. Therefore, to single out a purely vibrational effect on the stability one can use a dimensionless vibration parameter, $S = Ma/Ra_v^{1/2}$. The latter includes no temperature gradient and represents the dimensionless vibration rate, $b\Omega$ (in fact its inverse). Figure 4 presents the stability diagram in the variables Ma_* and S, where Ma_* denotes the threshold values of the Marangoni number obtained by minimization of Ma as a function of k, as in Fig. 1. The instability region lies above and to the left of the curves shown with prescribed value of Bi. The sequence 1, 2, 3 in Fig. 4 corresponds to the sequence 3, 2, 1 in Fig. 3.

Thus one may conclude that for suitable values of S the steady state is always stable, no matter how intense the heating. Accordingly, starting with such values of S, the stabilizing contribution of an increasing temperature gradient in the (vibrational) Rayleigh number prevails over its destabilizing role in the

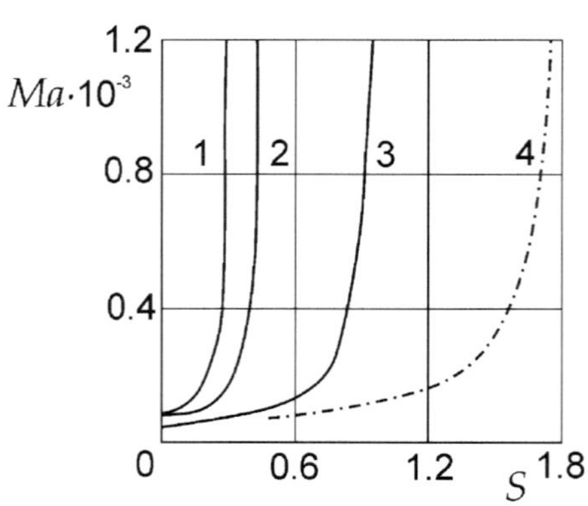

FIG. 4 Effect of high-frequency vibration on the surface tension gradient (Marangoni) driven instability of a free surface. Critical values of the Marangoni number vs. vibration parameter S at different values of the Biot number. The instability region lies above and to the left of the curves. The sequence 1, 2, 3 corresponds to sequence 3, 2, 1 in Fig. 3, curve 4 as 3, but calculated with the condition (i) in Eq. (10). (From Ref. 1.)

Marangoni number. These conclusions are equally valid for both models of free surface—undeformable or deformable with the pulsating motion—conditions (10) on the free surface. A quantitative difference in the results is illustrated in Fig. 4 when we compare curve 3 to curve 4, calculated with condition (i) from Eq. (10). Note also that with the increase of the Biot number, the influence of the Marangoni effect is reduced and therefore it is more easily suppressed by vibrations.

By analogy with the isothermal case, it would appear reasonable to suggest that Faraday ripple excitation sets a limit on the applicability of vibrations to suppress surface tension gradient (Marangoni) driven instability. Actually, the same high-frequency vibrations that suppress the abovementioned monotonic instabilities may be responsible for excitation of short waves. It can be shown that the threshold value of the vibration amplitude triggering wave motion depends on the Marangoni number. This result changes the region where equilibrium may be stable relative to the isothermal case.

II. EFFECT OF HIGH-FREQUENCY VIBRATION ON THE STABILITY OF AN UNDEFORMABLE INTERFACE

Let us now consider the influence of high-frequency transverse vibrations on the thermocapillary instability of a two-layer system. For a flat interface the vibrations alone are unable to cause instability. However, due to the presence of the averaged mass force the structure of motions initiated by the Marangoni effect changes and so does the stability threshold of the system.

A. Basic State and Equations for Disturbances in a Two-Layer System with an Inner Flat Interface

Let two layers of immiscible liquids of thickness h_1 and h_2 be confined between two solid surfaces held at temperatures Θ_1 and $-\Theta_2$ and let us assume that their interface, with its temperature taken at a reference value, remains flat (Fig. 5). The temperature gradients, A_1 and A_2, in the layers are related, $A_1\kappa_1 = A_2\kappa_2$, where κ_s

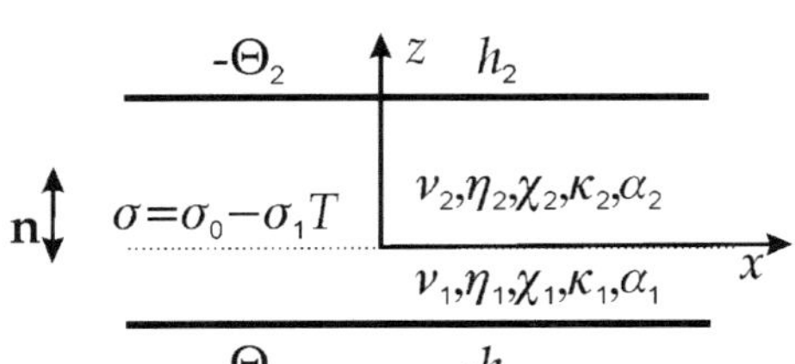

FIG. 5 Effect of high-frequency vibration on the stability of an interface between two liquids. Geometry of the problem and the coordinate system.

$(s = 1, 2)$ denotes liquid heat conductivity. Let the system as a whole be subjected to high-frequency vibrations of frequency Ω and amplitude b, and, as usual, $\sigma = \sigma_0 - \sigma_1 T$, with σ_1 positive.

The system can reach mechanical quasi-equilibrium with the zero averaged velocity and linear temperature distribution

$$\mathbf{v}_0 = 0, \qquad \mathbf{w}_0 = 0,$$

$$z < 0 : T_0 = -\Theta_1 \cdot \frac{z}{h_1} = -A_1 z \tag{11a}$$

$$z > 0 : T_0 = -\Theta_2 \cdot \frac{z}{h_2} = -A_2 z \tag{11b}$$

As in earlier cases studied, units and dimensionless parameters are defined by using two-layer quantities

$$d = h_1 + h_2, \qquad \nu = \nu_1 + \nu_2, \quad \eta = \eta_1 + \eta_2,$$
$$\chi = \chi_1 + \chi_2, \qquad \kappa = \kappa_1 + \kappa_2, \qquad \alpha = \alpha_1 + \alpha_2 \qquad \Theta = \Theta_1 + \Theta_2$$

where ν_s, η_s, χ_s, and α_s $(s = 1, 2)$ denote kinematic viscosity, dynamic (shear) viscosity, heat diffusivity, and thermal expansion coefficient, respectively. For such a choice the description of physical properties for each layer is symmetric and the relative characteristics of the layers, $\tilde{q}_s = q_s/q$ vary from 0 to 1. Then we set

$$\tilde{t} = \frac{t}{(d^2/\nu)}, \qquad \tilde{\mathrm{v}} = \frac{\mathrm{v}}{(\nu/h)}, \qquad (\tilde{T}, \tilde{\mathrm{w}}) = \frac{(T, \mathrm{w})}{\Theta},$$

$$\tilde{a}_1 = \frac{\tilde{\kappa}_2}{(\tilde{h}_1 \tilde{\kappa}_2 + \tilde{h}_2 \tilde{\kappa}_1)}, \qquad \tilde{a}_2 = \frac{\tilde{\kappa}_1}{(\tilde{h}_1 \tilde{\kappa}_2 + \tilde{h}_2 \tilde{\kappa}_1)},$$

$$Pr = \frac{\nu}{\chi}, \qquad Gr_{\mathrm{v}} = \frac{(b\Omega\alpha\Theta d)^2}{2\nu^2}, \qquad Ma = \frac{\sigma_1 \Theta d}{\eta \chi}$$

where t denotes time, v is the liquid velocity, T is the temperature disturbance, w is the amplitude of the vibrational component of the velocity, Pr is the Prandtl number, Gr_{v} is the vibrational analog of the Grashof number for heat, and Ma is the Marangoni number, which here is taken positive for a system heated on the side of the first layer, placed below.

With ω (real) denoting frequency, the neutral Fourier normal modes for oscillatory disturbances are

$$(\mathrm{v}_z, T, \mathrm{w}_z) = [\mathrm{v}(z), \theta(z), \mathrm{w}(z)] \exp[i(k_x x + k_y y - \omega t)]$$

are governed by the following equations

$$-i\omega\left(\frac{d^2v}{dz^2} - k^2v\right) = \nu_m\left(\frac{d^4v}{dz^4} - 2k^2\frac{d^2v}{dz^2} + k^4v\right) + G_m k^2 w \tag{12}$$

$$-i\omega\theta = Pr^{-1}\chi_m\left(\frac{d^2\theta}{dz^2} - k^2\theta\right) + a_m v \tag{13}$$

$$\frac{d^2w}{dz^2} - k^2w + k^2\theta = 0 \tag{14}$$

where

$$k^2 = k_x^2 + k_y^2, \qquad G_m = Gr_v\, a_m \alpha_m^2$$

and the boundary conditions are

$$z = -h_1, h_2 : v = \frac{dv}{dz} = \theta = w = 0;$$

$$z = 0 : v_1 = v_2 = 0, \qquad \frac{dv_1}{dz} = \frac{dv_2}{dz}, \qquad \theta_1 = \theta_2,$$

$$\kappa_1\left(\frac{d\theta_1}{dz}\right) = \kappa_2\left(\frac{d\theta_2}{dz}\right),$$

$$\eta_1\left(\frac{d^2v_1}{dz^2}\right) - \eta_2\left(\frac{d^2v_2}{dz^2}\right) = -k^2 Pr^{-1} Ma\,\theta_1,$$

$$w_1 = w_2 = 0 \tag{15}$$

The critical Marangoni number and the mode of neutral motions are determined by the boundary value problem of Eqs (13) and (14) that was solved by the Runge–Kutta method.

B. Solution of the Boundary-Value Problem

At high values of the vibrational Grashof number, Eqs (12)–(15) involve high-order derivatives with small parameter values. Therefore, to integrate this system numerically we need to ensure the linear independence of the particular solutions. An alternative is to construct solutions by the differential sweep method.

The scheme to construct the solution is based on a stepwise integration from the external layer boundaries, $z = -h_1$ and h_2, down to the liquid–liquid interface, $z = 0$. Let us introduce new variables, y_i, called boundary coefficients to relate the known functions of the corresponding boundary $z = -h_1$ and h_2 to the unknown functions:

$$
\begin{aligned}
\mathrm{v} &= y_1\mathrm{v}'' + y_2\mathrm{v}''' + y_3\theta' + y_4\mathrm{w}' \\
\mathrm{v}' &= y_5\mathrm{v}'' + y_6\mathrm{v}''' + y_7\theta' + y_8\mathrm{w}' \\
\theta &= y_9\mathrm{v}'' + y_{10}\mathrm{v}''' + y_{11}\theta' + y_{12}\mathrm{w}' \\
\mathrm{w} &= y_{13}\mathrm{v}'' + y_{14}\mathrm{v}''' + y_{15}\theta' + y_{16}\mathrm{w}' \qquad (16)
\end{aligned}
$$

To simplify notation for this subsection we shall use a prime ($'$) to denote differentiation with respect to $z(\mathrm{v}' = \mathrm{dv/dz}$, $\mathrm{v}'' = \mathrm{d^2v/dz^2}$, and so on).

Using Eq. (16), Eqs (12)–(15) lead to a system of ordinary differential equations for the new unknowns $y_i(z)$. We have

$$
\begin{aligned}
y_1' &= y_5 - y_2(p_1 + p_2y_1 + p_3y_{13}) - y_3(p_4y_9 + p_5y_1) + y_4p_6(y_9 - y_{13}) \\
y_2' &= y_6 - y_1 - y_2(p_2y_2 + p_3y_{14}) - y_3(p_4y_{10} + p_5y_2) + y_4p_6(y_{10} - y_{14}) \\
y_3' &= y_7 - y_2(p_2y_3 + p_3y_{15}) - y_3(p_4y_{11} + p_5y_3) + y_4p_6(y_{11} - y_{15}) \\
y_4' &= y_8 - y_2(p_2y_4 + p_3y_{16}) - y_3(p_4y_{12} + p_5y_4) + y_4p_6(y_{12} - y_{16}) \\
y_5' &= 1 - y_6(p_1 + p_2y_1 + p_3y_{13}) - y_7(p_4y_9 + p_5y_1) + y_8p_6(y_9 - y_{13}) \\
y_6' &= -y_5 - y_6(p_2y_2 + p_3y_{14}) - y_7(p_4y_{10} + p_5y_2) + y_8p_6(y_{10} - y_{14}) \\
y_7' &= -y_6(p_2y_3 + p_3y_{15}) - y_7(p_4y_{11} + p_5y_3) + y_8p_6(y_{11} - y_{15}) \\
y_8' &= -y_6(p_2y_4 + p_3y_{16}) - y_7(p_4y_{12} + p_5y_4) + y_8p_6(y_{12} - y_{16}) \\
y_9' &= -y_{10}(p_1 + p_2y_1 + p_3y_{13}) - y_{11}(p_4y_9 + p_5y_1) + y_{12}p_6(y_9 - y_{13}) \\
y_{10}' &= -y_9 - y_{10}(p_2y_2 + p_3y_{14}) - y_{11}(p_4y_{10} + p_5y_2) + y_{12}p_6(y_{10} - y_{14}) \\
y_{11}' &= 1 - y_{10}(p_2y_3 + p_3y_{15}) - y_{11}(p_4y_{11} + p_5y_3) + y_{12}p_6(y_{11} - y_{15}) \\
y_{12}' &= -y_{10}(p_2y_4 + p_3y_{16}) - y_{11}(p_4y_{12} + p_5y_4) + y_{12}p_6(y_{12} - y_{16}) \\
y_{13}' &= -y_{14}(p_1 + p_2y_1 + p_3y_{13}) - y_{15}(p_4y_9 + p_5y_1) + y_{16}p_6(y_9 - y_{13}) \\
y_{14}' &= -y_{13} - y_{14}(p_2y_2 + p_3y_{14}) - y_{15}(p_4y_{10} + p_5y_2) + y_{16}p_6(y_{10} - y_{14}) \\
y_{15}' &= -y_{14}(p_2y_3 + p_3y_{15}) - y_{15}(p_4y_{11} + p_5y_3) + y_{16}p_6(y_{11} - y_{15}) \\
y_{16}' &= 1 - y_{14}(p_2y_4 + p_3y_{16}) - y_{15}(p_4y_{12} + p_5y_4) + y_{16}p_6(y_{12} - y_{16}) \qquad (17)
\end{aligned}
$$

where

$$
p_1 = 2k^2 - \frac{i\omega}{\nu_m}, \qquad p_2 = -k^4 + \frac{i\omega k^2}{\nu_m}, \qquad p_3 = -\frac{G_m k^2}{\nu_m},
$$

$$
p_4 = k^2 - \frac{i\omega Pr}{\chi_m}, \qquad p_5 = -\frac{a_m Pr}{\chi_m}, \qquad p_6 = k^2
$$

The initial conditions for y_j are specified with the boundary conditions of Eq. (15) and in terms of Eq. (16) can be written as

$$z = -h_1, h_2 : y_j = 0, \text{ for all values of the subscript.} \tag{18}$$

The system of Eq. (17) with the initial conditions of Eq. (18) was integrated using the Runge–Kutta–Merson method to point $z = 0$. The boundary conditions of Eq. (15) lead to a homogeneous algebraic system of equations for the unknowns v''_-, v'''_-, θ'_-, w'_-, v''_+, v'''_+, θ'_+, w'_+ at point $z = 0$:

$$\begin{aligned}
&y_1^- v''_- + y_2^- v'''_- + y_3^- \theta'_- + y_4^- w'_- = 0\\
&y_1^+ v''_+ + y_2^+ v'''_+ + y_3^+ \theta'_+ + y_4^+ w'_+ = 0\\
&y_5^- v''_- + y_6^- v'''_- + y_7^- \theta'_- + y_8^- w'_- - y_5^+ v''_+ - y_6^+ v'''_+ - y_7^+ \theta'_+ - y_8^+ w'_+ = 0\\
&y_9^- v''_- + y_{10}^- v'''_- + y_{11}^- \theta'_- + y_{12}^- w'_- - y_9^+ v''_+ - y_{10}^+ v'''_+ - y_{11}^+ \theta'_+ - y_{12}^+ w'_+ = 0\\
&\kappa_1 \theta'_- - \kappa_2 \theta'_+ = 0\\
&\eta_1 v''_- - \eta_2 v''_+ + k^2 Pr^{-1} Ma(y_9^- v''_- + y_{10}^- v'''_- + y_{11}^- \theta'_- + y_{12}^- w'_-) = 0\\
&y_{13}^- v''_- + y_{14}^- v'''_- + y_{15}^- \theta'_- + y_{16}^- w'_- = 0\\
&y_{13}^+ v''_+ + y_{14}^+ v'''_+ + y_{15}^+ \theta'_+ + y_{16}^+ w'_+ = 0
\end{aligned} \tag{19}$$

where the subscripts "$-$" and "$+$" refer to liquid layers 1 and 2, respectively. The solvability condition for the system of Eq. (19) yields the critical Marangoni number. The requirement that the Marangoni number be real yields an equation for the disturbance frequency, ω, at onset. To find the velocity and temperature distribution in the layers one needs to perform a backward sweep procedure from the point $z = 0$. To do this, the system Eq. (17) is augmented with differential equations for the functions v'', v''', θ', and w', which are then used to define the functions v, v', θ, and w from relations of Eq. (17). Redefining the unknowns as

$$y_{17} = v'', \qquad y_{18} = v''', \qquad y_{19} = \theta', \qquad y_{20} = w' \tag{20}$$

we obtain from Eq. (12) and (17)

$$\begin{aligned}
y'_{17} &= y_{18}\\
y'_{18} &= p_1 y_{17} + p_2(y_1 y_{17} + y_2 y_{18} + y_3 y_{19} + y_4 y_{20})\\
&\quad + p_3(y_{13} y_{17} + y_{14} y_{18} + y_{15} y_{19} + y_{16} y_{20})\\
y'_{19} &= p_4(y_9 y_{17} + y_{10} y_{18} + y_{11} y_{19} + y_{12} y_{20})\\
&\quad + p_5(y_1 y_{17} + y_2 y_{18} + y_3 y_{19} + y_4 y_{20})\\
y'_{20} &= p_6((y_{13} - y_9) y_{17} + (y_{14} - y_{10}) y_{18} + (y_{15} - y_{11}) y_{19}\\
&\quad + (y_{16} - y_{12}) y_{20})
\end{aligned} \tag{21}$$

The initial conditions for the functions of Eq. (20) are determined by solving the system (19). The transverse and longitudinal liquid velocity and temperature distribution in the layer are then calculated by solving the system of Eqs. (17), (21), with (16).

C. Stability Analysis of the Model System

Let us now analyze the case when both liquids have the same volumetric expansion coefficient ($\alpha = 0.5$). In the absence of vibrations the lowest threshold for oscillatory disturbances is observed when $h_1 = 0.35$ and $Pr = 0.4$. For these parameter values Fig. 6 provides the neutral curves for $Gr_v = 0$ (lines 1) and $Gr_v = 10^6$ (lines 2) [2]. The broken lines correspond to oscillatory instability. Note the coexistence of oscillatory disturbances with monotonic disturbances at positive and negative values of the Marangoni number. Such a complex stability diagram of a two-layer system indicates that the layers interchange roles in the mechanism of thermocapillary instability.

Let us, for illustration, discuss the structure of neutral oscillations for disturbances with $k = \pi$ for $Ma = 22 \times 10^3$. Figure 7 illustrates the flow patterns

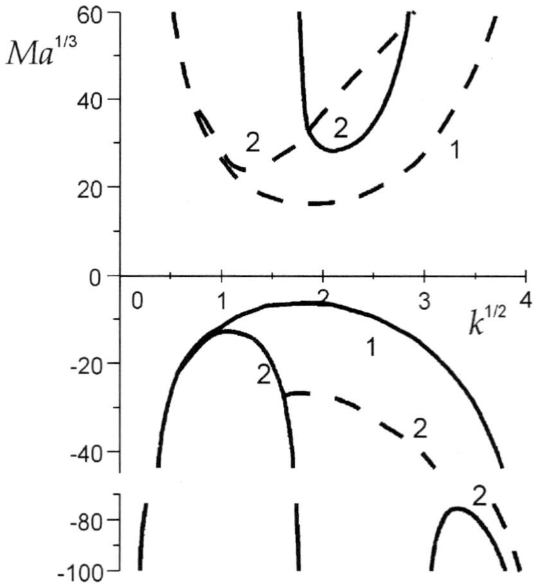

FIG. 6 Effect of high-frequency vibration on the stability of an interface between two liquids. Neutral curves for the model system with $h_1 = 0.35$, $Pr = 0.4$ without vibration (curves 1) and with vibration $Gr_v = 10^6$ (curves 2). Broken lines denote the neutral curves of oscillatory instability. (From Ref. 2.)

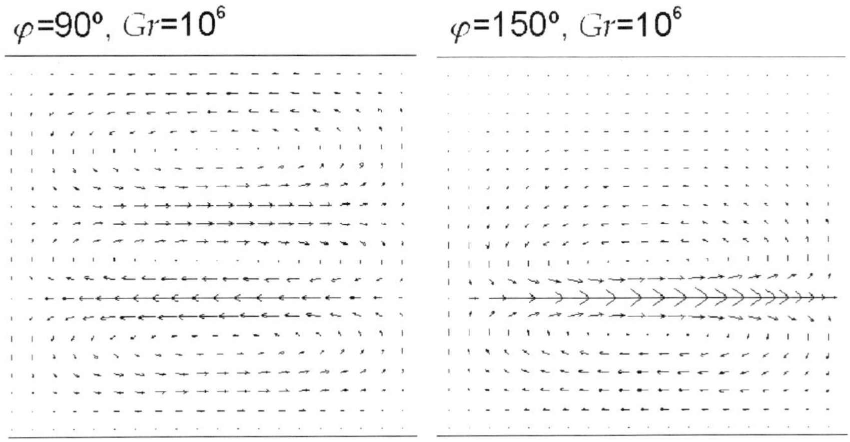

FIG. 7 Effect of high-frequency vibration on the stability of an interface between two liquids. Structure of flow at different phases of a standing wave. The size of an arrow is proportional to the actual velocity value at the corresponding point. (From Ref. 2.)

for a two-layer system in different phases of a standing wave. The first picture shows a convective cell in a phase of inertial flow ($\varphi = 90°$). The upper thicker layer involves two convective vortices, which reduce the efficiency of heat transfer through the layer and the variation of the longitudinal temperature gradient at the interface is dictated by the motion in the thinner layer. The flow pattern when $\varphi = 150°$ (second picture) indicates the reversal of the boundary flow and the generation of vortices rotating in opposite directions.

As in the vibration-free case the existence of undamped oscillations is related to a phase shift in the flow motion in the layers. Figure 8 shows the variation of the vortex intensity during oscillations in the first (line 1) and the second (line 2) layer. Note that the value of flow flux maximum in the thinner layer is higher than in the thicker one. As the direction of motion changes, the vortex near the interface decays faster in the thicker layer (recall that, as studied earlier, at vanishing vibrational Grashof number the vortex in the thinner layer decays faster). A small phase shift in the generation of vortices near the interface causes a marked increase in the critical Marangoni number. To illustrate the results predicted and to compare with experiments we shall now consider two different cases, a liquid-liquid interface and a liquid–air interface.

D. Liquid–Liquid Interface: Numerical Results for the Benzene–Water System

In Fig. 9 are given the neutral curves for the case of the benzene–water system when the water and benzene layers have equal thicknesses ($h_1 = 0.5$), for the

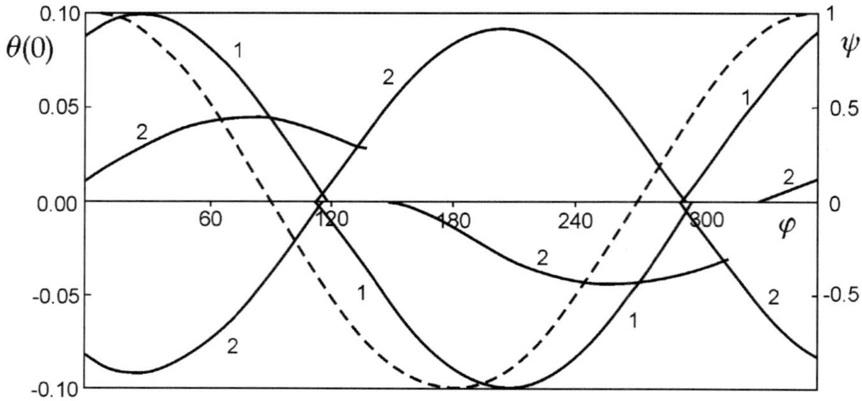

FIG. 8 Effect of high-frequency vibration on the stability of an interface between two liquids. Variation of the motion intensity (value of flow flux maximum) in the layers and the temperature gradient on the interface (broken line) for $Gr_v = 10^6$. (From Ref. 2.)

vibrational Grashof numbers $Gr_v = 0$, 10^3, and 10^6 (curves 1, 2, and 3, respectively) [3]. Monotonic instability is depicted by solid lines and overstability by broken lines.

Figure 9(b) provides the frequency of oscillatory disturbances. As mentioned earlier, for a system with equal layer thicknesses in the absence of vibrations, monotonic instability is achievable by heating on the side of benzene ($Ma < 0$, curve 1). High-frequency vibrations produce a strong stabilizing effect on the system in the region around the minimum of the neutral curve. At a certain value of the vibrational Grashof number the neutral curve breaks and there appears a region of wavenumbers where heating on the side of benzene does not excite monotonic instability ($Ma < 0$, locus between curves 2 and 3). A new region of oscillatory instability ($Ma < 0$, broken lines 2, 3) appears in this wavenumber range. At the same time, as the neutral curve for monotonic disturbances breaks at $Ma < 0$, then a region of monotonic instability appears when heating on the side of water ($Ma > 0$, solid lines 2, 3). With the increase of Gr_v the expected oscillatory instability, at $Ma > 0$, in the absence of vibrations is shifted to the region of high Marangoni numbers (broken lines 1, 2), the vibration effect on the dispersion relation being weak.

Using water layers of smaller thickness leads to the same instability modes for $d = 0.5$ and under high-frequency vibrations the neutral curves respond in a similar manner. As the benzene layer, responsible for temperature disturbances at the interface, is thicker, the generation of new instability regions is observed at higher values of the vibrational Grashof number. This effect is illustrated in Fig. 10

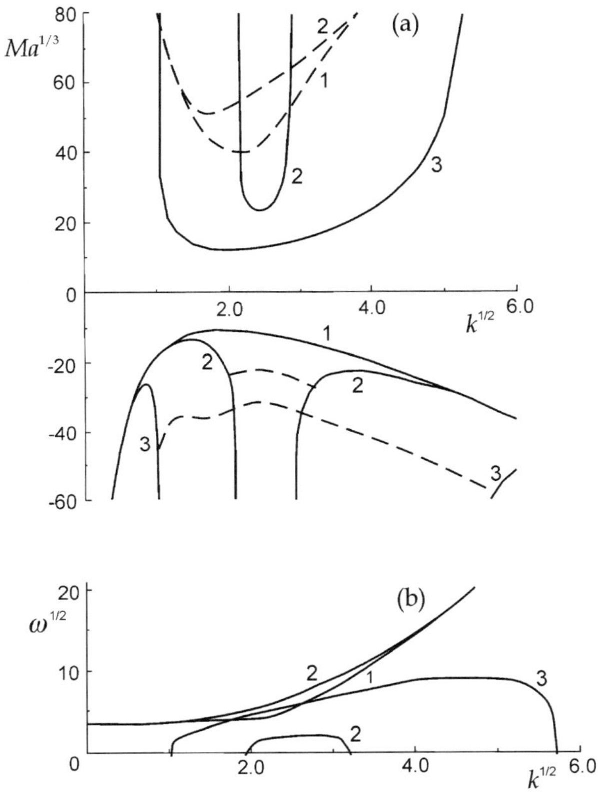

FIG. 9 Effect of high-frequency vibration on the stability of an interface between two liquids: (a) neutral curves of stability, and (b) frequency of oscillatory disturbances for the water–benzene system at $h_1 = 0.5$ and $Gr_v = 0$, 10^3, and 10^5 (curves 1, 2, and 3). Solid lines illustrate monotonic instability and broken lines, oscillatory instability. (From Ref. 3.)

where the neutral curves for monotonic and oscillatory disturbances are plotted for $Gr_v = 0$ and 10^5 with $h_1 = 0.2$. The neutral curves for the system with thicker water layer are shown in Fig. 11 ($h_1 = 0.8$). When vibrations are absent, heating on the side of the water layer ($Ma > 0$) gives rise to monotonic instability in the region of moderate wavenumbers ($k < 12$), while oscillatory instability appears when heating on the benzene layer side. Such behavior is typical of a system where the layers are drastically different in thickness but nearly similar in all other properties. However, the neutral curve at $Ma > 0$ has an asymptote and the appearance of an additional monotonic instability in the region of very short waves is caused by heating from the thinner layer (benzene) side.

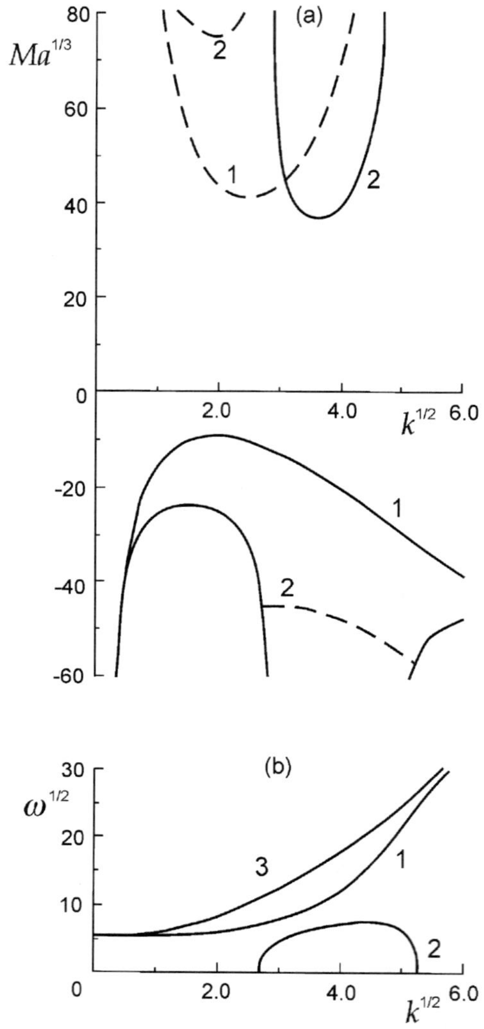

FIG. 10 Effect of high-frequency vibration on the stability of an interface between two liquids: (a) neutral curves of stability, and (b) frequency of oscillatory disturbances for the water–benzene system with $h_1 = 0.2$ for $Gr_v = 0$ and 10^5 (curves 1 and 2). (From Ref. 3.)

The situation changes radically in the presence of vibrations, which bring stabilization in the vicinity of the minima of the neutral curve and a widening of the wavenumber regions associated with the basic instability modes. It has been also found that vibrations suppress short-wave monotonic instability at $Ma < 0$.

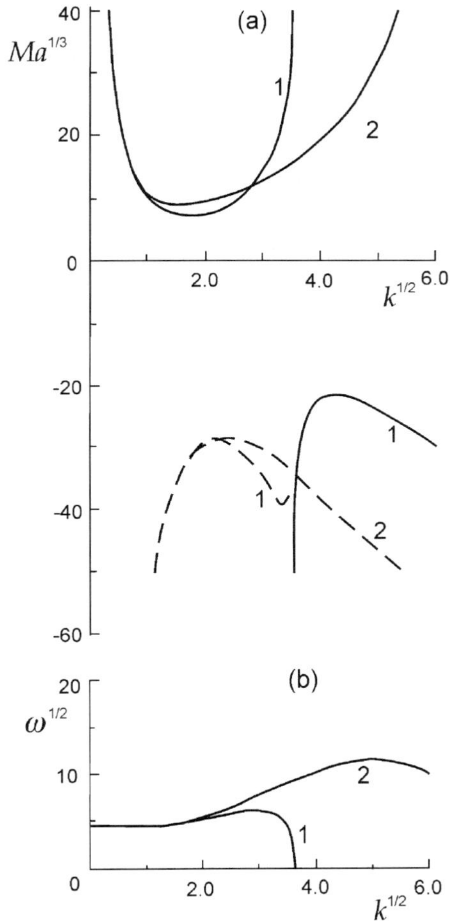

FIG. 11 Effect of high-frequency vibration on the stability of an interface between two liquids: (a) neutral curves of stability, and (b) frequency of oscillatory disturbances for the water–benzene system with $h_l = 0.8$ for $Gr_v = 0$ and 10^5 (curves 1 and 2). (From Ref. 3.)

E. Liquid–Gas Interface: Numerical Results for the *n*-Heptane–Air System

In the *n*-Heptane–air system the fluid layers have very different physical properties and therefore the influence of vibrations manifests itself differently than in a system with two similar liquid layers. Because the volumetric expansion coefficient of air is greater than that of *n*-heptane, at moderate values of the vibrational Grashof number the vibrations tend to suppress motion in the air layer, only. The stabilizing convective heat transfer from the interface on the air layer side diminishes, thus reducing the stability of the system [4]. With $h_1 = 0.4$ for $Gr_v = 10^5$ the critical Marangoni number reduces to 6.87 (using the lower layer parameter values we obtain $Ma = 81$). The Marangoni number actually attains the known value corresponding to a liquid layer thermally insulated from the air side ($Ma = 80$) [9].

Further increase of the vibrational Grashof number leads to the suppression of motion in the n-heptane layer and the stability is enhanced. No new regions of monotonic and oscillatory instability appear due to vibrations. There seems to be no reason to expect that layers with such drastically different properties could exchange their roles, leading to instability.

III. EFFECT OF HIGH-FREQUENCY VIBRATION ON THE STABILITY OF A TWO-LAYER SYSTEM WITH A DEFORMABLE INTERFACE

Let us now consider that the interface is deformable. We shall treat three different systems. In the first case all properties of the liquids are assumed to be similar while the layer thicknesses are different. In the second, the liquid layers are assumed to have the same thicknesses and different viscous properties. Finally, in the third problem, the asymmetry of the liquid layers arises from the layers having different thermal properties.

A. Two-Layer System with Deformable Interface and Transverse Vibrations

Let us consider a system with two immiscible liquid layers of thicknesses h_1 and h_2, respectively. The system is bounded by two solid surfaces, which are kept at constant temperatures, Θ_1 and $-\Theta_2$, to generate in both liquids the corresponding temperature gradients, A_1 and A_2. The temperature is calculated from the temperature of the liquid–liquid interface and as usual the surface tension of the interface depends linearly on the temperature, $\sigma = \sigma_0 - \sigma_1 T$, with σ_1 positive. Let the container and the liquids execute harmonic oscillations along the direction of the unit vector $\mathbf{n}$ with amplitude b and vibration frequency Ω (Fig. 5) [6,7].

In the frame of reference fixed to the vibrating cavity the heat and mass transfer equations are:

$$\operatorname{div} \mathbf{v}' = 0 \tag{22}$$

$$(1 - \beta_m T')\left[\frac{\partial \mathbf{v}'}{\partial t} + (\mathbf{v}' \cdot \nabla)\mathbf{v}'\right] = -\frac{1}{\rho_m}\nabla p' + \nu_m \Delta \mathbf{v}' + (1 - \alpha_m T') \times (\mathbf{g} - \mathbf{n} b \Omega^2 \cos \Omega t) \tag{23}$$

$$\frac{\partial T'}{\partial t} + \mathbf{v}' \cdot \nabla T' = \chi_m \Delta T' \tag{24}$$

where $\mathbf{v}'$, p', and T' denote relative velocity, total pressure, and temperature of a fluid, ν_m, χ_m, ρ_m, and α_m are the kinematic viscosity, thermal diffusivity, density, and thermal expansion coefficient of the fluid in layer $m = 1, 2$, $\mathbf{g}$ is the gravity vector pointing downwards. Note that primes ($'$) now refer to specific values of variables and not to differentiation. At variance with the BO approximation for natural convection in a heated layer, here the density variation with temperature appears in the inertial (convective) terms of the Navier–Stokes equations. This effect is expected to be of significance for a system with deformable boundary [7,8].

Owing to convective motion the interface may deviate from plane. For small deflections the boundary conditions on the interface, $z' = \zeta'(x', y', t)$, are

$$\eta_2\left(\frac{\partial v'_z}{\partial x}+\frac{\partial v'_x}{\partial z}\right)_2-\eta_1\left(\frac{\partial v'_z}{\partial x}+\frac{\partial v'_x}{\partial z}\right)_1=-\frac{\partial\sigma}{\partial x}$$
$$\eta_2\left(\frac{\partial v'_z}{\partial y}+\frac{\partial v'_y}{\partial z}\right)_2-\eta_1\left(\frac{\partial v'_z}{\partial y}+\frac{\partial v'_y}{\partial z}\right)_1=-\frac{\partial\sigma}{\partial y}$$
$$2\eta_2\left(\frac{\partial v'_z}{\partial z}\right)_2-2\eta_1\left(\frac{\partial v'_z}{\partial z}\right)_1-(p'_2-p'_1)=-\sigma\Delta_1\zeta' \qquad (25)$$

where

$$T'_1 = T'_2, \qquad \kappa_1\frac{\partial T'_1}{\partial l} = \kappa_2\frac{\partial T'_2}{\partial l},$$
$$\mathbf{v}'\cdot\mathbf{l}' = \frac{\partial\zeta'}{\partial t}, \qquad \mathbf{l}' = \left(-\frac{\partial\zeta'}{\partial x}, -\frac{\partial\zeta'}{\partial x}, 1\right)$$

where $\mathbf{l}'$ is vector normal to the interface from the first, "1", to the second, "2", fluid; and κ_m denotes fluid heat conductivity. On the solid boundaries we shall impose

$$z' = -h_1 : \mathbf{v} = 0,\ T_1 = \Theta_1$$
$$z' = h_2 : \mathbf{v} = 0,\ T_2 = -\Theta_2 \qquad (26)$$

If the vibration period is lower than both hydrodynamic and thermal times, but the Mach number is small, then the velocity, temperature, and pressure fields can be taken to be composed of slowly varying quantities with time (the triplet $\{\mathbf{v}, T, p\}$ accounts for the average motion) plus a pulsational component, $\{\tilde{\mathbf{v}}, \tilde{T}, \tilde{p}\}$, rapidly oscillating in the vicinity of zero. Thus two time scales can be introduced [8]: a slow time t and a fast time $\tau = \Omega t$. Then all quantities are split accordingly. We set

$$\mathbf{v}' = \mathbf{v}(t) + \tilde{\mathbf{v}}(t, \tau), \qquad T' = T(t) + \Omega^{-1}\tilde{T}(t, \tau), \qquad p' = p(t) + \Omega\tilde{p}(t, \tau),$$
$$\zeta' = \zeta(t) + \Omega^{-1}\tilde{\zeta}(t, \tau) \tag{27}$$

Substituting these expressions to the first order in the Navier–Stokes equations and boundary condition for pressure, and to the zeroth order in the remaining equations, we obtain the following problem for the pulsational component of the motion, which obeys the following equations:

$$\operatorname{div} \tilde{\mathbf{v}} = 0 \tag{28}$$

$$(1 - \alpha_m T)\frac{\partial \tilde{\mathbf{v}}}{\partial \tau} = -\frac{1}{\rho_m}\nabla\tilde{p} + (1 - \alpha_m T)\mathbf{n}b\Omega\cos\tau \tag{29}$$

$$\frac{\partial \tilde{T}}{\partial \tau} + \tilde{\mathbf{v}} \cdot \nabla T = 0 \tag{30}$$

where

$$z = \zeta(x, y, t) : \tilde{p}_1 = \tilde{p}_2, \qquad \tilde{\mathbf{v}}_1 \cdot \mathbf{l} = \tilde{\mathbf{v}}_2 \cdot \mathbf{l} = \frac{\partial \tilde{\zeta}}{\partial \tau}$$

$$z = -h_1, h_2 : \tilde{\mathrm{v}}_z = 0$$

Note that the fast component motion problem includes values of the slow one T, and $\mathbf{l}$, which upon integration yield a constant. Owing to the smallness of the amplitude pulsation component of the motion, the boundary conditions on the fluid interface are taken on an interface defined by the average motion.

The solution for the fast component can be sought in the form

$$\tilde{\mathbf{v}} = \mathbf{w}(x, y, z, t)b\Omega\sin\tau, \qquad \tilde{p} = \Phi(x, y, z, t)\rho_m b\Omega\sin\tau,$$
$$\tilde{T} = (\mathbf{w} \cdot \nabla T)b\Omega\cos\tau, \qquad \tilde{\zeta} = -(\mathbf{w} \cdot \mathbf{l})b\Omega\cos\tau \tag{31}$$

The slowly varying functions $\mathbf{w}(x, y, z, t)$ and $\Phi(x, y, z, t)$ obey the equations:

$$\operatorname{div} \mathbf{w} = 0 \tag{32}$$

$$(1 - \alpha_m T)\mathbf{w} = -\nabla\Phi - (1 - \alpha_m T)\mathbf{s} \tag{33}$$

$$z = \zeta : \rho_1\Phi_1 = \rho_2\Phi_2, \qquad \mathbf{w}_1 \cdot \mathbf{l} = \mathbf{w}_2 \cdot \mathbf{l}$$

$$z = -h_1, h_2 : \mathbf{w}_z = 0$$

To obtain the equations for the slow motion component, we introduce the expressions (27) and the solution (31) in the initial system (22)–(26) and average them over the vibration period. Then, together with Eqs (32) and (33) we have

$$\operatorname{div} \mathbf{v} = 0 \tag{34}$$

$$(1-\alpha_m T)\left[\frac{\partial \mathbf{v}}{\partial t} + (\mathbf{v}\cdot\nabla)\mathbf{v}\right] = -\frac{1}{\rho_m}\nabla p + \nu_m \Delta \mathbf{v} + (1-\alpha_m T)\mathbf{g}$$
$$-(1-\alpha_m T)\frac{b^2\Omega^2}{2}(\mathbf{w}\cdot\nabla)\mathbf{w} + \alpha_m \frac{b^2\Omega^2}{2}(\mathbf{w}\cdot\nabla T)(\mathbf{w}+\mathbf{n}) \tag{35}$$

$$\frac{\partial T}{\partial t} + \mathbf{v}\cdot\nabla T = \chi_m \Delta T \tag{36}$$

$$z = \zeta : \mathbf{v}_1 = \mathbf{v}_2, \qquad (\mathbf{v}\cdot\mathbf{l}) = \frac{\partial \zeta}{\partial t}$$

$$\eta_2\left(\frac{\partial v_z}{\partial x} + \frac{\partial v_x}{\partial z}\right)_2 - \eta_1\left(\frac{\partial v_z}{\partial x} + \frac{\partial v_x}{\partial z}\right)_1 = -\frac{\partial \sigma}{\partial x}$$

$$\eta_2\left(\frac{\partial v_z}{\partial y} + \frac{\partial v_y}{\partial z}\right)_2 - \eta_1\left(\frac{\partial v_z}{\partial y} + \frac{\partial v_y}{\partial z}\right)_1 = -\frac{\partial \sigma}{\partial y}$$

$$2\eta_2\left(\frac{\partial v_z}{\partial z}\right)_2 - 2\eta_1\left(\frac{\partial v_z}{\partial z}\right)_1 - (p_2 - p_1)$$
$$+\frac{b^2\Omega^2}{2}\left[\rho_2 \frac{\partial \Phi_2}{\partial z}(\mathbf{w}_2\cdot\mathbf{l}) - \rho_1 \frac{\partial \Phi_1}{\partial z}(\mathbf{w}_1\cdot\mathbf{l})\right] = -\sigma \Delta_1 \zeta$$

$$T_1 = T_2, \qquad \kappa_1 \frac{\partial T_1}{\partial l} = \kappa_2 \frac{\partial T_2}{\partial l}$$

B. Base State and Equations for Perturbations

As mechanical quasi-equilibrium, with zero average velocity, and linear distribution of temperature is possible, we can set

$$\mathbf{v}_0 = 0, \qquad \mathbf{w}_0 = 0$$

$$z < 0 : T_{0,1} = -\Theta_1 \cdot \frac{z}{h_1} = -A_1 z, \qquad \Phi_{0,1} = -z - \frac{\alpha_1 A_1}{h_1}\frac{z^2}{2} \tag{37}$$

$$z > 0 : T_{0,2} = -\Theta_2 \cdot \frac{z}{h_2} = -A_2 z, \qquad \Phi_{0,2} = -z - \frac{\alpha_2 A_2}{h_2}\frac{z^2}{2}$$

where A_1 and A_2 denote the gradients of temperature in the layers obeying the relation $\kappa_1 A_1 = \kappa_2 A_2$.

Let us now consider the stability of the state of Eq. (37). The equations for small perturbations can be obtained from Eqs (31)–(36) by linearization around Eq. (37). Using for the description of perturbations the same symbols as for the complete value of the slow component of the motion, the dimensionless equations for small perturbations become

$$\operatorname{div}\,\mathbf{v} = 0, \qquad \operatorname{div}\,\mathbf{w} = 0 \tag{38}$$

$$(1 - \varepsilon\alpha_m T_{0,m})\left(\frac{\partial \mathbf{v}}{\partial t}\right) = -\frac{\nabla p}{\rho_m} + \nu_m \Delta \mathbf{v} + \alpha_m Gr_{\mathrm{v}}(\mathbf{w}\nabla T_{0,m})\mathbf{n} \tag{39}$$

$$\frac{\partial T}{\partial t} + \mathbf{v}\cdot\nabla T_{0,m} = \chi_m Pr^{-1}\Delta T \tag{40}$$

$$(1 - \varepsilon\alpha_m T_{0,m})\mathbf{w} = -\nabla\Phi + \alpha_m T\mathbf{n}$$

$$Pr = \frac{\nu}{\chi}, \qquad Gr_{\mathrm{v}} = \frac{(\alpha b\Omega\Theta h)^2}{2\nu^2}, \qquad \varepsilon = \alpha\Theta$$

where we have neglected gravity. We have introduced:

$$d = h_1 + h_2, \qquad \nu = \nu_1 + \nu_2, \qquad \eta = \eta_1 + \eta_2,$$

$$\chi = \chi_1 + \chi_2, \qquad \kappa = \kappa_1 + \kappa_2, \qquad \alpha = \alpha_1 + \alpha_2, \qquad \Theta = \Theta_1 + \Theta_2$$

and we have taken as units of time, velocity, pressure, and temperature, d^2/ν, ν/d, $\eta\nu/d^2$, and Θ, respectively. For the functions w and Φ the scales used are $\alpha\Theta$ and $\alpha\Theta d$, respectively. Then, due to symmetry, the physical properties in each layer, and hence all relative characteristics of layers vary from 0 to 1.

The boundary conditions on the deformable interface can be defined at $z = 0$:

$$\mathbf{v}_1 = \mathbf{v}_2, \qquad \mathrm{v}_z = \frac{\partial\zeta}{\partial t}$$

$$\eta_2\left(\frac{\partial \mathrm{v}_z}{\partial x} + \frac{\partial \mathrm{v}_x}{\partial z}\right)_2 - \eta_1\left(\frac{\partial \mathrm{v}_z}{\partial x} + \frac{\partial \mathrm{v}_x}{\partial z}\right)_1 = Ma\ Pr^{-1}\left(\frac{\partial T_1}{\partial x} + \frac{\partial T_{0,1}}{\partial z}\frac{\partial\zeta}{\partial x}\right)$$

$$\eta_2\left(\frac{\partial \mathrm{v}_z}{\partial y} + \frac{\partial \mathrm{v}_y}{\partial z}\right)_2 - \eta_1\left(\frac{\partial \mathrm{v}_z}{\partial y} + \frac{\partial \mathrm{v}_y}{\partial z}\right)_1 = Ma\ Pr^{-1}\left(\frac{\partial T_1}{\partial y} + \frac{\partial T_{0,1}}{\partial z}\frac{\partial\zeta}{\partial y}\right) \tag{41}$$

$$2\eta_2\left(\frac{\partial \mathrm{v}_z}{\partial z}\right)_2 - 2\eta_1\left(\frac{\partial \mathrm{v}_z}{\partial z}\right)_1 - (p_2 - p_1)$$

$$+ Gr\left(\rho_2\frac{\partial\Phi_{0,2}}{\partial z}w_{z,2} - \rho_1\frac{\partial\Phi_{0,1}}{\partial z}w_{z,1}\right) = -Ca\Delta_1\zeta$$

$$T_1 + \frac{\partial T_{0,1}}{\partial z}\zeta = T_2 + \frac{\partial T_{0,2}}{\partial z}\zeta, \qquad \kappa_1\frac{\partial T_1}{\partial l} = \kappa_2\frac{\partial T_2}{\partial l}$$

$$\rho_1\left(\Phi_1 + \zeta\frac{\mathrm{d}\Phi_{0,1}}{\mathrm{d}z}\right) = \rho_2\left(\Phi_2 + \zeta\frac{\mathrm{d}\Phi_{0,2}}{\mathrm{d}z_2}\right), \qquad w_{z,1} = w_{z,2},$$

$$Ma = \frac{\sigma_1\Theta d}{\eta\chi}, \qquad Ca = \frac{\sigma_0 d}{\eta\nu}$$

On the solid boundaries, we assume that

$$z = -h_1, h_2 : \mathbf{v} = 0, \qquad T = 0, \qquad w_z = 0 \tag{42}$$

Five dimensionless parameters appear in the problem: the earlier defined vibration analog of the Grashof number, Gr_v; the Marangoni number, Ma, which here is taken positive when heating from the liquid layer "1"; the capillary number, Ca; the Prandtl number, Pr; and the smallness parameter ε.

Once more we consider Fourier normal modes, $\exp(-\lambda t + ik_x x + ik_y y)$. Eliminating pressure, the longitudinal components of the velocity, the vector $\mathbf{w}$, and the function Φ, then the amplitudes of perturbations $\{\mathrm{v}(z), \theta(z), \mathrm{w}(z), \xi\}$ of the cross-velocity component, temperature, the z-component of $\mathbf{w}$, and the interface deflection obey the following boundary-value problem:

$$-\lambda\left(\frac{d^2 \mathrm{v}}{dz^2} - k^2 \mathrm{v}\right) = \nu_m\left(\frac{d^4 \mathrm{v}}{dz^4} - 2k^2\frac{d^2 \mathrm{v}}{dz^2} + k^4 \mathrm{v}\right) - G_m k^2 \mathrm{w} \tag{43}$$

$$-\lambda\theta = Pr^{-1}\chi_m\left(\frac{d^2\theta}{dz^2} - k^2\theta\right) + a_m \mathrm{v} \tag{44}$$

$$\frac{d^2 \mathrm{w}}{dz^2} - k^2 \mathrm{w} + k^2 \alpha_m \theta = 0 \tag{45}$$

$$k^2 = k_x^2 + k_y^2, \qquad G_m = Gr_v a_m \alpha_m^2$$

$$a_1 = \frac{\kappa_2}{(h_1\kappa_2 + h_2\kappa_1)}, \qquad a_2 = \frac{\kappa_1}{(h_1\kappa_2 + h_2\kappa_1)}$$

$$z = -h_1, h_2 : \mathrm{v} = \frac{d\mathrm{v}}{dz} = \theta = \mathrm{w} = 0$$

$$z = 0 : \mathrm{v}_1 = \mathrm{v}_2 = -\lambda\xi, \qquad \frac{d\mathrm{v}_1}{dz} = \frac{d\mathrm{v}_2}{dz} \tag{46}$$

$$\eta_1\left(\frac{d^2\mathrm{v}_1}{dz^2} + k^2\mathrm{v}_1\right) - \eta_2\left(\frac{d^2\mathrm{v}_2}{dz^2} + k^2\mathrm{v}_2\right) = -k^2 Pr^{-1} Ma(\theta_1 - a_1\xi) \tag{47}$$

$$\lambda\left(\rho_1\frac{d\mathrm{v}_1}{dz} - \rho_2\frac{d\mathrm{v}_2}{dz}\right) + \eta_1\left(\frac{d^3\mathrm{v}_1}{dz^3} - 3k^2\frac{d\mathrm{v}_1}{dz}\right) - \eta_2\left(\frac{d^3\mathrm{v}_2}{dz^3} - 3k^2\frac{d\mathrm{v}_2}{dz}\right) + k^2 G\alpha_1 \mathrm{w}_1\frac{(\rho_1 - \rho_2)}{\varepsilon} = k^4 Ca\xi \tag{48}$$

$$\theta_1 - a_1\xi = \theta_2 - a_2\xi, \qquad \kappa_1\frac{d\theta_1}{dz} = \kappa_2\frac{d\theta_2}{dz}, \qquad \mathrm{w}_1 = \mathrm{w}_2,$$

$$\rho_1\frac{d\mathrm{w}_1}{dz} = \rho_2\frac{d\mathrm{w}_2}{dz} - k^2\xi\frac{(\rho_1 - \rho_2)}{\varepsilon}$$

The boundary-value problem of Eqs (45)–(47), defining critical Marangoni numbers and disturbing frequency, $\omega = \mathrm{Im}\ \lambda$, was solved by a differential sweep method.

C. Instability Threshold Dependence on the Asymmetry in Layer Thicknesses

Let us first consider the case where both liquids have similar properties and hence the only dimensionless parameter specifying the difference between the two layers is their layer thickness. All other two-layer parameter values are equal to 0.5. Let us consider also that $h_1 < h_2$, thus positive Marangoni numbers here correspond to heating the system from the side of the thicker layer.

Figure 12 illustrates the neutral curves and dispersion relations for the system with $h_1 = 0.3$, $Pr = 0.1$, and $Ca = 10^4$ when Gr_v is set to zero [7]. The solid line accounts for monotonic instability, and dashed lines for overstability.

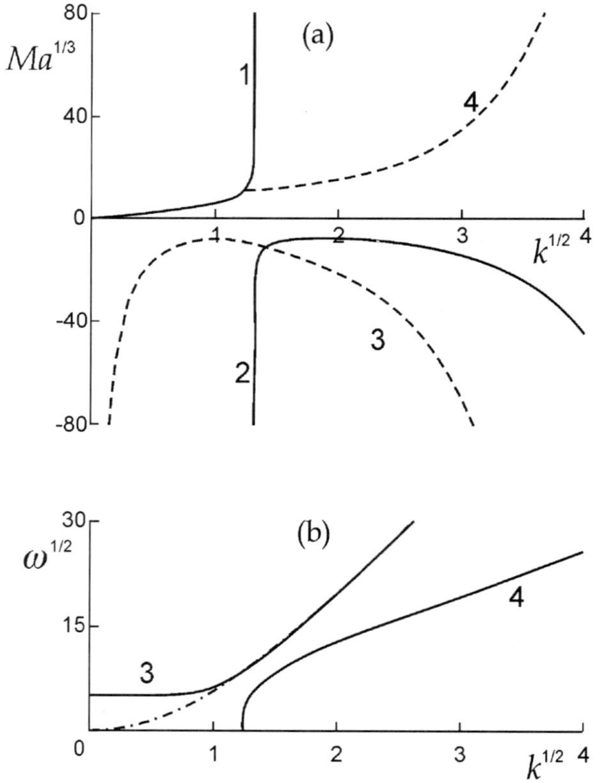

FIG. 12 Effect of high-frequency vibration on the stability of an interface between two liquids: (a) curves of neutral monotonic (solid lines) and oscillatory (dashed lines) instability, and (b) dispersion relations for the system with $h_1 = 0.3$, $Pr = 0.1$, and $Ca = 10^4$ at $Gr_v = 0$. The dash-dot line gives the dispersion relations for capillary waves on the interface of isothermal inviscid liquid. (From Ref. 7.)

Recall that, for an undeformable interface, monotonic instability is found when heating on the side of the thicker layer, and oscillatory otherwise. In the latter case, the dispersion curve, in the long-wave range, describes an oscillation frequency practically independent of the wavelength, hence waves in such a region exhibit no dispersion. When the interface is deformable various instability modes are possible: a long-wavelength branch of monotonic instability excited when heating on the thinner layer side (solid line 1); an oscillatory mode when heating on the side of the thicker layer in the region of medium and short waves, whose dispersion relation follows very closely the dispersion curve for capillary waves in an inviscid fluid (broken line 3); simultaneously, the long-wave region of the neutral curves for modes in the absence of deformation is truncated, thus allowing for a short-wave branch of monotonic instability (lower branch of curve 1); and, finally, overstability when heating on the side of the thinner layer (curve 2).

Let us now consider the effect of transverse vibrations on the above discussed instabilities. High-frequency vibrations are not capable of causing instability in the system because we have assumed that the frequency of vibrations is higher than the driving frequency for parametric oscillations of the interface. However, because of the availability of average body force the structure of motions caused by the Marangoni effect and the instability threshold of the system can be altered.

Let us, in particular, consider the case when the volumetric expansion coefficient in both fluids is identical ($\alpha = 0.5$). In the absence of vibrations the threshold for oscillatory perturbations appears when $h_1 = 0.3$ and $Pr = 0.1$. For these parameter values and $Gr_v = 10^6$, the neutral curves (a) for monotonic (solid line) and oscillatory (dashed lines) instability and the dispersion relations (b) are plotted in Fig. 13. Here and in the following figures, the dashed-dotted line gives the dispersion relation for capillary waves on the interface of an isothermal inviscid liquid. Comparison of these curves with those of Fig. 12 shows a marked stabilizing effect of the vibration.

Particularly noteworthy is the essential change of a neutral curve featuring the Pearson mechanism of instability described in Chapter 5 (a short-wave branch by curve 1) [9], and the occurrence of neutral oscillatory perturbations at negative values of the Marangoni number and monotonic perturbations at positive values. Such stability variation of the system is bound to change the role of layers in the mechanism of thermocapillary instability. The analysis of the structure of the flow shows that the vibrations reduce fluid flow along the direction of vibrations. Indeed, with intense vibrating action the motion in the thicker layer becomes a two-vortex flow. The weak convective heat transfer of this motion is responsible for the exchange of contributions between the thicker and thinner layers, leading to the instability mechanism, and the monotonic instability studied by Pearson [9] occurs when heating on the side of the thinner layer.

The oscillatory instability in this region of wavenumbers is a result of a phase shift of the motion in the layers and arises when heating from the opposite

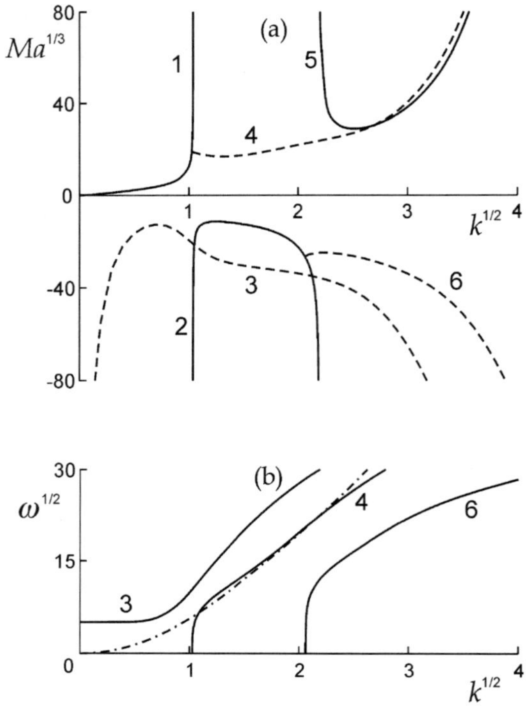

FIG. 13 Effect of high-frequency vibration on the stability of an interface between two liquids: (a) curves of neutral monotonic (solid lines) and oscillatory (dashed lines) instability, and (b) dispersion relations for the system with $h_1 = 0.3$, $Pr = 0.1$, and $Ca = 10^4$ at $Gr_v = 5 \times 10^6$. The dash-dot line gives the dispersion relations for capillary waves on the interface of isothermal inviscid liquids. (From Ref. 7.)

side. The small phase shift at a stopping of vortexes near to a interface causes a noticeable rise of the critical Marangoni number for oscillatory perturbations.

D. Instability Threshold Dependence on the Difference in the Viscosity of the Liquids

Let us now consider a geometrically symmetric two-layer system ($h_1 = h_2 = 0.5$) with fluids having identical thermal properties, but largely different kinematic and dynamic viscosities.

As discussed earlier, in two-layer systems with a flat undeformable interface, monotonic instability due to the Marangoni effect either comes from unequal layer depths or from the disparity in thermal diffusivities [5]. Heating on the side of a layer with higher value of kinematic viscosity, there is the possibility of oscillatory instability. The dynamic (shear) viscosities of the layers also exert influence on the growth of perturbations in the system, but, in contrast to the kinematic viscosity their inequality does not bring instability.

In a system with a deformable interface and subjected to vibrations, density difference matters. However, we shall consider the simplest case, $\rho_1 = \rho_2 = 1$.

The neutral curves and frequencies of neutral disturbances when $\eta_1 = \nu_1 = 0.45$, $Pr = 0.1$, $Ca = 2 \times 10^{-4}$ without vibrations $(Gr_v = 0)$ and with vibrations $(Gr_v = 5 \times 10^4)$ are given in Fig. 14. The neutral curves of oscillatory disturbances (curves 2 and 3) and the corresponding frequency appear insensitive to vibrations. There is monotonic instability in the system with deformable interface when heating on the side of the liquid with higher viscosity. In the absence of vibrations the instability domain is determined by curve 1. Vibrations (curve 4) induce a decrease in the stability of system in the short-wave region ($k > 3$). Apparently, short-wave disturbances add a significant

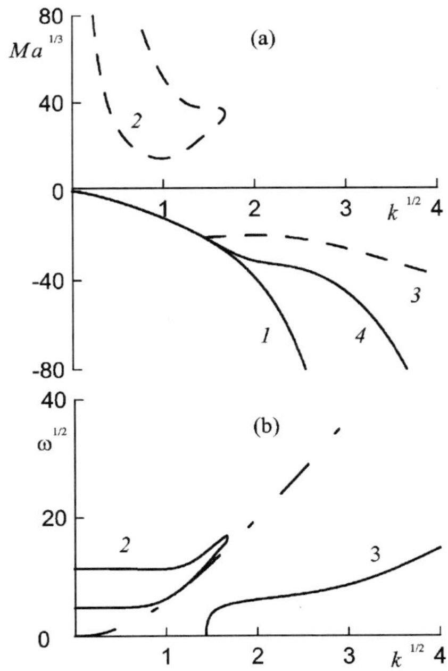

FIG. 14 Effect of high-frequency vibration on the stability of an interface between two liquids: (a) curves of neutral monotonic (solid lines) and oscillatory (dashed lines) instability, and (b) dispersion relations for the system with $\eta_1 = \nu_1 = 0.45$, $Pr = 0.1$, and $Ca = 2 \times 10^4$ in the absence of vibrations ($Gr_v = 0$) and with vibrations ($Gr_v = 5 \times 10^4$, line 4 in place of 1). The dash-dot line gives the dispersion relation for capillary waves on the interface of isothermal inviscid liquids. (From Ref. 7.)

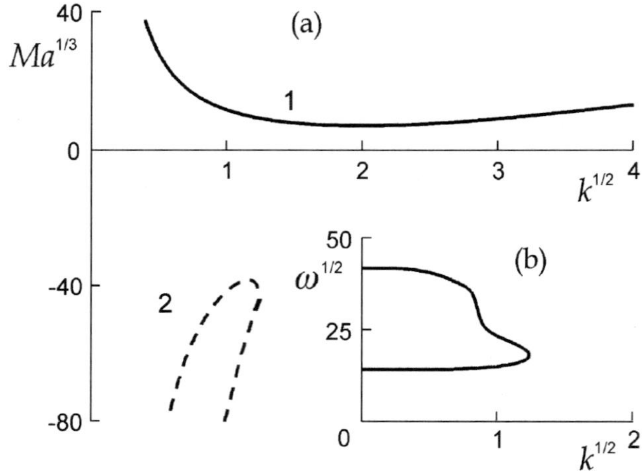

FIG. 15 Effect of high-frequency vibration on the stability of an interface between two liquids: (a) curves of neutral monotonic (solid lines) and oscillatory (dashed lines) instability, and (b) dispersion relation for the system with $\chi_1 = 0.3$, $Pr = 0.1$, and $Ca = 10^4$ at $Gr_v = 0$. (From Ref. 7.)

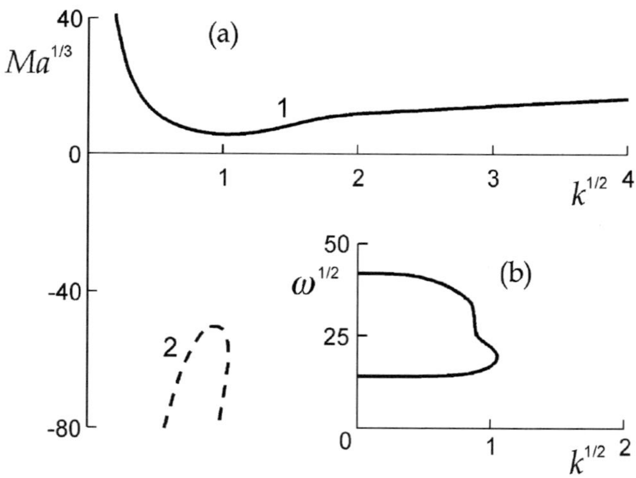

FIG. 16 Effect of high-frequency vibration on the stability of an interface between two liquids: (a) curves of neutral monotonic (solid lines) and oscillatory (dashed lines) instability, and (b) dispersion relation for the system with $\chi_1 = 0.3$, $Pr = 0.1$, and $Ca = 10^4$ at $Gr_v = 5 \times 10^6$. (From Ref. 7.)

component to the cross-velocity, which is suppressed by vibrating forces. In the liquid with lower viscosity, as may be seen from the Navier–Stokes equation, the vibrating body force is higher and flow in this layer is more strongly suppressed.

E. Instability Threshold Dependence on the Difference in Thermal Transport Properties

Finally, let us consider a geometrically symmetrical two-layer system ($h_1 = h_2 = 0.5$) when the layers differ from each other only in their heat diffusions. In the absence of vibrations, monotonic instability appears when heating on the side of the layer with lower χ (curve 1 in Fig. 15), while oscillations arise when cooling

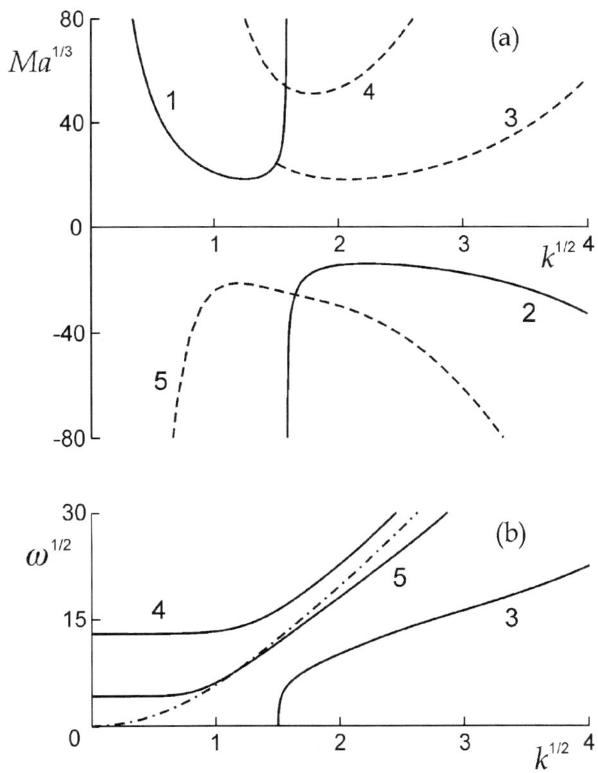

FIG. 17 Effect of high-frequency vibration on the stability of an interface between two liquids: (a) curves of neutral monotonic (solid lines) and oscillatory (dashed lines) instability, and (b) dispersion relations for the system with $\kappa_1 = 0.3$, $Pr = 0.1$, and $Ca = 10^4$ at $Gr_v = 5 \times 10^5$. The dash-dot line gives the dispersion relation for capillary waves on the interface of isothermal inviscid liquids. (From Ref. 7.)

the system from this side (curve 2). The latter arises at considerably higher (in absolute value) Marangoni number than for the monotonic one. The heat conductivity of the fluids also exerts influence on the fate of disturbances in the system, but at $\chi_1 = \chi_2$ there is stability whatever the values of κ_1 and κ_2 (hence specific heats and/or densities play a significant role). The effect of high-frequency vibrations on oscillatory and monotonic modes of instability in a system with different thermal diffusivities can be seen by comparing Figs 15 and 16. Furthermore, vibration adds new modes of monotonic and oscillatory instability directly related to the difference in heat conductivities. Neutral curves and dispersion relations are shown in Fig. 17 for $Gr_v = 5 \times 10^5$.

Finally, it appears that, at variance with systems with asymmetry in layer depths or in liquid viscosities, for a two-layer system with only unequal heat properties the deformation of the interface does not play a significant role. Instability happens to occur identically for two-layer liquid layers with or without interfacial deformability.

REFERENCES

1. Birikh, R.V., Briskman, V.A., Zuev, A.L., Chernatynsky, V.I., Yakushin, V.I. Interaction of the thermovibrational and thermocapillary convection mechanisms. Fluid Dyn. **1994**, *29* (5), 681–692.
2. Birikh, R.V., Rudakov, R.N. Mechanism of thermocapillary oscillation in a system with interface. In *Vibrational Effects in Hydrodynamics*; Perm State University: Perm, 1998; 38–48 (in Russian).
3. Birikh, R.V., Briskman, V.A., Rudakov, R.N., Velarde, M.G. Interface Marangoni instability in the high-frequency vibration field. Proceedings of the Joint Xth Europ. and VIth Russian Symp. on "Physical Sciences in Microgravity". St. Petersburg, Russia, 1997, Vol. 1, 205–212.
4. Birikh, R.V., Briskman, V.A., Rudakov, R.N., Velarde, M.G. Thermocapillary instability of real liquids interfaces. In *Convection in Immiscible Liquid–Liquid Systems*; UB RAS: Ekaterinburg, 1999; 25–44 (in Russian).
5. Smith, K.A. On convective instability induced by surface-tension gradients. J. Fluid Mech. **1966**, *24*, 401–414.
6. Birikh, R.V., Boushoueba, S.V. Thermocapillary instability in a two-layer system with a deformable interface. Izv. Akad. Nauk SSSR, Mech. Zhidk. Gaza **2001**, *3*, 13–20 (in Russian).
7. Birikh, R.V., Briskman, V.A., Boushoueva, S.V., Rudakov, R.N. Thermocapillary and vibrational instability in a two-layer system with a deformable interface. In *Concentration and Thermocapillary Effects in Complex Systems*; Birikh, R.V., ed; UrO RAN: Ekaterinburg, 2002 (in Russian).
8. Gershuni, G.Z., Lyubimov, D.V. *Thermal Vibrational Convection*; Wiley: New York, 1998.
9. Pearson, J.R.A. On convective cells induced by surface tension. J. Fluid Mech. **1958**, *4*, 489–500.

Index

Naren